Introduction to Physiology and Homeostasis

Chapter Overview

The cells, the fundamental units of life, are organized into tissues, organs, and systems, which in turn maintain homeostasis. This homeostasis is essential for the survival of the cells. The multi-cellular organism provides this discrete internal environment through its structural organization. The functions of tissues, organs, and systems are specializations of specific cellular capabilities. A cyclic phenomenon exists: The organism, its tissues, organs, and systems, is composed of cells maintained through the homeostasis of an internal environment that is controlled by body systems. This is the basic foundation of physiology.

Chapter Outline

INTRODUCTION TO PHYSIOLOGY
- Physiology is the study of the functions of the body or how the body works. Physiology is closely related with anatomy, the study of the structure of the body. There are two approaches to explain physiological events, one emphasizing the purpose of the function the other emphasizing why it happens. Physiologists focus on the latter.

Structure and function are inseparable.
- Physiology is closely related with anatomy.

LEVELS OF ORGANIZATION IN THE BODY
The chemical level: Various atoms and molecules make up the body.
- Atoms form molecules.

The cellular level: Cells are the basic units of life.
- The smallest unit capable of performing the basic functions essential for life is the cell.
- Whether they exist as solitary cells or as part of a multi-cellular organism, all cells perform certain basic functions essential for survival of the organism. Organisms are independent living entities. As a result of cell differentiation, your body is made up of many different specialized cell types.

Basic cell functions.
- These functions include: (1) obtaining nutrients and oxygen, (2) performing certain chemical reactions that use nutrients and oxygen to provide energy, (3) eliminating from the cell carbon dioxide and other by-products or wastes, (4) synthesizing proteins and other components, (5) controlling the exchange of material between the cell and its environment, (6) moving materials within the cell and even the cell itself, (7) being sensitive to changes in the environment, and (8) reproducing.

Specialized cell functions.
- Each cell also performs a specialized function such as protein synthesis, muscle contraction, and the transmission of information.

The tissue level: Tissues are groups of cells of similar specialization.

- Cells of similar origin, structure, and function are referred to as tissues.
- The four primary types are muscle, nervous, epithelial, and connective tissue.
- There are three types of muscle tissue: skeletal, cardiac, and smooth muscle. Muscle cells are specialized for contraction which generates tension and produces movements.
- Nervous tissue consists of cells specialized for initiation and transmission of electrical impulses, sometimes over long distances.
- These impulses are transmitted to the brain and subsequently to muscles or glands for the appropriate response.
- Epithelial tissue lines or covers the body and its various organs, vessels, and cavities.
- The specialization of epithelial tissue is the exchange of materials.
- Highly specialized epithelial cells form glands that provide essential secretions.
- Glands may be either exocrine or endocrine.
- Connective tissue, as its name implies, connects, supports, and anchors the various body parts.

The organ level: An organ is a unit made up of several tissue types.

- Organs are composed of two or more types of primary tissues organized to perform some specific function within the body.
- Organs are further organized into body systems, each of which is a collection of organs that perform related functions and interact to accomplish a common activity.

The body system level: A body system is a collection of related organs.

- The total body is composed of various organ systems linked together as an entity that is separate from the external environment.

The organism level: The body systems are packaged together into a functional whole body.

- The human body is made up of cells organized into life-sustaining systems.

CONCEPT OF HOMEOSTASIS
Body cells are in contact with a privately maintained internal environment.

- The cells in a multi-cellular organism must contribute to the survival of the organism as a whole; they cannot live and function without contributions from the other body cells because the vast majority of cells are not in direct contact with the external environment.
- The external environment is the surrounding environment in which the organism lives.
- The internal environment is the fluid that surrounds the cells and through which they make life-sustaining exchanges. .
- The fluid contained with all body cells is called the intracellular fluid (ICF) and the fluid outside the cells is called the extracellular fluid (ECF)
- The ECF is made up of plasma, and interstitial fluid, which surrounds and bathes the cells.
- Various body systems accomplish exchanges between the external environment and the internal environment.
- Body systems perform the necessary functions to maintain homeostasis.

Body systems maintain homeostasis, a dynamic, steady state in the internal environment.

- Homeostasis is essential for the survival of each cell, and each cell, through its specialized activities, contributes as part of a body system to the maintenance of the internal environment shared by all cells.
- To maintain homeostasis, the body must be able to detect deviations in the internal environment factors that need to be held within narrow limits, and it must be able to control the various body systems responsible for adjusting these factors.

Introduction

The physiology professor has selected an excellent text: *Fundamentals of Human Physiology*, Forth Edition, by Lauralee Sherwood. It is a state-of-the-art book designed to adequately prepare the student in the knowledge of human physiology. Most of the ancillaries, for any text, aid the instructor in the presentation of the material. This *Study Guide* is designed solely for the student.

The goal of this study guide is to help the student gain a better understanding of the material covered in the textbook. Throughout, the concepts, ideas, and phenomena are restated in various forms to help the student understand human physiology. From the student's perspective, this guide is designed to improve test grades and elevate the grade point average. With proper use of this guide, these interrelated goals should become a reality.

The **Chapter Overview** links the chapter to the main fundamental phenomenon of physiology: homeostasis. It mentions broad general statements to relate the chapter to other chapters. Several specifics may be noted or listed to indicate how the physiology covered by the chapter is achieved. In short, the Chapter Overview is a very brief summary. It identifies some of the important concepts that should be understood upon completion of the chapter.

The **Chapter Outline** is a synopsis of the text chapter. The format follows that of the textbook. In each division and subdivision of a chapter, the most important and succinct sentences have been taken directly from the text. The Chapter Outline should be used as a list of important entries that must be thoroughly understood to have a good working knowledge of the material in the chapter. It should be remembered that this is an outline. Additional explanation and examples are found in the textbook.

The **Key Terms** section is a reiteration of the important ideas, concepts, and structures in a one-word format. This should be used for learning by association. A definition for each key term should be prepared and the role of the term, in the physiology depicted in the chapter, should be duly noted.

The questions found in the **Review Exercises** are not intended to indicate the manner in which a student will be tested on particular information. Between the textbook and the *Study Guide*, most of the material has been presented in test form. This is another method to check for an accurate understanding of the information presented in the chapter.

The information found under **Points to Ponder** may or may not be answered with the knowledge gained from the current chapter. Each entry is designed to stimulate the use of certain material presented in the chapter. In some instances additional information may be required to answer the questions, whereas, in others it may involve a social or cultural perspective. Each Point is included as a possible application of the new material. These Points to Ponder are just that: something to think about.

In **Clinical Perspectives**, the application of the new material must be employed. Unlike Points to Ponder, these questions can be answered with the material gained from the chapter. Usually, these examples are common applications of the physiology under consideration.

The **Experiment of the Day** section utilizes another method for learning a concept or phenomenon. While most of the science learned today is the result of experimentation, it is not always easy to understand the knowledge gained from such tests. Frequently, a simple activity can aid in the learning process. Although these activities do not qualify as experiments in the purest sense of the word, they encourage the use of curiosity to help understand the physiological material.

Study Guide

Fundamentals of Human Physiology

FOURTH EDITION

Lauralee Sherwood
West Virginia University

Prepared by

Mark A. Paternostro
West Virginia University

BROOKS/COLE
CENGAGE Learning

Australia • Brazil • Japan • Korea • Mexico • Singapore • Spain • United Kingdom • United States

For product information and technology assistance, contact us at **Cengage Learning Customer & Sales Support, 1-800-354-9706**

For permission to use material from this text or product, submit all requests online at **www.cengage.com/permissions** Further permissions questions can be emailed to **permissionrequest@cengage.com**

ISBN-13: 978-1-111-42786-3
ISBN-10: 1-111-42786-0

Brooks/Cole
20 Davis Drive
Belmont, CA 94002-3098
USA

Cengage Learning is a leading provider of customized learning solutions with office locations around the globe, including Singapore, the United Kingdom, Australia, Mexico, Brazil, and Japan. Locate your local office at: **www.cengage.com/global**

Cengage Learning products are represented in Canada by Nelson Education, Ltd.

To learn more about Brooks/Cole, visit **www.cengage.com/brookscole**

Purchase any of our products at your local college store or at our preferred online store **www.cengagebrain.com**

Printed in the United States of America
1 2 3 4 5 6 7 14 13 12 11 10

Contents

HOMEOSTATIC CONTROL SYSTEMS
Homeostatic control systems may operate locally or body wide.

- Control systems that operate to maintain homeostasis can be grouped into two classes: intrinsic and extrinsic controls.
- Intrinsic, or local, controls are built in or inherent to an organ.
- Extrinsic, or systemic, controls are regulatory mechanisms initiated outside an organ. Most factors are maintained by extrinsic controls.
- Extrinsic control of the various organs and systems is accomplished by the nervous and endocrine systems.

Negative feedback opposes an initial change and is widely used to maintain homeostasis.

- The body's homeostatic control mechanisms primarily operate on the principle of negative feedback.

- Negative feedback occurs when a change in a controlled variable triggers a response that opposes the change, driving the variable in the opposite direction of the initial change.

Positive feedback amplifies an initial change.

- Positive feedback occurs less frequently in the body.
- With positive feedback, the output is continually enhanced so that the controlled variable continues to be moved in the direction of the initial change.

Disruptions in homeostasis can lead to illness and death.

- When one or more of the body's systems fail to function properly, homeostasis is disrupted, and all cells suffer.
- Various pathophysiological states ensue.

Key Terms

Anatomy
Body system
Cell
Cell differentiation
Connective tissue
Effector
Endocrine glands
Epithelial tissue
Exocrine glands
Extracellular fluid
Extrinsic controls
Feedback
Glands
Homeostasis
Homeostatic control system
Internal environment
Interstitial fluid

Intracellular fluid
Intrinsic (local) controls
Extrinsic (systemic) controls
Lumen
Muscle tissue
Negative feedback
Nervous tissue
Organisms
Organs
Pathophysiology
Physiology
Plasma
Positive feedback
Secretion
Set point
Tissues

Review Exercises
Answers are in the appendix.

True/False-if an answer is false, underline the incorrect portion of the statement and rewrite it to make the statement true.

_____ 1. Tissues perform functions that are specializations of the normal function of individual cells.

_____ 2. Homeostasis maintains body systems to enhance the survival of cells.

_____ 3. Each body system is composed of different organs.

_____ 4. The endocrine system is a major control system.

_____ 5. The nervous system is an example of intrinsic control.

_____ 6. Positive feedback mechanisms are frequently used in the reproductive system.

_____ 7. All definitive actions initiated by nervous tissue are either muscle contractions or glandular secretions.

_____ 8. When physiologists study functions of the body they focus on the purpose of the function rather than emphasizing why it happens.

_____ 9. Cells require homeostasis, body systems maintain homeostasis, and cells make up body systems.

_____ 10. With positive feedback, a control system's input and output continue to enhance each other.

_____ 11. Cells within the body make life-sustaining exchanges directly with the external environment in which we live.

_____ 12. Glands are formed during embryonic development by pockets of epithelial tissue that dip inward from the surface.

_____ 13. Most human cells can be cultured; that is, when removed from the body, they will continue to thrive and reproduce in laboratory dishes when supplied with appropriate nutrients and other supportive materials.

_____ 14. Endocrine glands have ducts.

_____ 15. Control systems that operate to maintain homeostasis can be grouped into two classes— intrinsic and extrinsic.

_____ 16. Humans are multi-cellular organisms.

_____ 17. Negative feedback principles underlie the concept of homeostasis.

_____ 18. Connective tissue is distinguished by having relatively few cells dispersed within an abundance of extra-cellular material.

_____ 19. The uterus is in the pelvic cavity.

_____ 20. In the human body, there are five primary tissue types.

_____ 21. Defense against microorganisms is one of the functions of the immune system.

_____ 22. The nervous and endocrine systems play a major role in integration and coordination within the human body.

_____ 23. There are four major types of muscle tissue found within the human body.

_____ 24. The integumentary system plays a role in protecting the body.

_____ 25. Connective tissue may also be called "supporting tissue."

Fill in the Blank

26. _____ glands have ducts.

27. A(n) _____ is the cavity within the interior of hollow organs and tubes.

28. Intrinsic control mechanisms are also known as _____ controls.

29. _____ is essential for the survival of cells in the body.

30. In addition to plasma _____ fluid is part of the extracellular fluid.

31. The _____ system transfers water and electrolytes from the external environment into the internal environment.

32. _____ is a common regulatory mechanism for maintaining homeostasis.

33. The smallest unit capable of carrying out the processes associated with life is the _____.

34. The body cells are in direct contact with, and make life-sustaining exchanges with, the _____.

35. The _____ is the portion of the internal environment that surrounds and bathes all cells.

36. _____ exists when a change in a regulated variable triggers a response that opposes the change.

37. _____ refers to the abnormal functioning of the body associated with disease.

38. The basic unit of both structure and function is the _____.

39. In addition to the nervous system, this system is essential for communication within the body:_____.

40. _____ are epithelial-tissue derivatives that are specialized for secretion.

41. Organs are further organized into _____, each of which is a collection of organs that perform related functions and interact to accomplish a common activity that is essential for survival of the whole body.

42. There are eleven major body systems. List them: (1) _____,
    (2) _____ (3) _____, (4) _____,
    (5) _____, (6) _____, (7) _____,
    (8) _____, (9) _____, (10) _____,
    (11) _____.

43. _____ is the study of all of the vital functions within the human body.

44. The fluid outside of the body cells is called _____ fluid.

45. When an initial stimulus is reinforced, this is a good example of _____ feedback.

46. The internal and external forces that create imbalances within the body are called
    _____.

47. _____ tissue may also be called supporting tissue.

48. The lymphatic system is made up of many small vessels that carry _____.

49. The fluid inside the cell is called _____.

50. The tonsils are part of the _____ system.

Matching
Match the following organ or body part to the correct body system.

_____ 51. Skin   a. Reproductive

_____ 52. Joints   b. Circulatory system

_____ 53. Trachea   c. Digestive system

_____ 54. Thyroid   d. Immune system

_____ 55. Special sense organs   e. Integumentary system

_____ 56. Uterus   f. Muscular system

_____ 57. Tonsils   g. Skeletal system

_____ 58. Skeletal muscles   h. Endocrine system

_____ 59. Blood   i. Urinary system

_____ 60. Urethra   j. Nervous system

_____ 61. Liver   k. Respiratory system

Match the system to the function.

62. _____ Circulatory system   a. Transports food, vitamins, minerals, and other

                                               nutrients

63. _____ Respiratory system   b. The chemical and mechanical breakdown of food

64. _____ Digestive system   c. Responsible for continuation of the human race

65. _____ Urinary system   d. Excretes various wastes

66. _____ Reproductive system   e. Exchange of oxygen and carbon dioxide

Multiple Choice

67. These cells, when destroyed by trauma or disease, cannot be replaced.
    a. glandular tissue
    b. epithelial tissue
    c. muscles and nerves
    d. connective tissue
    e. none of the above

68. Which system is important in maintaining the proper pH of the internal environment?
    a. circulatory
    b. digestive
    c. muscular
    d. respiratory
    e. immune

69. This tissue lines the stomach.
    a. connective
    b. muscle
    c. epithelial
    d. nervous
    e. mesodermal

70. Which type of control is exemplified by an air conditioner, its thermostatic device, and their electrical connections which function together to keep a room at a constant temperature?
    a. positive feedback
    b. feed-forward
    c. biofeedback
    d. cybernetic
    e. none of the above

71. Which muscle type moves the contents of food through the intestines?
    a. cardiac
    b. skeletal
    c. smooth
    d. extrinsic

72. Physiologist use which approach when studying the body? a. a microscopic approach

b. why something happens

c. how something happens

d. all of the above

e. none of the above

73. Which is not a basic cell function?
    a. Maintain a storage reservoir for potassium, which will maintain a relatively constant concentration of potassium in the extra-cellular fluid.
    b. Perform various chemical reactions that use nutrients and oxygen to provide energy for the cell.
    c. Obtain food and oxygen from the environment surrounding the cell.
    d. Be sensitive and responsive to changes in the environment surrounding the cell.
    e. Control to a large extent the exchange of materials between the cell and its surrounding environment.

74. Which type of gland loses its cellular connection during fetal development?
    a. holocrine
    b. endocrine
    c. exocrine
    d. mesocrine
    e. endoderm

75. Which type of tissue utilizes the inherent capability of cells to produce intracellular movement?
    a. epithelial
    b. muscle
    c. nervous
    d. connective
    e. endoderm

76. Which type of tissue utilizes the inherent capability to synthesize digestive enzymes?
    a. epithelial
    b. muscle
    c. nervous
    d. connective
    e. glandular

77. During the minute that it will take to answer this question, all but one of these activities will occur.
    a. Your body will use about 2 calories of energy.
    b. Your body will exchange 6 liters of air between the atmosphere and your lungs.
    c. More than 10 liters of blood will flow through your kidneys.
    d. Your heart will beat 70 times, pumping 5 liters of blood to your lungs and another 5 liters to the rest of your body.
    e. Your brain will process 325 afferent impulses.

78. Which of the following best describes embryonic stem cells?
    a. the "mother cells" resulting from the early division of the egg
    b. all of the cells in the embryo
    c. the last cells to develop in an embryo
    d. the implantation of cells into embryos

79. Factors of an internal environment that must be homeostatically maintained include:

a. its pH and temperature
b. its concentration of water, salt, and other electrolytes
c. its volume and pressure
d. all of the above
e. none of the above

80. Which of the following is a component of the immune system?
a. appendix
b. adrenals
c. bulbourethral glands
d. hypothalamus
e. none of the above

81. Which system controls and coordinates bodily activities that require swift responses?
a. skeletal system
b. muscular system
c. reproductive system
d. nervous system
e. endocrine system

82. What feedback continually enhances the output so that the controlled variable continues to be moved in the direction of the initial change?
a. negative
b. positive
c. feed-forward
d. biofeedback
e. cybernetic

83. What controls are built in or inherent to an organ?
a. intrinsic
b. extrinsic
c. endocrine
d. exocrine
e. holocrine

84. What controls are regulatory mechanisms initiated outside of an organ to alter the activity of the organ?
a. intrinsic
b. extrinsic
c. mesocrine
d. epicrine
e. holocrine

85. What controls use the nervous and endocrine systems to control various organs and systems?
a. intrinsic
b. extrinsic
c. mesocrine
d. holocrine
e. exocrine

86. The medical branch that studies the structure of the human body is called:
    a. physiology
    b. organology
    c. histology
    d. cytology
    e. anatomy

87. The thoracic cavity contains:
    a. the heart and lungs
    b. the testes, ovaries, and other reproductive organs
    c. just the lungs
    d. the diaphragm and heart
    e. all of the above

88. The smallest functional unit in the human body from a physiologic perspective is the:
    a. nucleus
    b. atom
    c. cell
    d. tissue
    e. organ

89. When blood glucose levels fall below normal, the body will respond by increasing blood glucose levels. This is an example of:
    a. positive feedback only
    b. negative feedback only
    c. exocrine action
    d. both a and b only
    e. both a and c only

90. Which of the following systems does not play a role in protecting the human body?
    a. integumentary
    b. lymphatic
    c. circulatory
    d. digestive
    e. nervous

91. All of the following are examples of membranes that line body cavities except:
    a. mucous
    b. serous
    c. synovial
    d. pleura
    e. viscera

92. Which of the following is not a type of connective tissue?
    a. bone
    b. cartilage
    c. blood
    d. tendon
    e. muscle

93. Which of the following is a major transport system in the human body?
    a. muscular
    b. nervous
    c. circulatory
    d. integumentary
    e. urinary

94. Which of the following systems plays a role in integration and coordination?
    a. nervous
    b. skeletal
    c. urinary
    d. respiratory
    e. integumentary

95. Which of the following systems sends electrical signals?
    a. nervous
    b. respiratory
    c. circulatory
    d. muscular
    e. digestive

96. Bone is considered which type of tissue?
    a. connective
    b. epithelial
    c. both a and b
    d. none of the above

97. All glands are composed of which of the following tissue types?
    a. connective
    b. muscular
    c. epithelial
    d. a and b
    e. b and c

98. How many body systems are present in your body?
    a. five
    b. six
    c. seven
    d. ten
    e. eleven

99. In your body, you would find a lot of elastin in:
    a. connective tissue
    b. muscle tissue
    c. nervous tissue
    d. skeletal tissue
    e. all of the above

100. The respiratory system contains all of the following except:
    a. nose
    b. pharynx
    c. larynx
    d. bronchi
    e. hair

Points to Ponder

1. Why does the heart rate of a well-trained athlete begin to increase a few moments before the competitive event occurs?

2. As you sit in class you start yawning. Looking around, you see other students doing the same. What do you think is happening, aside from the obvious?

3. How would you rate homeostasis as a diagnostic tool?

4. After studying this chapter on homeostasis, why do you think it is important for people to have thorough examinations by their physicians?

5. How are the different structural levels of the human body related?

6. If a bullet entered the right side of the body just below the armpit and exited at the same point on the other side of the body, which cavities would be pierced (in order), and which organs would be probably damaged?

7. Using insulin as an example, explain how the secretion of this hormone is controlled by the effects of insulin's action in various places in the body.

8. What are some reasons artificial organs are so difficult to develop?

9. What does exercise physiology tell us about normal body functions?

10. Do illness and aging involve the same pathophysiological process? Explain your answer.

11. Why are strokes so damaging to the human body?

12. What are some similarities and differences between endocrine and exocrine glands?

13. How does each cell in the body contribute to homeostasis?

14. What are some of the major hurdles to the use of embryonic stem cells?

15. What are some pros and cons with respect to tissue engineering?

Clinical Perspectives

1. What does your heart rate (in beats per minute) indicate when you are resting versus when you are exercising?

2. Why do health personnel take one's temperature? What does this tell you about normal body function?

3. Why is positive feedback necessary for childbirth to occur? What mechanisms, hormones, and systems are involved?

4. What is one of the first effects of an exercise program on the heart?

Cell Physiology

Chapter Overview

The basic processes associated with life require cellular functions at the molecular level. These molecular activities are performed by various organelles of the cell. Some organelles such as mitochondria and lysosomes act directly in the basic life processes, while cytoskeletal structures, polyribosomal clusters, and others function in areas of structural integrity, maintenance, and movement. The molecular activity of the organelles and other cellular components is essential for the proper functioning of tissues, organs, and body systems in the maintenance of homeostasis. Energy produced in organelles drives all the body systems. The role of a body system in homeostasis is inseparable from the basic molecular functions of its cells.

Chapter Outline

INTRODUCTION
- Cells are the bridge between molecules and humans.
- Even though the chemicals comprising a cell have been thoroughly analyzed, researchers have not yet organized these chemicals into a cell in the laboratory.
- Cells are the living building blocks for the whole body.
- All body functions depend on the activities of individual cells.

OBSERVATION OF CELLS
- Human cells average about 10 to 20 micrometers in diameter. They are too small to be seen by the unaided eye.
- Lined up side by side, about 100 average-sized cells would occupy a distance of only 1 millimeter.
- Cells were unknown until the middle of the seventeenth century.
- The electron microscope has revealed the great diversity and complexity of the, internal structure of the cell.
- Through the availability of better microscopy, biochemical techniques, cell-culture technology, and genetic engineering, the cell has emerged as a complex, highly organized, compartmentalized structure.

AN OVERVIEW OF CELL STRUCTURE
- The trillions of cells in a human body are classified into about 200 different types based on specific variations in structure and function
- Most cells are subdivided into the plasma membrane, nucleus, and cytoplasm.

The plasma membrane bounds the cell.
- The thin membranous structure that encloses the cell is the plasma membrane, or cell membrane, which serves as the boundary of the cell.
- The plasma membrane is composed mostly of lipid molecules and studded with proteins.
- The fluid within the cell is the "intracellular fluid" (ICF), while fluids outside the cell are referred to as "extra-cellular fluid" (ECF).

- The plasma membrane selectively controls the movement of molecules between the ICF and ECF.

The nucleus contains the DNA.
- The nucleus, surrounded by the double-layered nuclear membrane (the nuclear envelope) is the largest single organized cellular component.
- Deoxyribonucleic acid (DNA) is found in the nucleus.
- DNA controls protein synthesis.
- Since most cellular activities involve structural or enzymatic proteins, DNA provides the instructions and thus the control of these activities.
- As a genetic blueprint DNA ensures cellular reproduction and organismal reproduction.
- Three types of ribonucleic acid (RNA) play a role in protein synthesis.
- The DNA's genetic code for a particular protein is transcribed into a messenger RNA molecule.
- Within the cytoplasm, messenger RNA (mRNA) delivers the coded message to ribosomal RNA.
- Ribosomal RNA (rRNA) "reads" the code and translates it into the appropriate amino acid sequence.
- Transfer RNA (tRNA) transfer the appropriate amino acid within the cytoplasm to their designated site in the protein under construction.

The cytoplasm consists of various organelles and the cytosol.
- That part of the cell outside the nucleus is called the cytoplasm and is composed of a gel-like mass known as "cytosol." Organelles are distinct, highly organized structures that perform specialized functions within the cell.
- There are six structures or organelles dispersed in the cytosol: endoplasmic reticulum, Golgi complex, lysosomes, peroxisomes, mitochondria, and vaults.
- The cytoskeleton is an interconnected system of protein fibers and tubes that extends throughout the cytosol.

- The cytoskeleton provides shape to the cell, internal organization, and internal movement.
- The portion of the cytoplasm not occupied by organelles and cytoskeleton consists of a semi-liquid gel-like mass known as the cytosol.

ENDOPLASMIC RETICULUM AND SEGREGATED SYNTHESIS
- There are two types of endoplasmic reticulum (ER): smooth and rough.
- The ER is one continuous membranous organelle with many interconnected channels.
- The rough ER derives its name from dark-staining particles, ribosomes, which are attached to the ER membrane.
- Ribosomes are composed of ribosomal RNA-protein complexes.

The rough endoplasmic reticulum synthesizes proteins for secretion and membrane construction.
- The rough ER and its associated ribosome synthesize proteins and release these products into the ER lumen.
- Some of these new proteins are destined for export to the exterior of the cell as a secretion.
- Other new proteins are used within the cell as structural proteins.

The smooth endoplasmic reticulum packages new proteins in transport vesicles.
- Newly synthesized proteins are passed to the smooth ER where they "bud off" and become membrane enclosed transport vesicles.
- These transport vesicles move to the Golgi complex for further processing.
- Smooth and rough ER membranes contain enzymes for lipid synthesis.
- In liver cells the smooth ER has detoxifying enzymes.
- In muscles the smooth ER becomes modified into the sarcoplasmic reticulum and stores calcium.

GOLGI COMPLEX AND EXOCYTOSIS
- The Golgi complex is a refining plant and directs molecular traffic.
- The Golgi complex consists of membranous, flattened, slightly curved sacs stacked in layers called "Golgi stacks."

Transport vesicles carry their cargo to the Golgi complex for further processing.
- Transport vesicles enter the Golgi stacks at a point nearest the center of the cell.
- The Golgi complex performs two important interrelated functions: processing the raw materials into finished products, and sorting and directing the finished products to their final destination as coated vesicles or secretory vesicles.

The Golgi complex packages secretory vesicles for release by exocytosis.
- Vesicles with their selected cargo destined for different sites are wrapped in membranes containing different surface proteins. .
- These surface proteins allow the vesicles to "dock" and unload their cargo at the proper location within the cell. Once a vesicle has docked at its target it opens and empties its contentSecretory vesicles release their contents out of the cell in a process called exocytosis.

LYSOSOMES AND ENDOCYTOSIS
Lysosomes serve as the intracellular digestive system.
- Lysosomes are membrane-enclosed sacs containing 40 different powerful hydrolytic enzymes.

Extracellular material is brought into the cell by phagocytosis for attack by lysosomal enzymes.
- In lysosomes the organic molecules are cell debris and foreign material that have been brought into the cell Endocytosis can be accomplished in one of two ways: pinocytosis and phagocytosis.

Pinocytosis.
- The nonselective uptake of a droplet of extracellular fluids is pinocytosis. The plasma membrane invaginates, forming a pouch that contains a small bit of extracellular fluid.
- The plasma membrane then seals at the surface of the pouch, forming a small, intracellular membrane-enclosed vesicle.

Receptor-mediated endocytosis.
- Receptor-mediated endocytosis is a highly selective process.
- This process is triggered by the binding of a specific protein to a surface membrane receptor site.
- The plasma membrane then seals at the surface of the pouch, forming a small, intracellular membrane-enclosed vesicle.

Phagocytosis.
- White blood cells use a special form of endocytosis called "phagocytosis" to trap multimolecular particles.
- Lysosomes fuse to the internalized vesicle and release their hydrolytic enzymes into the vesicle.
- Lysosomes become residual bodies upon completion of their digestive activities and are eliminated by exocytosis.

Lysosomes remove useless but not useful parts of the cell.
- Lysosomes can fuse with aged or damaged organelles to remove them from the cell.
- Serious diseases may occur if an enzyme is missing in the lysosome.
- The result is massive accumulation within the lysosomes of the specific compound that is normally digested by the missing enzyme.
- Among these so-called storage diseases is Tay-Sachs disease.
- It is characterized by abnormal accumulations of molecules which bring about progressive nervous system degeneration.

PEROXISOMES
Peroxisomes house oxidative enzymes that detoxify various wastes.

- **Peroxisomes** are membranous organelles that produce and decompose hydrogen peroxide in the process of degrading potentially toxic molecules
- About one-third to one-half the size of lysosomes, peroxisomes are membrane-bound sacs containing oxidative enzymes that strip hydrogen from toxic wastes.
- Peroxisomes form hydrogen peroxide (H_2O_2) from molecular oxygen and the hydrogen atoms stripped from the toxic molecule.
- Peroxisomes also contain catalase, an antioxidant, which breaks down hydrogen peroxide to water and molecular oxygen.

MITOCHONDRIA AND ATP PRODUCTION
Mitochondria, the energy organelles, are enclosed by a double membrane.

- The rod-shaped mitochondria are the energy organelles of the cell.
- Mitochondria have smooth outer membranes and inner membranes that have infoldings, or cristae, into the gel-like matrix.
- Cristae are the site of the electron transport chain activity.
- The citric acid cycle occurs in the matrix.

Mitochondria play a major role in generating ATP.

- To obtain immediate usable energy, cells split the terminal phosphate bond of adenosine triphosphate (ATP), which yields adenosine diphosphate (ADP) plus inorganic phosphate plus energy.
- Cellular respiration refers collectively to the intracellular reactions in which energy-rich molecules are broken down to form ATP.
- Cellular respiration involves three separate stages: glycolysis, the citric acid cycle, and oxidative phosphorylation.

Glycolysis.

- Glucose from the digestive system via the blood is transported across the plasma membrane into the cytosol, where it is converted to pyruvate and two molecules of ATP and NADH are produced. .

Citric acid cycle.

- Pyruvate is converted to acetyl coenzyme A in the matrix of the mitochondria. Acetyl coenzyme A joins oxaloacetic acid in the citric acid cycle, which produces two molecules of carbon dioxide, oxaloacetic acid, and one molecule of ATP. Hydrogen is removed using carrier molecules nicotinamide adenine dinucleotide (NAD) and flavine-adenine dinucleotide (FAD).
- One molecule of glucose produces two acetic acid molecules and two turns of the citric acid cycle yielding two molecules of ATP.

Oxidative Phosphorylation

- Oxidative phosphorylation refers to the process by which ATP is synthesized using the energy released by electrons as they are transferred to O_2.
- The process involves two groups of proteins located at the inner mitochondrial membrane: the electron transport chain system and ATP synthase.
- The hydrogen carrier molecules enter the electron transport chain, which produces far more energy than the citric acid cycle itself.
- As electrons move through the chain of reactions they release energy that is used to synthesize ATP.
- These reactions produce 28 molecules of ATP
- At three sites in the electron transport chain ATP is produced using oxygen in a process known as "oxidative phosphorylation."

- The total number of ATP produced for each glucose utilized is 32 (2 during glycolysis, 2 during citric acid cycle reactions and 28 through oxidative phosphorylation.

The cell generates more energy in aerobic than in anaerobic conditions.
- Glycolysis is anaerobic.
- The citric acid cycle and the electron transport chain are aerobic.
- An oxygen insufficiency will not impede glycolysis thereby producing excessive pyruvic acid.
- Excess pyruvic acid is converted to lactic acid through a reversible reaction.

The energy stored within ATP is used for synthesis, transport, and mechanical work.
- Cellular activities that require energy expenditure fall into three main categories:
(1) synthesis of new compounds such as protein synthesis by the ER,
(2) membrane transport such as selective transport in the kidney, and
(3) mechanical work such as muscular contraction.

CYTOSOL: CELL GEL
The cytosol is important in intermediary metabolism, ribosomal protein synthesis, and nutrient storage.
- Intermediary metabolism refers to the large set of intracellular chemical reactions that involve degradation, synthesis, and transformation of small organic molecules such as simple sugars, amino acids, and fatty acids.

Ribosome protein synthesis.
- Throughout the cytosol, free ribosomes are dispersed that synthesize proteins for use in the cytosol.

Storage of fat, glycogen, and secretory vesicles.
- Glycogen and fat are stored in the cytosol as inclusions.

CYTOSKELETON: CELL "BONE AND MUSCLE"
- The cytoskeleton is found in the cytosol and is composed of three distinct elements: microtubules, microfilaments, intermediate filaments

Microtubules help maintain asymmetrical cell shapes and play a role in complex cell movements.
- Microtubules are long, slender tubes composed of the small globular protein, tubulin.
- Microtubules are the dominant structure and functional components of cilia and flagella.
- Cilia are numerous tiny protrusions.
- Flagella are long whiplike structures.
- Both have microtubules arranged in a ring of nine pairs or doublets, with one pair in the center.
- Molecular motors are transporters. These proteins attach to the particle to be transported and move along the microtubule
- Dynein and kinesin are two such proteins. Microtubules are also important structural components of cilia and flagella.
- Humans have cilia on the cells lining the trachea, with cilia also lining the oviducts in females.
- Sperm are the only human cells that contain flagella.

Microfilaments are important to cellular contractile systems and as mechanical stiffeners.
- The major events associated with cell division, mitosis and cytokinesis depend upon the action of microfilaments.
- Microfilaments are composed of actin in most cells and also myosin in muscle cells.
- Actin microfilaments are two twisted strands of the globular protein actin.

Microfilaments in cellular contractile systems.

- In muscles the contractile force is generated by actin microfilaments sliding past myosin microfilaments.
- Actin microfilaments provide the constricting force to separate the cytoplasm into two daughter cells in cytokinesis
- Actin microfilaments are used to produce amoebid movement necessary for locomotion of fibroblasts, skin cells, and white blood cells.

Microfilaments as mechanical stiffeners.

- Microfilaments serve as mechanical stiffeners in the microvilli, which line the free edge of epithelial cells lining the intestine and kidney tubules.
- These microvilli increase the absorptive surface area.
- The hair cells of the inner ear are specialized microvilli.

Intermediate filaments are important in cell regions subject to mechanical stress.

- These filaments derive their name by being between microtubules and microfilaments in size.
- Intermediate filaments are found as neurofilaments in axons and provide strength to these elongated structures.
- In skeletal muscle, intermediate filaments hold actin and myosin contractile units in proper alignment.
- Intermediate filaments, made of the protein keratin, form irregular networks that connect to extracellular filaments to strengthen and waterproof the skin.

The cytoskeleton functions as an integrated whole and links other parts of the cell together.

- This lattice of exceedingly fine filaments serves to suspend organelles, microtubules, and free ribosomes.
- These filaments extend from the inner layer of the plasma membrane throughout the cytoplasm.

Key Terms

Actin	Flagellum
Adenosine diphosphate (ADP)	Glycogen
Adenosine triphosphate (ATP)	Glycolysis
Adipose tissue	Golgi complex
Aerobic	Hydrolytic enzymes
Amoeboid movement	Inclusions
Anaerobic	Intermediary metabolism
ATP synthase	Intermediate filaments
Cell membrane	Keratin
Cellular respiration	Kinesin
Cilia	Krebs cycle
Citric-acid cycle	Lysosomes
Cytoplasm	Matrix
Cytokinesis	Messenger RNA
Cytoskeleton	Microfilaments
Cytosol	Microtubules
Deoxynucleic acid (DNA)	Microvilli
Deoxyribonucleic	Mitochondria
Endocytosis	Myosin
Endocytic vesicle	Nicotinamide adenine dinucleotide (NAD)
Endoplasmic reticulum (ER)	Nucleus
Exocytosis	Organelle

Oxidative enzymes
Oxidative phosphorylation
Peroxisomes
Phagocytosis
Pinocytosis
Plasma membrane
Pseudopods
Receptor-mediated endocytosis
Ribonucleic acid (RNA)
Ribosomal RNA
Ribosomes

Rough ER
Secretion
Secretory vesicles
Smooth ER
Tay-Sacs disease
Transfer RNA
Transport vesicles
Tubulin
Vesicle

Review Exercises

Answers are in the appendix.

True/False- if an answer is false, underline the incorrect portion of the statement and rewrite it to make the statement true.

_____ 1. Tay-Sachs is a disease associated with abnormal mitochondrial function.

_____ 2. Glycolysis reactions are considered anaerobic and produce 2 ATP per glucose utilized.

_____ 3. Ketoglutaric acid is part of the citric acid cycle.

_____ 4. Anaerobic reactions are associated with oxidative phosphorylation.

_____ 5. The chemiosmotic mechanism is another name for the tricarboxylic acid cycle.

_____ 6. Carbon dioxide produced by the Krebs cycle is released into the atmosphere.

_____ 7. Dynein is an example of a molecular motor protein.

_____ 8. "Life" is due to the complex organization and interaction of chemical molecules within the cells.

_____ 9. The cells of the human body are surrounded by a cell wall.

_____ 10. All body functions ultimately depend on the activities of the individual cells that compose the body.

_____ 11. About 100 average-sized cells lined up side by side would stretch a distance of only 1 mm.

_____ 12. The plasma membrane does not merely serve as a mechanical barrier to hold in the contents of the cell; it has the ability to selectively control movement of molecules between the ICF and ECF.

_____ 13. The nucleus is the largest multicellular component usually located near the center of the cell.

_____ 14. Within the nucleus is the cell's genetic material, RNA.

_____ 15. Organelles are like intracellular "specialty shops."

_____ 16. Protein synthesis is carried out by lysosomes.

_____ 17. All cells contain the same number of mitochondria.

_____ 18. The smooth ER contains ribosomes.

_____ 19. In liver cells, the smooth ER has a special capability. It contains enzymes that are involved in detoxifying harmful substances produced within the body by metabolism, or substances that enter the body from the outside in the form of drugs or other foreign compounds.

_____ 20. The Golgi complex consists of sets of flattened, slightly curved membrane-bound sacs, or cisternae, stacked in layers.

_____ 21. Cell respiration refers to the collection of reactions in which nutrient molecules are broken down to produce ATP.

_____ 22. Release of the contents of the secretory vesicle constitutes the process of secretion.

_____ 23. Residual bodies are lysosomes that have completed their digestive activities.

_____ 24. Each mitochondrion is enclosed by a single membrane.

_____ 25. Energy derived from food is stored in the mitochondria.

_____ 26. In glycolysis, one molecule of glucose has a net yield of only 32 molecules of ATP.

_____ 27. During citric acid cycle reactions, the pyruvate produced by glycolysis in the cytosol is selectively transported into the mitochondrial matrix.

_____ 28. The cytosol is the semiliquid portion of the cytoplasm that surrounds the organelles.

_____ 29. The endoplasmic reticulum is one continuous organelle consisting of many tubules and cisternae.

_____ 30. Coated vesicles enclose a representative mixture of proteins present in the Golgi sac before budding off.

_____ 31. Sometimes healthy organelles and/or cells are attacked by lysosomal enzymes.

_____ 32. Mitochondrial reactions are both anaerobic and aerobic.

_____ 33. Centrioles and basal bodies are identical in structure and may be interconvertible.

_____ 34. The protective, waterproof outer layer of skin is formed by the tough skeleton of the microtrabecular lattice, which persists after the surface skin cells die.

_____ 35. Each ribosome can synthesize only one specific type of protein.

_____ 36. Only one ribosome can translate the mRNA message at any given time.

Fill in the Blank

37. The _____ "bud off" from the smooth ER.

38. The _____ appears to suspend microtubules and microfilaments, as well as various organelles.

39. The _____ occupies about 55% of the total cell volume.

40. _____ is a transport molecule.

41. NAD and FAD function to carry _____.

42. Oxidative enzymes are found in the _____.

43. A cell is made up of three major subdivisions: (1) _____, (2) _____, and (3) _____.

44. The cytosol is pervaded by a protein scaffolding called _____, which serves as the "bone and muscle" of the cell.

45. The ribosomes of the rough ER synthesize _____, whereas its membranous walls contain enzymes essential for the synthesis of _____.

46. Endocytosis can be accomplished in three ways—_____, _____ and _____.

47. The _____ ER is the central packaging and discharge site for molecules to be transported from the ER.

48. The process in which a secretory vesicle fuses with the plasma membrane, then opens up and extrudes its contents to the exterior, is known as _____.

49. Lysosomes contain (what type of) _____ enzymes.

50. Foreign material to be attacked by lysosomal enzymes is brought into the cell by the process of _____.

51. _____, an enzyme found in peroxisomes, decomposes potentially toxic hydrogen peroxide.

52. The nucleus of the cell contains _____, the cell's genetic material.

53. _____ is the enzyme associated with the production of ATP during oxidative phosphorylation.

54. _____ help maintain asymmetric cell shapes.

Matching

Match the component or reaction to the organelle or structure in which it occurs.

_____ 55. Chemiosmosis	a. cytosol
_____ 56. Catalase	b. Golgi complex
_____ 57. Microtubules	c. microvilli
_____ 58. Glycolysis	d. mitochondrial matrix
_____ 59. Hydrolytic enzymes	e. ribosome
_____ 60.. mRNA formation	f. pseudopoda
_____ 61. Secretory vesicles	g. mitochondrial inner membrane
_____ 62. Microfilaments	h. nucleus
_____ 63. Plasmasol	i. lysosomes
_____ 64. protein synthesis	j. flagella
_____ 65. Citric acid cycle	k. peroxisomes

Multiple Choice

66. Which of the following is involved in cytokinesis?
    a. clathrin
    b. actin
    c. kinesin
    d. malic acid
    e. dynein

67. Which is the storage form of glucose?
    a. citric acid
    b. galactose
    c. myosin
    d. guanosine triphosphate
    e. glycogen

68. Tay-Sachs disease is a storage disease involving dysfunction of which organelle?.
    a. lysosome
    b. peroxisome
    c. nucleus
    d. ribosome
    e. mitochondria

69. Which is the correct order for the citric acid cycle?
    a. oxaloacetic acid—citric acid—isocitric acid—ketoglutaric acid—fumaric acid—succinyl CoA—succinic acid—malic acid
    b. oxaloacetic acid—citric acid—isocitric acid—ketoglutaric acid—succinyl CoA—succinic acid—fumaric acid—malic acid
    c. citric acid—isocitric acid—oxaloacetic acid—ketoglutaric acid—succinyl acid—succinic acid—malic acid—fumaric acid
    d. oxaloacetic acid—isocitric acid—citric acid—succinyl CoA—succinic acid—ketoglutaric acid—fumaric acid—malic acid
    e. oxaloacetic acid—citric acid—isocitric acid—ketoglutaric acid—succinic acid—succinyl CoA—fumaric acid—malic acid

70. Which compound is used as mechanical stiffeners in microvilli?
    a. actin
    b. kinesin
    c. dynein
    d. myosin
    e. clathrin

71. Which is composed of RNA and proteins?
    a. chromosomes
    b. peroxisomes
    c. lysosomes
    d. ribosomes
    e. cortisone

72. Which of the following functions as the modifier, packager, and distributor for newly synthesized proteins?
    a. nucleus
    b. Golgi complex
    c. secretory vesicles
    d. endoplasmic reticulum
    e. mitochondria

73. Which of the following serves as the cell's "bone and muscle?"
    a. cytoskeleton
    b. Golgi complex
    c. residual bodies
    d. free ribosomes
    e. mitochondria

74. Which component holds the actin-myosin units in proper alignment in skeletal muscle?
    a. microtubules
    b. microfilaments
    c. microtrabecular lattice
    d. spindle fibers
    e. intermediate filaments

75. Dynein is found in which of the following?
    a. cilia and flagella
    b. cilia and mitotic spindles
    c. flagella and centrioles
    d. mitotic spindles and flagella
    e. none of the above

76. Which of the following processes acetyl CoA?
    a. electron transport chain
    b. citric acid cycle
    c. glycolysis
    d. respiratory chain
    e. fermentation

77. Which of the following converts glucose into two pyruvate molecules?
    a. electron transport chain
    b. glycolysis
    c. citric acid cycle
    d. respiratory chain
    e. transition reaction

78. Which cellular structure organizes the glycolytic enzymes in sequential alignment?
    a. microtubules
    b. microfilaments
    c. intermediate filaments
    d. microtrabecular lattice
    e. vaults

79. Which of the following are the smallest elements visible with a conventional electron microscope?
    a. microtubules
    b. microfilaments
    c. intermediate filaments
    d. microtrabecular lattice
    e. vaults

80. Which ribosome synthesizes proteins used to construct new cell membranes?
    a. free ribosome
    b. rough ER bound ribosome
    c. smooth ER
    d. all of the above
    e. none of the above

81. Which ribosome synthesizes proteins used intracellularly within the cytosol?
    a. free ribosome
    b. rough ER bound ribosome
    c. neither
    d. both a and b
    e. none of the above

82. Which of the following takes place in the mitochondrial matrix?
    a. glycolysis
    b. citric acid cycle
    c. electron transport chain
    d. fermentation
    e. cytolysis

83. Which reaction produces water as a by-product?
    a. electron transport chain
    b. citric acid cycle
    c. glycolysis
    d. Calvin reaction
    e. fermentation

84. Which of the following is not characteristic of the cytoskeleton?
    a. supports the plasma membrane and is responsible for the particular shape, rigidity, and
       spatial geometry of each different cell type
    b. plays a role in regulating cell growth and division
    c. elements are all rigid, permanent structures
    d. is responsible for cell contraction and cell movement
    e. supports and organizes the ribosomes, mitochondria, and lysosomes

85. This structure has hairlike motile protrusions.
    a. flagella
    b. cilia
    c. actin bundles
    d. sperm
    e. fermentation

86. In the plasma membrane, which of the following provides a selective barrier to the
    movement of molecules and materials?
    a. water
    b. lipids
    c. proteins
    d. surface glycoproteins
    e. internal carbohydrates

87. The Golgi complex is:
    a. located in both the cytosol and nucleoplasm of the cell
    b. never associated with the endoplasmic reticulum
    c. always associated with ribosomes
    d. where some proteins are concentrated before being secreted from the cell
    e. involved in cell reproduction and the movement of the chromosome during cell division

88. Which of the following best describes a lysosome?
    a. a membrane sac containing hydrolytic enzymes
    b. an organelle formed in the smooth endoplasmic reticulum
    c. the site of protein synthesis for the cell
    d. a type of microfilament found in the cytoplasm
    e. a vesicle containing substances for export

89. Which of the following best describes the endoplasmic reticulum of a cell?
    a. It is the site of ATP synthesis.
    b. It is the site for the attachment of all ribosomes.
    c. It is the major store for carbohydrates, proteins, and fats.
    d. It helps a variety of materials circulate throughout the cytoplasm.
    e. It is physically connected to the plasma membrane but not to the nuclear envelope.

90. Which of the following is the most accurate statement concerning the cell's cytoskeleton?
    a. The major structural component is chromatin.
    b. It is an inflexible, rigid structure.
    c. It is present throughout the cell, including the nucleus.
    d. It is a latticed framework that extends throughout the cytoplasm to help organelles in place.
    e. Its shape and structure are not altered according to the activities of the cell.

91. Which term refers to release to the cell's exterior, on appropriate stimulation, of a product produced by the cell?
    a. pinocytosis
    b phagocytosisc. secretion.
    d. endocytosis.
    e. none of the above.

92. Which of the following structures provides a clue concerning the physiological relationship between the Golgi apparatus and the endoplasmic reticulum?
    a. ribosomes
    b. vesicles
    c. nuclear pores
    d. mitochondria
    e. various cytoplasmic inclusions

93. During exocytosis, the membrane of a vacuole fuses with the membrane of:
    a. the cell
    b. another vacuole
    c. the Golgi apparatus
    d. ribosomes
    e. the nuclear envelope

94. The nucleolus is best described by which of the following statement?
    a. It has a thick membrane.
    b. It is the site of ribosomal RNA synthesis.
    c. It specifies the chemical structure of enzymes.
    d. It has fewer secretory cells than nonsecretory cells.
    e. It contains ATP.

95. mRNA must pass through these openings to reach the cytoplasm.:
    a. nuclear pores
    b. ribosomes
    c. polyribosomes
    d. nucleosomes
    e. vacuoles

96. Peroxisomes contain:
    a. oxidative enzymes
    b. DNA
    c. RNA
    d. mRNA
    e. all of the above

97. Which of the following organelles destroys bacteria?
    a. peroxisomes
    b. lysosomes
    c. vacuoles
    d. Golgi complex
    e. mitochondria

98. How many different types of endoplasmic reticula are found within the cell?
    a. one
    b. two
    c. three
    d. four
    e. five

Points to Ponder

1. What conditions would prompt primitive bacteria or early mitochondria to invade primitive cells?

2. Aside from energy production, what is the advantage in having the citric acid cycle?

3. How do you think your cells replace nicotinamide adenine dinucleotide?

4. Would it be possible to calculate the ATP production knowing how much carbon dioxide was produced?

5. Why is the plasma membrane considered a semi-permeable structure?

6. Why is the cytoskeleton considered a skeleton?

7. Why are lysosomes and peroxisomes vital to the life of a cell?

8. How does the endoplasmic reticulum differ from the Golgi apparatus?

9. In what way does the meaning of the term "cellular respiration" differ from that of carbohydrate metabolism?

10. Is all the energy liberated from the catabolism of one molecule of sugar trapped in ATP? Explain.

Clinical Perspectives

1. In carbon monoxide poisoning, the gas carbon monoxide readily combines with hemoglobin. How does this phenomenon produce such fatigue that the victim may not be able to escape the gas?

2. Release of lysosomal enzymes from white blood cells that phagocytose bacteria can contribute to the symptoms of inflammation. Suppose, to alleviate inflammation, you develop a drug that destroys lysosomes. What would be the negative side effects of this drug?

3. A pathologist examines a liver biopsy sample. The cells contain extensive amounts of smooth endoplasmic reticulum. The patient admits to substance abuse. What is the relationship between the drug abuse and the increased ER?

4. Based on what you have learned in this chapter, which is better for a healthy young person—aerobic exercises or lifting weights?

5. What is the relationship between neurofilaments and amyotrophic lateral sclerosis?

3

The Plasma Membrane and Membrane Potential

Chapter Overview

All cells are enveloped by a plasma membrane, a thin, flexible lipid barrier that separates the contents of the cell from its surroundings. The plasma membrane is a lipid bilayer with membrane proteins, membrane carbohydrates, and cholesterol arranged as a fluid mosaic model. Some proteins form channels across the lipid bilayer, others serve as carrier molecules, and still others are receptor sites on the outer surface of the cell. Using these proteins and several enzymes, molecules can enter the cell. Cells can be held together using fibrous proteins: (1) collagen, (2) elastin, and (3) fibronectin. Cells can be directly linked together using (1) desmosomes, (2) tight junctions, and (3) gap junctions. Two forces are used to transport molecules across a membrane: (1) passive and (2) active. Diffusion is passive down a concentration gradient. Ions may use an electrical gradient for transport. Osmosis is water diffusing down its own concentration gradient. Larger molecules use carrier-mediated transport. The sodium-potassium pump is such a mechanism that performs several functions for the cell. Vesicular transport is a special situation to gain, secrete, rebuild, or destroy various entities. A membrane potential exists due to separated charged ions, primarily sodium, potassium, and large anions. The membrane potential and the various transport mechanisms demonstrate the importance of the plasma membrane in maintaining homeostasis.

Chapter Outline

MEMBRANE STRUCTURE AND FUNCTIONS
The plasma membrane is a fluid bilayer embedded with proteins.
- Each cell type is unique in its intracellular contents.
- The plasma membrane is a thin layer of lipids, proteins and small amounts of carbohydrates forming the outer boundary of the cell.
- This membrane maintains the specific composition of the intracellular fluid by selectively permitting specific substances to pass between the cell and its environment.
- The lipids are predominately phospholipid with some cholesterol.
- Phospholipids vary from most lipids by having two fatty acids and a phosphate.
- The polar head is hydrophilic and negatively charged.
- Consistent with the properties of fatty acids, the tails are hydrophobic.
- The two lipid layers are arranged such that the outer polar heads are in contact with the extracellular fluid, and the inner polar heads are in contact with the intracellular fluid.
- The hydrophobic tails of each lipid layer are toward the inside of the plasma membrane.
- The fluid nature of the membrane is maintained by the lack of bonding between phospholipids.
- Cholesterol molecules between the phospholipid molecules prevent the fatty acid chains from packing and crystallizing

The molecular arrangement of a plasma membrane is called the "fluid mosaic model."

The lipid bilayer forms the structural barrier that encloses the cell.
- The lipid bilayer provides three important functions: (1) It forms the basic structure of the membrane. (2) It prevents passage of water-soluble substances between the ICF and ECF. (3) It provides fluidity to the membrane.

The membrane proteins perform various specific membrane functions.
- Membrane proteins may extend through the thickness of the membrane while others are found only on the inner and outer surfaces.
- Membrane proteins provide seven specialized functions: (1) Proteins spanning the membrane form highly selective channels that allow or prevent passage of small particles such as ions. (2) Other proteins serve as carrier molecules to transport substances between the ECF and ICF. (3) Proteins on the inner surface can bind with proteins on secretory vesicles. (4) Proteins located on either the inner or outer surface also function as membrane-bound enzymes to control surface chemical reactions. (5) Proteins on the outer surface may serve as receptor sites for specific molecules in the ECF. n. (6) Additional proteins function on the outer surface as cell adhesion molecules (CAMs) and are used to attach cells and tissues together. (7) Still other proteins, functioning with special carbohydrates, serve to recognize themselves and other cells.

The membrane carbohydrates serve as self-identity markers.
- The short sugar chains on the outer-membrane surface serves as self-identity markers that enable cells to identify and interact with each other in the following ways: (1) These carbohydrate chains play an important role cells recognizing other cells of the same types. (2) Carbohydrate containing surface markers also appear to be involved in tissue growth, which is normally held within certain limits of cell density.

CELL-TO-CELL ADHESIONS
The extra-cellular matrix serves as the biological "glue."
- The plasma membrane functions in cell-to-cell adhesions that permit groups of cells to bind together into tissues and organize further into organs.
- Once arranged, cells are held together by three different means: (1) CAMS, (2) the extracellular matrix, and (3) specialized cell junctions.
- The extra-cellular matrix is a meshwork of fibrous proteins in a watery gel composed of complex carbohydrates.
- The protein fibers are collagen, elastin, and fibronectin.
- The extra-cellular matrix is secreted by fibroblasts.

Some cells are directly linked together by specialized cell junctions.
- Some tissue cohesion is provided by cell adhesion molecules.
- There are three types of specialized cell junctions (1) desmesomes, (2) tight junctions, (3) gap junctions.

Desmosomes.
- At a desmosome, filaments extend between the plasma membranes of two closely adjacent but non touching cells to anchor the cells together.

Tight junctions.
- Sheets of epithelial tissue are joined by tight junctions.
- Adjacent cells are joined together in a tight seal, which is impermeable and thereby prevents materials from passing between cells.
- Selected substances must pass through the cells.

Gap junctions.
- The third type of junction between cells is a communicating junction known as a "gap junction."
- A gap exists between adjacent cells that are linked by small connecting tunnels known as "connexons."
- Gap junctions allow particles such as ions to pass between cells whereas larger molecules are blocked.

OVERVIEW OF MEMBRANE TRANSPORT
- The membrane is permeable to the substance if the substance can cross the membrane and impermeable if the substance cannot cross.
- Selectively permeable membranes permit some particles to pass but exclude others.
- Two properties of particles determine whether they can cross the membrane without assistance: (1) Particles must be soluble in lipids. (2) Particles must be small.
- Forces are required to move particles across the membrane.
- Passive forces do not require the cell to expend energy to produce movement.
- In conditions where the cell must expend energy, the movement is achieved by active forces.

UNASSISTED MEMBRANE TRANSPORT
Particles that can permeate the membrane passively diffuse down their concentration gradient.
- Simple diffusion uses passive forces to move particles down a concentration gradient.
- While diffusion involves movement of molecules in all directions, the net movement of a substance is from an area where the concentration of the substance is greater to an area of lesser concentration.
- When the molecules of the substance are uniformly distributed, a steady state exists.
- If the membrane is impermeable to a substance no diffusion can take place even if a concentration gradient exists.

Fick's law of diffusion.
- Fick's law of diffusion demonstrates the effects of several factors on rate of net diffusion: (1) The greater the difference in concentration, the faster the rate of net diffusion. (2) The more permeable the membrane is to a substance, the more rapidly the substance can diffuse down its concentration gradient. (3) The larger the available surface area of the membrane, the greater rate of diffusion it can accommodate. (4) Lighter molecules diffuse more rapidly. (5) The greater the distance through which diffusion must take place, the slower the rate of diffusion.

Ions that can permeate the membrane also passively move along their electrical gradient.
- Diffusion also moves substances along an electrical gradient.
- Positively charged ions (cations) move toward negatively charged areas, whereas negatively charged ions (anions) move toward positively charged areas.
- When an electrical gradient (difference in charge) exists between the ICF and ECF, and the membrane is permeable to the ions in question, then these ions will diffuse along this electrical gradient.

Osmosis is the net diffusion of water down its own concentration gradient.
- Osmosis is a special type of diffusion involving a water-soluble solute and water.
- Water molecules can either pass directly through the lipid bilayer (slow) or pass more easily through water channels known as aquaporins.
- If both solute and water can easily pass through the membrane then both will cross the membrane until each has the same concentration on both sides.
- The resulting volume will be the same as at the onset of the diffusion.

Movement of water when a selective permeable membrane separates pure water from a solution of non-penetrating solutes

- If a nonpenetrating solute is present on side two and pure water is present on side one, osmosis occurs from side one to side two.
- The net diffusion of water is known as osmosis.
- Osmosis occurs from side 1 to side 2, but the concentrations between the two compartments can never be equal.
- As the volume expands in compartment 2, the pressure exerted by the water on side 2 increases and tends to push water back to side 1. This pressure is known as hydrostatic pressure.
- Osmotic pressure of a solution is a measure of the tendency for osmotic flow of water into that solution because of the relative concentration of non-penetrating solutes and water.

Movement of water and solutes when a membrane separates unequal solutions of a penetrating solute.

-
- The solute will move its concentration gradient in the opposite direction of water. The movements continue until both the solute and water are evenly distributed across the membrane.
- Solutes that can penetrate the membrane do not contribute to osmotic differences between the ECF and ICF.

Movement of water when a membrane separates equal and unequal solutions of nonpenetrating solute.

- If solutions of equal concentration of nonpenetrating solutes are separated by a membrane no concentration gradients for water exist, so not net osmosis occurs.
- If solutions of unequal concentration of nonpenetrating solutes are separated by a membrane the side with the higher solute concentration will gain water.

Tonicity.

- The tonicity of a solution refers to the effect the solution has on cell volume.
- The tonicity of a solution is determined by its concentration of nonpenetrating solutes.
- An isotonic solution has the same concentration of nonpenetrating solutes as normal body cells do. In this case no water enters or leaves the cell.
- A cell placed in a hypertonic solution will lose water to the solution whereas a cell placed in a hypotonic solution will gain water.

ASSISTED MEMBRANE TRANSPORT

- Carrier-mediated transport is used for large, poorly lipid-soluble molecules such as glucose, proteins, and amino acids, which cannot cross the plasma membrane.
- This impermeability ensures that large essential molecules cannot escape from the cell.

Carrier-mediated transport is accomplished by a membrane carrier flipping its shape.

- The carrier proteins span the thickness of the plasma membrane.
- These proteins change shape to expose the binding sites to each surface of the membrane alternately.
- The substance to be transported attaches to a carrier protein that is specific for one substance or at most a few closely related compounds.
- The transport maximum of a substance is the upper limit that can be transported per unit time.
- This transport maximum occurs when all the carriers are being used, thus saturation.
- This saturation condition can be overcome by changing the affinity of the binding site for the passenger or by increasing the number of binding sites.
- Closely related chemical compounds compete for binding sites.

Facilitated diffusion is passive Carrier-mediated transport.

- Carrier-mediated transport takes two forms: (1) Facilitated diffusion goes down a concentration gradient and is passive. (2) Active transport goes against the concentration gradient.
- These active transport mechanisms are called "pumps."
- The most notable example of facilitated diffusion is the transport of glucose into cells.
- Since glucose is higher in the blood than in the tissues, a gradient exists for the net diffusion of glucose into the cells by way of a transport protein.

Active transport is carrier-mediated and uses energy to move a substance against its concentration gradient.

- Active transport involves the use of a carrier protein to transfer a specific substance
  across the membrane against its concentrations gradient.
- ATP energy is needed for this transport.
- Active transport comes in two forms: (1) primary active transport in which energy is directly required to move a substance; (2) secondary active transport in which energy is required but not directly used to move the substance.
- These active transport mechanisms are called "pumps," analogous to water pumps that require energy to lift water against the downward pull of gravity.

Na+-K+ PUMP.

- The Na+-K+ pump moves three Na+ out of the cell for every two K+ it pumps in.
- The sodium-potassium pump performs three important roles: (1) It establishes sodium and potassium gradients across the plasma membrane of all cells. (2) It helps regulate cell volume. (3) The energy used to run the sodium potassium pump also indirectly serves as the energy source for secondary active transport.

Secondary Active Transport

- The movement of Na^+ into the cell down its concentration drives the uphill transport of another solute by a secondary active-transport carrier.
- Secondary active-transport carriers have two binding sites: one for the solute being moved and one for Na^+.
- Secondary active transport is called symport when the Na+ and solute are moved in the same direction and antiport when they are moved in the opposite direction.
- The sodium and glucose cotransporter or SGLT,is an important cotransport system within the intestines.

With vesicular transport, material is moved into or out of the cell wrapped in a membrane.

- Vesicular transport occurs when either a material is brought into the cell (endocytosis) or a substance leaves the cell (exocytosis) in a membrane-enclosed vesicle.

Endocytosis.

- Endocytosis is the formation of a vesicle by pinocytosis (cell drinking) or phagocytosis (cell eating).
- Such a vesicle will either be degraded by a lysosome or pass across the cell and be released.

Exocytosis.

- In exocytosis A membrane-enclosed secretory vesicle formed within the cell fuses with the plasma membrane, then opens up and releases its contents to the exterior.

MEMBRANE POTENTIAL
Membrane potential is a separation of opposite charges across the plasma membrane.

- The unequal distribution of a few key ions between the ECF and ICF and their selective movement through the plasma membrane are responsible for the electrical properties of the membrane.

- Membrane potential refers to the separation of charges across the membrane.
- Because separated charges have the "potential" to do work, a separation of charges across the membrane is referred to as a "membrane potential."
- The membrane potential is relatively low therefore the unit used is the millivolt.
- These separated charges represent only a small fraction of the total number of charged particles present in the ICF and ECF.
- The magnitude of the potential depends on the degree of separation of the opposite charges: the greater the number of charges separated, the larger the potential.

Membrane potential is due to differences in the concentration and permeability of key ions.

- All living cells have a membrane potential characterized by a slight excess of positive charges outside and a corresponding slight excess of negative charges on the inside.
- The cells of excitable tissues (nerve cells and muscle cells) have the ability to produce rapid, transient changes in their membrane potential when excited. These brief fluctuations in potential serve as electrical signals
- The ions primarily responsible for the generation of membrane potential are sodium, potassium, and anionic intracellular proteins.
- These large, negatively charged proteins are found only inside the cell.

Effect of the sodium-potassium pump on membrane potential.

- The Na+-K+ pump transports three Na+ out of the cell for every two K+ it moves into the cell.
- This results in an unequal distribution of positive charges across the membrane with inside being more negative compared to the outside.

- The potential difference generated this way is small relative to the overall membrane potential.
- The majority of the membrane potential results from the passive movements of Na+ and K+ down their concentration gradients.
- These gradients are established and maintained by the Na+-K+ pump.

Effect of movement of potassium alone on membrane potential: Potassium equilibrium potential.

- The membrane is permeable to K+ so K+ diffuses out of the cell down its concentration gradient. As K+ leaves the cell, the inside becomes more negative.
- At the same time, the negative charges associated with the large intracellular proteins (A-) would be left behind because they are too large to diffuse out of the cell.
- Two opposing forces are acting on K+ movements. The concentration gradient moving K+ out of the cell and the electrical gradient tending to move the ion into the cell.
- The process would continue until the two opposing forces exactly balanced each other.
- The potential that would exist at this point is known as the equilibrium potential for K+.
- Under normal conditions, this potential would be -90mV.

Effect of the movement of sodium ions alone on membrane potential: the sodium equilibrium potential.

- The membrane is permeable to Na+ and Na+ will enter the cell down its concentration gradient.
- This produces a buildup of positive charges inside the cell and leaves the negatively charged Cl- outside the cell in the ECF.
- Net diffusion of Na+ into the cell would continue until the negatively charged ECF exactly balances the inward movement of Na+ down its concentration gradient.

- At this point, the equilibrium potential for Na+ would be reached and would be measured at +60mV.

Concurrent potassium and sodium effects on membrane potential.
- Because neither K+ or Na+ exist alone in body fluids, their equilibrium potentials do not exist.
- In living cells the effects of both Na+ and K+ must be taken into account.
- Because the membrane is 25-30 times more permeable to K+ than Na+. the normal resting membrane potential is much closer to the equilibrium potential for K+.
- Normal resting potential is -70mV.

Balance of passive leaks and active pumping at resting membrane potential.
- Both ions leak across the membrane.

- The sodium-potassium pump counterbalances the rate of leakage.
- The pump transports back into the cell essentially the same number of potassium ions that have leaked out and simultaneously transports to the outside the sodium ions that have leaked in.
- Not only is the sodium-potassium pump initially responsible for the sodium and potassium concentration differences across the membrane, but it also maintains these differences.

Specialized use of membrane potential in nerve and muscle cells.
- Nerve and muscle cells have developed a specialized use for membrane potential.
- This causes nerve impulses and muscles to contract.

Key Terms

Active transport
Aquaporins
Carrier-mediated transport
Carrier molecules
Channels
Chemical gradient
Cholesterol
Collagen
Competition
Concentration gradient
Connexons
Desmosome
Diffusion
Elastin
Electrical gradient
Electrochemical gradient
Endocytosis

Exocytosis
Extracellular matrix (ECM)
Facilitated diffusion
Fibroblasts
Fibronectin
Fick's law of diffusion
Fluid mosaic model
Gap junction
Hydrostatic fluid pressure
Hypertonic solution
Hypotonic solution
Impermeable
Isotonic solution
Lipid bilayer
Membrane-bound enzymes
Membrane carbohydrate
Membrane potential
Membrane proteins
Millivolt (mV)

Net diffusion
Osmosis
Osmotic pressure
Permeable
Phospholipids
Plasma membrane
Pumps
Receptor sites (receptors)
Resting membrane potential

Saturation
Selectively permeable
Sodium-potassium pump
Specificity
Steady state
Tight junction
Tonicity
Transport maximum
Vesicular transport

Review Exercises

Answers are in the appendix.

True/False-if an answer is false, underline the incorrect portion of the statement and rewrite it to make the statement true.

_____ 1. Cholesterol is found within the lipid bilayer.

_____ 2. Most cells have active transport for the chloride ion.

_____ 3. Large anions are in the ICF.

_____ 4. Pinocytosis is the process whereby the white blood cell eats bacteria.

_____ 5. The proteins on the inner and outer surface of the plasma membrane are different in composition and function.

_____ 6. Phospholipids have hydrophilic heads that can interact with water.

_____ 7. When equilibrium is achieved and no net diffusion is taking place, there is no movement of molecules.

_____ 8. Phosphorylation of a carrier can alter the affinity of its binding sites.

_____ 9. Under normal circumstances, the equilibrium potential for K+ is more negative than the equilibrium potential for Na+.

_____ 10. At resting membrane potential, passive and active forces exactly balance each other so there is no net movement of ions across the membrane.

_____ 11. The survival of every cell depends on the maintenance of intracellular contents unique for that cell type despite the remarkably different composition of the extracellular fluid surrounding it.

_____ 12. All plasma membranes consist mostly of lipids and proteins plus small amounts of carbohydrates.

_____ 13. The small amount of membrane carbohydrate is located on the outer surface.

_____ 14. The lipid bilayer forms the basic structure of the membrane.

_____ 15. It is not possible for a given channel to be open or closed to its specific ion as a result of changes in channel shape in response to a controlling mechanism.

_____ 16. In a solution of salt water, salt is the solvent.

_____ 17. Elastin is a rubberlike protein fiber, most abundant in tissues that must be capable of stretching.

_____ 18. The extracellular matrix is secreted by local cells, most commonly by fibronectin.

_____ 19. Sheets of epithelial tissue are joined by gap junctions.

_____ 20. Tight junctions restrict the movement of membrane proteins to one of two cell surfaces: The luminal border or basolateral border.

_____ 21. Connexons are formed by the joining of proteins that extend outward from each of the adjacent plasma membranes.

_____ 22. Gap junctions are found in skeletal muscle.

_____ 23. Two properties of particles influence whether they can permeate the plasma membrane without any assistance: the relative solubility of the particle in lipid and the polarity of the particle.

_____ 24. If 10 molecules move from A to B while two molecules simultaneously move from B to A, the net diffusion is 12 molecules moving from A to B.

_____ 25. According to Fick's Law, the greater the difference in concentration of a substance across a membrane the slower the rate of net diffusion.

_____ 26. The more permeable the membrane is to a substance, the more rapidly the substance can diffuse down its concentration gradient.

_____ 27. The sodium and glucose cotransporter (SGLT) is an example of secondary active transport.

_____ 28. The Na+-K+ pump maintains the resting membrane potential.

_____ 29. At rest, a typical cell is losing more positive charges than it is gaining.

_____ 30. Opposite charges attract each other.

_____ 31. A solution with a high solute concentration exerts greater osmotic pressure than does a solution with a lower solute concentration.

_____ 32. Cells utilize two different mechanisms to accomplish these selective-transport processes: carrier-mediated transport and vesicular transport.

_____ 33. If fluid is internalized by endocytosis, the process is termed "pinocytosis."

_____ 34. The separation of like charges across the plasma membrane is referred to as a "membrane potential."

_____ 35. All living cells have a membrane potential characterized by a slight excess of positive charges outside and a correspondingly slight excess of negative charges on the inside.

_____ 36. The larger the concentration gradient for an ion, the lesser the ion's equilibrium potential.

Fill in the Blank

37. _____ serves to prevent phospholipid crystallization in the membrane

38. Once arranged, cells are held together by three different means. They are (1) _____, (2) _____, and (3) _____.

39. _____ are adhering junctions that anchor cells together in tissues subject to considerable stretching.

40. List the three major types of protein fibers: (1) _____, (2) _____, and (3) _____.

41. The extracellular matrix is secreted by local cells, most commonly by _____ present in the matrix.

42. Some cells are directly linked together by one of three types of specialized cell junctions, which are: (1) _____, (2) _____, and (3) _____.

43. Tight junctions restrict the movement of membrane proteins to one of two cell surfaces. Name the two cell surfaces: (1) _____ and (2) _____.

44. Sheets of epithelial tissue are joined by _____.

45. _____ form(s) cablelike fibers or sheets that provide tensile strength.

46. All cells are enveloped by a(n) _____, a thin, flexible lipid barrier that separates the contents of the cell from its surroundings.

47. Cells have a(n) _____, which refers to a slight excess of _____ (+/-) charges lined up along the inside of the membrane and separated from a slight excess of positive charges on the outside.

48. _____ have a polar head containing a negatively charged phosphate group and two nonpolar fatty acid tails.

49. A polar end that interacts with water molecules is referred to as _____, and a nonpolar end that will not mix with water is called _____.

50. _____ are the membrane component that exhibit a variety of diverse functions.

51. The _____ forms the primary barrier to diffusion within the membrane.,

52. Some proteins serve as _____ that transfer substances unable to cross the membrane on their own. Many of the proteins on the outer surface serve as _____ that "recognize" and bind with specific molecules in the environment of the cell.

53. _____ is a rubberlike protein fiber most abundant in tissues that must be capable of easily stretching and then recoiling after the stretching force is removed.

54. _____ are communicating junctions consisting of _____, small connecting tunnels that permit movement of charge-carrying ions between two adjacent cells.

55. _____ promotes cell adhesion and holds cells in position.

56. The equilibrium potential for _____ is -90mV.

57. _____ refers to the difference between two opposing movements.

58. When movement of molecules from A to B are exactly matched by movement of molecules from B to A, this situation is known as a(n)_____ or _____.

59. If a substance is unable to pass, the membrane is _____ to it.

60. All _____ proteins span the thickness of the plasma membrane and are able to undergo reversible changes in shape so that specific binding sites can alternately be exposed at either side of the membrane.

61. There is a limit to the amount of a substance that can be transported across the membrane via a carrier in a given time. This limit is referred to as _____.

62. With _____, energy is directly required to move a substance uphill.

63. With _____, energy is required in the entire process, but it is not directly required to run the pump.

64. In _____, the plasma membrane surrounds the substance to be ingested, then fuses over the surface, pinching off a membrane-bound vesicle that encloses the engulfed material within the cell.

65. If large multimolecular particles are engulfed, such as bacteria or cellular debris, the process is called _____.

66. _____ refers to a separation of charges across the membrane or to a difference in the relative number of cations and anions in the ICF and ECF.

67. If a relative difference in charge exists between two adjacent areas, the positive charged ions, called _____, tend to move toward the more _____ charged area, whereas the negatively charged ions, called _____, tend to move toward the more _____ charged area.

68. A difference in charge between two adjacent areas thus produces a(n) _____ that passively induces the movement of ions.

69. The simultaneous existence of an electrical gradient and concentration (chemical) gradient for a particular ion is referred to as a(n) _____.

70. The term _____ refers to the density of the solute (dissolved substance) in a given volume of water.

71. The net diffusion of water is known as _____.

72. The magnitude of opposing pressure necessary to completely stop osmosis is equal to the _____ or hydrostatic pressure.

Matching

Match substance to method of transport.

_____ 73. sodium ions	a. diffusion through lipid bilayer
_____ 74. water	b. endocytosis
_____ 75. carbon dioxide	c. exocytosis
_____ 76. calcium ions	d. diffusion through protein channel
_____ 77. potassium ions	e. osmosis
_____ 78. oxygen	f. facilitated diffusion
_____ 79. bacteria	
_____ 80. chloride ions	
_____ 81. glucose	
_____ 82. fatty acids	

Multiple Choice

83. Which of the following acts as "spot rivets" to anchor cells together?
    a. tight junctions
    b. gap junctions
    c. desmosomes
    d. ribosomes
    e. archaeosomes

84. Which type of movement is the net diffusion of water down its own concentration gradient?
    a. carrier-mediated
    b. osmosis
    c. facilitated diffusion
    d. secondary active transport
    e. exocytosis

85. Which process is used to recycle portions of the plasma membrane?
    a. primary active transport
    b. exocytosis
    c. dephosphorylation
    d. facilitated diffusion
    e. endocytosis

86. Which of the following types of cell junctions involve connexons?
    a. tight junctions
    b. gap junctions
    c. desmosomes
    d. all the above
    e. none of the above

87. Cholesterol contributes to the fluidity and stability of the membrane. Between which molecules is cholesterol found?
    a. membrane proteins
    b. membrane carbohydrates
    c. cell adhesions molecules
    d. phospholipids
    e. G proteins

88. Phagocytosis is an example of which of the following?
    a. endocytosis
    b. mesocytosis
    c. ectocytosis
    d. hydrocytosis
    e. none of the above

89. Membrane potential is measured in these units.
    a. volts
    b. nanovolts
    c. kilovolts
    d. millivolts
    e. decivolts

90. The negative sign of the membrane potential applies to which of the following?
    a. a truly negative potential
    b. the polarity of excess charge on the inside of the membrane
    c. the polarity of excess charge on the outside of the membrane
    d. the concentrations of large anions in the ECF
    e. none of the above

91. What is the ratio of sodium pumped out to potassium pumped in by the active transport mechanism?
    a. 9:1
    b. 3:1
    c. 5:2
    d. 4:1
    e. 3:2

92. Which of the following ions is most responsible for the resting membrane potential of a typical cell?
    a. sodium
    b. chloride
    c. potassium
    d. calcium e. both sodium and potassium

93. What is the function of carbohydrates in a plasma membrane?
    a. form a barrier to diffusion
    b. perform specific membrane functions
    c. play an important role in "self" recognition
    d. both a and b
    e. both b and c

94. Dense plaque and strong glycoprotein filaments are found in:
    a. tight junctions
    b. desmosomes
    c. gap junctions
    d. both a and b
    e. none of the above

95. Cellular communication occurs most often in which of the following types of junctions?
    a. tight junctions
    b. desmosomes
    c. gap junctions
    d. both a and b
    e. none of the above

96. How many types of gated channels are there?
    a. one
    b. two
    c. three
    d. four
    e. five

97. Which ion is normally used to provide the energy utilized in secondary active transport?

    a. Sodium
    b. Potassium
    c. Calcium
    d. Chloride
    e. None of the above

98. Which statement is true?

    a. The membrane is normally more permeable to sodium than to potassium
    b. Large intracellular proteins (A-) cannot diffuse out of the cell
    c. Cl- are typically found in the ICF
    d. Sodium is at equilibrium across the membrane
    e. The Na+-K+ pump moves the same number of Na+ and K+ across the membrane

Points to Ponder

1. After a long hard day you soak in the tub for several minutes. It feels good to just relax in the water. As you dry your body you notice that your fingers and toes are very wrinkled. What has happened?

2. As you watch people sitting in the smoking section of a restaurant you notice that shortly after one patron lights up a cigarette others soon follow. What is happening?

3. How would you explain that many different molecules totally man-made, such as some drugs, are transported across membranes? Even carcinogenic molecules, both large and small, enter the cell. How is this possible?

4. Why can't humans drink salt water? What happens when they do?

5. If a person has an abnormally low concentration of proteins in his/her plasma, excessive accumulation of fluid in the tissues (edema) will occur. Why?

Clinical Perspectives

1. Mannitol is a sugar that does not cross the blood-brain barrier. It is a particle that is osmotically active. During head trauma, the brain swells. How might administering mannitol to a head trauma patient prevent this swelling?

2. A physiology student was diagnosed as having hyperkalemia (high potassium) and an abnormal electrocardiogram (EKG). Based on what you know about the membrane potential, what effect would a high potassium level have on the membrane potential of the heart to produce an abnormal EKG?

3. Severe diarrhea is responsible for almost half of all deaths worldwide in children under the age of 4. Because intravenous rehydration therapy is not economically feasible, oral rehydration therapy is used. The concept behind oral rehydration therapy is to cause sodium to move into the body and then water follows. Based on the principles of osmosis, what is the rationale for this therapy?

4. If a diabetic patient takes too much insulin, this can lead to hypoglycemia, unconsciousness, and death due to a coma. The rate of facilitated diffusion into the brain depends directly on the plasma glucose concentration. Based on what you know about facilitated diffusion, why does a coma occur?

5. Based on the concepts of isotonic, hypertonic, and hypotonic, why is Ringer's lactate or 5% dextrose used for IV fluids?

Experiments of the Day

1. Materials: Two glasses
                  One face cloth
                  Water
                  Cooking oil

Place one glass half filled with water next to the other one-fourth filled with cooking oil. Carefully insert one end of the dry face cloth into the water glass and the other end into the glass with oil. Be sure both ends are in the respective fluids. Mark the fluid levels. Explain what happens over the next several days.

2. Take a diet coke and a regular coke. Place them in a tub or bucket of water. One can will sink and one can will float. Based on what you have learned in this chapter, you should be able to explain why this occurs.

Principles of Neural and Hormonal Communication

Chapter Overview

Nerve cells, or neurons, make up the nervous system, one of the two major regulatory systems of the body. As changes occur in the external and internal environments of the body, the nerve cells receive this information and transmit it to other cells. Eventually nerve cells initiate the necessary responses to such changes. Therefore, neurons are essential in maintaining homeostasis. The process of transmission of information within a neuron involves electrical signals such as graded and action potentials. Through changes in permeability the nerve cell undergoes depolarization, hyperpolarization, and repolarization. These phenomena transmit the information. The speed of transmission depends on the diameter of the fiber and whether or not the neuron is myelinated. When the action potential reaches the synaptic knob a neurotransmitter is released that diffuses across the synaptic cleft and changes the permeability of the postsynaptic neuron. Through temporal or spatial summation an action potential is initiated in the axon hillock of the postsynaptic neuron. Synapses may be excitatory or inhibitory. Drugs and diseases may alter synaptic effectiveness.

The second major integrating and coordinating system is the endocrine system. Hormonal communication is accomplished by hormones, which are long distance chemical messengers secreted by the endocrine glands into the blood stream. The blood stream carries the hormones to distant target sites where they regulate processes that require duration rather than speed, such as metabolic activities, water and electrolyte balance, reproduction, and growth.

Chapter Outline

INTRODUCTION TO NEURAL COMMUNICATION
Nerve and muscle are excitable tissues.
- All cells of the body process a membrane potential related to the distribution and permeability of sodium ions, potassium ions, and large intracellular anions.
- Nerve and muscle are excitable tissues, capable of producing electrical signals when stimulated.

Membrane potential becomes less negative during depolarization and more negative during hyperpolarization.
- Charges are separated across the plasma membrane so that the membrane has potential.
- When depolarized, the membrane is less polarized than at resting potential. The inside becomes less negative
- During repolarization, the membrane returns to a resting potential.
- When a membrane is hyperpolarized, it is more polarized (more negative) that at a resting potential.

Electrical signals are produced by changes in ion movement across the plasma membrane.

- Changes in ionic movement across the membrane are responsible for the changes in potential.
- Changes in ionic movement in turn are brought about by changes in the permeability of the membrane to specific ions.
- There are two types of channels: leak channels and gated channels.
- Specifically there are four types of gated channels: (1) voltage-gated; (2) chemically-gated; (3) mechanically-gated; (4) thermally-gated. The differences relate to the factor that causes the channel to change its shape.

GRADED POTENTIALS

The stronger a triggering event, the larger the resultant graded potential.

- Graded potentials are local changes in membrane potential that occur in varying grades or degrees of magnitude; that is the stronger the triggering event, the larger the graded potential.
- The triggering event might be: (1) a stimulus, (2) an interaction of a chemical messenger, or (3) a spontaneous change of potential caused by imbalances in the leak-pump cycle.
- When a graded potential occurs in excitable tissue a different potential exists in this area than in the remainder of the membrane, which is still at resting potential.
- The magnitude of the graded potential is proportional to the strength of the triggering event.

Graded potentials spread by passive current flow.

- Because opposite charges attract each other, current flows from this active area to the neighboring inactive areas.
- Any reduction in resting potential will produce a current flow.
- Since these areas on excitable tissues are not insulated the current flow leaks into the ECF.

Graded potentials die out over short distances.

- The graded potential dies out after a short distance.
- Graded potentials are critically important to the body as: (1) postsynaptic potentials, (2) receptor potentials, (3) end-plate potentials, (4) pacemaker potentials, and (5) slow-wave potentials.

ACTION POTENTIALS

- The rapid reversals of membrane potential in excitable tissues are known as action potentials.
  This entire rapid change in potential, from threshold potential to depolarization and back, is called the "action potential."

During an action potential, the membrane potential rapidly, transiently reverses.

- During an action potential, marked changes in membrane permeability of sodium and potassium take place, permitting rapid fluxes of these ions.
- Action potentials are initiated when a depolarizing potential is sufficiently strong enough to depolarize the cell to a critical level known as threshold potential.
- At threshold potential an explosive depolarization takes place.
- The rising phase of the action potential (depolarization) is due to a sodium ion influx induced by an explosive increase in sodium ion permeability at threshold.
- The falling phase (repolarization) is brought about by a potassium ion efflux caused by a marked increase in potassium ion permeability occurring simultaneously with an inactivation of sodium channels at the peak of the action potential.
- After the action potential, the ion distribution has been altered slightly.
- Sodium enters the cell during the rising phase.
- Potassium exits the cell during the falling phase.

- The sodium-potassium pump restores the normal distribution of these ions.

Marked changes in membrane permeability and ion movement lead to an action potential.

- During an action potential, depolarization of the membrane to threshold potential triggers sequential changes in permeability, caused by conformational changes in voltage-gated sodium and potassium channels.

Voltage-gated sodium and potassium channels.

- Voltage-gated sodium and potassium channels are the ones involved in action potentials.
- The voltage-gated sodium channel has two gates: an activation gate and an inactivation gate.

Changes in permeability and ion movement during an action potential.

- Permeability changes bring about a brief reversal of membrane potential, with sodium influx being responsible for the rising phase, followed by potassium efflux during the falling phase.

The sodium-potassium pump gradually restores the concentration gradients disrupted by action potentials.

- At the completion of an action potential, the membrane potential has been restored to its resting condition by the sodium-potassium pump.

Action potentials are propagated from the axon hillock to the axon terminals.

- A single action potential involves only a small patch of the total surface membrane of an excitable cell.
- The nerve cell, or neuron, has three basic parts: (1) the cell body, (2) the dendrites, and (3) the axon.
- The nucleus and organelles are in the cell body.
- The dendrites project as antennae to receive signals from other neurons.
- The dendrites carry signals toward the cell body.

- The axon carries the signal away from the cell body. The first portion of the axon is the axon hillock and is the site where action potentials are initiated.

Once initiated, action potentials are conducted throughout a nerve fiber.

- Once an action potential is triggered nothing further is necessary to activate the remainder of the nerve fiber.
- The impulse carried by an axon ends at the axon terminal where it stimulates the release of chemical messengers, which excite other cells.
- Impulse conduction is by local current flow called contiguous conduction or saltatory conduction.

The refractory period ensures unidirectional propagation of the action potential and limits the frequency of action potentials.

- The refractory period has two components: (1) the absolute refractory period and (2) the relative refractory period.
- The absolute refractory period is the time when a recently activated patch of membrane is completely incapable of initiating another action potential no matter how strongly it is stimulated.
- The relative refractory period is the time during which a second action potential can be produced only by a stimulus considerably stronger than is usually necessary.
- The refractory period ensures the unidirectional propagation of the action potential down the axon away from the initial site of activation.
- An excitable membrane either responds to a stimulus with a maximal action potential that spreads nondecrementally throughout the membrane, or it does not respond with an action potential at all.

Action potentials occur in all-or-none fashion.

- An excitable membrane responds to a triggering event with a maximal action potential that spreads nondecrementally throughout the membrane, or it does not respond with an action potential at all.

- This property is called the "all-or-none law."

The strength of a stimulus is coded by the frequency of action potentials.

- Weak and strong stimuli evoke the same action potential but stronger stimuli initiate more action potentials.
- Once initiated, the velocity or speed with which an action potential travels down the axon depends on whether the axon is myelinated.
- The fast method of conduction known as "salutatory conduction" occurs in myelinated fibers.

Myelination increases the speed of conduction of action potentials.

- The velocity with which an action potential travels down the axon depends on two factors: (1) whether or not the fiber is myelinated and (2) the diameter of the fiber.
- Myelin is composed primarily of lipids.
- Water-soluble ions cannot permeate this thick lipid barrier.
- Myelin-forming cells wrap around the axon in jelly-roll fashion.
- In the central nervous system, oligodendrocytes are the myelin-forming cells, while Schwann cells perform this function in the peripheral nervous system.
- Between the myelinated regions, the axonal membrane is bare and exposed to the ECF.
- These bare areas are known as "nodes of Ranvier."
- The distance between nodes (about 1 mm) is short enough that local current from an active node can reach an adjacent node.
- This is saltatory conduction.
- Myelinated fibers conduct impulses about fifty times faster than unmyelinated fibers of comparable diameter.
- Since ion fluxes are confined to nodes, energy is conserved.

- The larger the diameter of the nerve fiber, the faster it can propagate an action potential.
- Myelinated fibers conduct impulses 50X faster than unmyelinated fibers of comparable size.
- Multiple sclerosis is a pathophysiological condition in which demyelination of nerve fibers occurs in various locations of the nervous system.

SYNAPSES AND NEURONAL INTEGRATION

Synapses are junctions between presynaptic and postsynaptic neurons.

- The junction between two neurons is known as the synapse.
- The axon terminal of the presynaptic neuron is referred to as a "synaptic knob."
- The synaptic knob contains synaptic vesicles, which store a neurotransmitter.
- The space between the presynaptic and postsynaptic membrane is called the "synaptic cleft."

A neurotransmitter carries the signal across a synapse.

- The action potential in the presynaptic neuron induces the release of the neurotransmitter by opening calcium ion channels in the synaptic knob.
- The released neurotransmitter binds to specific sites on the subsynaptic membrane.
- This binding triggers the opening of specific ion channels, thereby altering the permeability of the postsynaptic neuron.
- Synapses operate in one direction only.

Some synapses excite whereas others inhibit the postsynaptic neuron.

- At an excitatory synapse, the permeability of the sodium and potassium ions is increased by the opening of the respective channels. The result is a net movement of positive ions into the postsynaptic neuron or a small depolarization.

- The potential across the membrane is now closer to the threshold potential.
- This postsynaptic potential change is called an "excitatory postsynaptic potential" or EPSP.
- At an inhibitory synapse, the permeability of the potassium or chloride ion is increased.
- The resulting ion movements bring about a small hyperpolarization.
- The inside of the postsynaptic neuron becomes more negative.
- The difference between this new potential and the threshold potential is greater.
- This postsynaptic potential change is called an "inhibitory postsynaptic potential" or IPSP.
- EPSPs and IPSPs are produced by opening of chemically gated channels, unlike action potentials, which are produced by opening of voltage-gated channels.
- A given synapse is either always excitatory or always inhibitory.

Neurotransmitters are quickly removed from the synaptic cleft.
- The transmitter may be inactivated by specific enzymes or may be actively taken back up into the axon terminal

The grand postsynaptic potential depends on the sum of the activities of all presynaptic inputs.
- EPSPs and IPSPs are graded potentials and do not follow the all-or-none rule. Which means they can be summed on top of each other.
- The grand postsynaptic potential is the total potential produced by the presynaptic neurons.
- The summing of several postsynaptic potentials occurring very close together in time because of successive firing of a single presynaptic neuron is known as "temporal summation."
- A second method to elicit a grand postsynaptic potential is through concurrent activation of several excitatory inputs located at different positions. This is known as spatial summation.

Temporal summation.
- The summing of several EPSPs or IPSPs occurring very close together in time because of successive firing of a single presynaptic neuron is called "temporal summation."

Spatial summation.
- The summation of EPSPs or IPSPs originating simultaneously from several different presynaptic inputs is known as "spatial summation."

Cancellation of concurrent EPSPs and IPSPs.
- If an excitatory and inhibitory input are simultaneously activated, they cancel each other out.

Importance of postsynaptic integration.
- The magnitude of GPSP depends on the sum of activity in all presynaptic inputs and in turn determines whether or not the neuron will undergo an action potential to pass information on to the cells at which the neuron terminates.
- Some neurons secrete neuromodulators in addition to neurotransmitters. In addition to secreting neurotransmitters, some neurons also secret larger chemicals known as neuropeptides.
- Neuropeptides acts as neuromodulators.
- Neuropeptides bind to neuronal receptors at nonsynaptic sites and bring about long-term changes that subtly depress or enhance synaptic effectiveness.
- Whereas neurotransmitters serve as chemical messengers for rapid communication between neurons, neuromodulators are involved with more long-term neural events such as learning.

Drugs and disease can modify synaptic transmission.
- Synaptic drugs may block an undesirable effect or enhance a desirable effect.

- Possible drug actions include: (1) altering the synthesis, axonal transport, storage, or release of a neurotransmitter; (2) modifying neurotransmitter receptor; (3) influencing neurotransmitter reuptake or destruction; or (4) replacing a deficient neurotransmitter with a substitute transmitter.

Neurons are linked through complex converging and diverging pathways.

- Through convergence, a single cell is influenced by thousands of other cells.
- Divergence refers to the branching of axon terminals so that a single cell synapses with many other cells.
- A single neuron may be connected to between 5000 and 10,000 other neurons.

INTERCELLULAR COMMUNICATION AND SIGNAL TRANSDUCTION

- Coordination of the diverse activities of cells throughout the body to accomplish life-sustaining and other desired activities depends on the ability of the cells to communicate with each other.

Communication between cells is largely orchestrated by extracellular chemical messengers.

- There are three types of intercellular communication: (1) through gap junctions, (2) through transient direct linkup of surface markers (3) indirectly through extracellular chemical messengers.
- There are four types of extracellular chemical messengers molecules: (1) paracrines are local chemical messengers whose effect is exerted on neighboring cells in the immediate environment (2) neurotransmitters are very short-range chemical messengers in response to electrical signals; (3) hormones are long-range chemical messengers that are produced by endocrine glands in response to an appropriate signal; (4) neurohormones are hormones released into the blood by neurosecretory neurons.

Extracellular chemical messengers bring about cell responses primarily by signal transduction.

- The term "signal transduction" refers to the process by which incoming signals are conveyed to the target cell's interior for executionDuring signal transduction, the extracellular signal is transduced, or changed into a form necessary to modify intracellular activities to accomplish the desired outcome
- Signal transduction occurs by different mechanisms, depending on the messenger and receptor type.

Some extracellular chemical messengers open chemically gated receptor-channels.

- Some chemical messengers carry out the assigned task by opening or closing specifically gated receptor-channels to regulate the movement of a particular ion into or out of the cell.

Many extracellular chemical messengers activate second-messenger pathways via G-protein-coupled receptors.

- The first messenger issues an order by triggering a pass-it-on process.
- Binding of the first messenger to the receptor actives the G protein which alters the activity of a nearby membrane protein called an effector protein. .
- Once activated the effector protein leads to an increased concentration of the second messenger.
- The second messenger by influencing a series of biochemical pathways causes a change in shape and function of designated proteins. These proteins accomplish the cellular response.

HORMONAL COMMUNICATION

- Endocrinology is the study of homeostatic chemical adjustments and other activities accomplished by hormones, which are secreted into the blood by endocrine glands.
- Hormones are long distance chemical messengers secreted by the endocrine glands into the blood, which transports them to specific target sites where they

control a particular function by altering protein activity within target cells.

Hormones are classified chemically as hydrophilic or lipophilic.
- Hydrophilic water-soluble hormones include the peptides (the majority of hormones) and the catecholamines (secreted by the adrenal medulla).
- The lipophilic (lipid-soluble) hormones include the thyroid hormone and the steroid hormones (the sex hormones and those secreted by the adrenal cortex).

The mechanisms of synthesis, storage, and secretion of hormones vary according to their chemical differences.
- Hydrophilic peptide hormones are synthesized and packaged for export by the endoplasmic reticulum/Golgi complex, stored in secretory vesicles, and release by exocytosis on appropriate stimulation.

Hydrophilic hormones dissolve in the plasma; lipophilic hormones are transported by plasma proteins.
- Hydrophilic hormones bind to receptors on the surface of the cell, whereas lipophilic hormones bind to receptors inside the cell.

Hormones generally produce their effect by altering intracellular proteins.
- At their target cells, hydrophilic hormones bind with surface membrane receptors. On binding, these hormones trigger a chain of intracellular events by means of second messenger systems that ultimately alter pre-existing cellular proteins, usually enzymes, which exert the effect leading to the target cell's response to the hormone.

Locations of receptors for hydrophilic and lipophilic hormones
- Hydrophilic hormones cannot pass through the plasma membrane and bind with receptors located on the outer plasma membrane surface.

- Lipophilic hormones easily pass through the membrane to bind with receptors located inside the cell..

Hydrophilic hormones alter preexisting proteins via second-messenger systems.
- The activation of G-protein-coupled surface membrane receptors leads to the production of second messengers inside the cell.
- Cyclic AMP is the most widely used second messenger. .
- Second messenger systems take advantage of cascading or multiplying effects to produce signal amplification.

By stimulating genes lipophilic hormones promote synthesis of new proteins.
- Lipophilic hormones readily enter through the lipid membrane barriers of their target cells and bind with nuclear receptors.
- The hormone-receptor complex binds to DNA at a specific attachment site known as the hormone response element.
- Hormonal binding activates a specific gene within the target cell leading to production of a new protein that carry out the hormones' effect on the target cell.

COMPARISON OF THE NERVOUS AND ENDOCRINE SYSTEMS
The nervous system is "wired" and the endocrine system is "wireless."
- Nerves go to their target cells whereas hormones are carried by the blood.

Neural specificity is due to anatomic proximity and endocrine specificity to receptor specialization.
- Specificity of neural action depends on the anatomic proximity of the neurotransmitter-releasing neuronal terminal to its target organ.
- Specificity of endocrine action depends on specialization of target-cell receptors for a specific circulating hormone.

The nervous and endocrine systems have their own realms of authority but interact functionally.

- In general, the nervous system coordinates rapid responses, whereas the endocrine system regulates activities that require duration rather than speed.

Key Terms

Absolute refractory period
Action potential
Adenyl cyclae
All-or-none law
Axon hillock
Cascading
Cell body
Chemically gated channels
Continuous conduction
Convergence
Current
Cyclic adenosine monophosphate (cAMP)
Dendrites
Depolarization
Effector protein
Endocrine system
Endocrinology
Excitable tissue
Excitatory synapse
First messenger
Graded potential
Grand postsynaptic potential
Hormonal communication
Hormone response element (HRE)
Hormones
Hydrophilic hormones
Hyperpolarization
Inhibitory synapse
IPSP
Leak channels
Lipophilic hormones
Mechanically gated channels
Multiple sclerosis
Myelin
Myelinated fibers

Nervous system
Neurocommunication
Neurohormones
Neuromodulators
Neuron
Neuropeptides
Neurotransmitters
Nodes of Ranvier
Oligodendrocytes
Overshoot
Paracrines
Polarization
Postsynaptic neuron
Presynaptic inhibition
Presynaptic neuron
Refractor period
Relative refractory period
Repolarization
Resistance
Saltatory conduction
Schwann cell
Second messenger
Signal transduction
Spatial summation
Synapse
Synaptic cleft
Synaptic knob
Synaptic vesicle
Target cells
Temporal summation
Tetanus toxin
Thermally gated channels
Threshold potential
Voltage-gated channels

Review Exercises

Answers are in the appendix.

True/False-if an answer is false, underline the incorrect portion of the statement and rewrite it to make the statement true.

True/False

_____ 1. The only way in which a neurotransmitter receptor combination can influence the postsynaptic cell is to directly alter its permeability to specific ions.

_____ 2. Synapses permit two-way transmission of signals between two neurons.

_____ 3. A given synapse may can be either excitatory or inhibitory.

_____ 4. A balance of IPSPs and EPSPs will negate each other so that the grand postsynaptic potential is essentially unaltered.

_____ 5. Neurons are specialized for rapid electrical and chemical signaling.

_____ 6. The greater the magnitude of a triggering event, the smaller the graded potential.

_____ 7. Repolarization means that the membrane returns to resting potential after having been depolarized.

_____ 8. When a neuron is driven to threshold potential voltage-gated potassium channels rapidly open.

_____ 9. The threshold potential is reached typically between -50 mV and - 55 mV.

_____ 10. Lipophilic hormones activate second messenger systems when interacting with target cells.

_____ 11. Membrane channels are formed by proteins that span the thickness of the membrane.

_____ 12. Action potentials can serve as faithful long-distance signals.

_____ 13. Once an action potential is initiated in one part of a nerve cell membrane, a self-perpetuating cycle is initiated so that the action potential is propagated throughout the rest of the fiber automatically.

_____ 14. Myelinated fibers conduct impulses about 80 times faster than nonmyelinated fibers of comparable size.

_____ 15. As a general rule, the most urgent types of information are transmitted via unmyelinated fibers, whereas the nervous pathways carrying less urgent information are myelinated.

_____ 16. Damaged neuronal fibers in the brain and spinal cord often regenerate.

_____ 17. The refractory period ensures the unidirectional propagation of the action potential down the axon from the initial site of activation.

_____ 18. When a neuron is at rest, its voltage-gated sodium channels are s in their "closed but capable of opening" conformation.

_____ 19. Some neurotransmitters function through intracellular second messenger systems rather than by directly altering membrane permeability.

_____ 20. The highest threshold is present at the axon hillock because this region has an abundance of voltage-gated sodium channels.

_____ 21. The hormone CCK has been found in synaptic vesicles in diffuse areas of the brain.

_____ 22. Drugs like cocaine increase the amount of neurotransmitter in the synapse by inhibiting neurotransmitter reuptake.

_____ 23. There are an estimated 100 billion neurons in the brain alone.

Fill in the Blank

24. The constant membrane potential that exists when an excitable tissue cell is not displaying rapid changes in potential is referred to as the _____ _____ _____.

25. _____ carry signals toward the cell body while _____ or _____ carry action potentials away from the cell body.

26. The intervening unmyelinated regions of an axon are known as Nodes of _____.

27. Name three neuropeptides that start with the letter G. (1) _____, (2) _____, and (3) _____.

28. The site of synthesis in neurotransmitters is _____ and in neuromodulators is _____.

29. _____ means the membrane has potential; there is a separation of opposite charges.

30. A faster method of propagation, _____, takes place in myelinated fibers.

31. The space between the presynaptic and postsynaptic neurons is called the _____.

32. _____ is an membrane enzyme that when activated produces cAMP.

33. The postsynaptic neuron can be brought to threshold in two ways: (1)_____ and (2) _____.

34. There are three kinds of channels: (1) _____, (2)_____, and (3) _____.

35. The first portion of the axon plus the region of the cell body from which the axon leaves is known as the _____.

36. These myelin-forming cells are _____ in the central nervous system and _____ in the peripheral nervous system.

37. The portion of DNA where the hormone-receptor complex binds is known as the _____.

38. Name three neurotransmitters that start with the letter G: (1) _____, (2) _____, and (3) _____.

39. Large precursor proteins, or _____, are synthesized by ribosomes on the rough endoplasmic reticulum (ER).

40. The nucleus and organelles are housed in the _____, from which numerous extensions known as _____ typically project like antennae to increase the surface area available for receiving signals from other nerve cells.

41. Myelin is composed primarily of _____.

42. An excitable membrane either responds to a stimulus with a maximal action potential that spreads nondecrementally throughout the membrane, or it does not respond with an action potential at all. This is called the _____.

43. In _____.myelin is destroyed leading to a variety of neurological deficits.

44. The entire system of motor neurons that supply skeletal muscles all use the same neurotransmitter _____.

45. _____ refers to the branching of axon terminals so that a single cell synapses with many other cells.

Matching

_____ 46. Gastrin   a. neurotransmitters

_____ 47. Epinephrine   b. neuropeptides

_____ 48. Histamine

_____ 49. Neurotensin

_____ 50. Insulin

_____ 51. Somatostatin

_____ 52. Glutamate

_____ 53. Glucagon

_____ 54. Oxytocin

_____ 55. Aspartate

_____ 56. Small, rapid-acting molecules

_____ 57. Large, slow-acting molecules

Multiple Choice

58. Multiple sclerosis is a pathophysiological condition in which:
   a. The release of the neurotransmitter is blocked.
   b. The demyelination of nerve fibers occurs.
   c. There is a deficiency of dopamine in a particular region of the brain.
   d. The release of gamma-aminobutyric acid is prevented.
   e. The viscosity of mucus increases.

59. Which of the following neurotransmitters is not a small, rapid-acting molecule?
   a. insulin
   b. dopamine
   c. acetylcholine
   d. epinephrine
   e. serotonin

60. Which sequence best describes the events in an action potential?
   a. polarization—hyperpolarization—depolarization—repolarization
   b. hyperpolarization—repolarization—depolarization—polarization
   c. polarization—hyperpolarization—repolarization—depolarization
   d. polarization—depolarization—hyperpolarization—repolarization
   e. depolarization—hyperpolarization—polarization—repolarization

61. Which of the following releases the neurotransmitters?
   a. axon hillock
   b. dendrite
   c. synaptic vesicles
   d. subsynaptic membrane
   e. synaptic cleft

62. Which of the following are the myelin-forming cells in the central nervous system?
   a. presynaptic neurons
   b. postsynaptic cells
   c. oligodendrocytes
   d. Renshaw cells
   e. Schwann cells

63. Which part of the neuron has the lowest threshold potential?
   a. cell body
   b. axon hillock
   c. collaterals
   d. dendrites
   e. axon terminal

64. Which part of the neuron directs the signal toward the cell body?
   a. nodes of Ranvier
   b. axon hillock
   c. collaterals
   d. dendrites
   e. axon

65. Which best describes a postsynaptic neuron? The neuron whose:
    a. Action potentials are propagated toward the synapse.
    b. Myelin is formed by oligodendrocytes.
    c. Action potentials are propagated away from the synapse.
    d. Axons are unmyelinated.
    e. Dendrites are unmyelinated.

66. Which cells form a regeneration tube in peripheral nerves?
    a. neurons
    b. Schwann cells
    c. oligodendrocytes
    d. nodes of Ranvier
    e. Golgi cells

67. Cocaine acts on which neurotransmitter system? a. serotonin b. acetylcholine c. dopamine
    d. thyroid hormone e. GABA

68. A neuron is undergoing an EPSP. Which of the following events is likely taking place?
    a. sodium is entering the cell
    b. potassium is leaving the cell
    c. chloride is entering the cell
    d. voltage-gated channels are opening
    e. the membrane potential moves from -70mV to -90mV

69. Which of the following moves by way of an ATP dependent pump?
    a. only sodium ions
    b. only potassium ions
    c. only chloride ions
    d. both a and b above
    e. all of the above

70. All of the following are a major group of neurotransmitters except one. Which one isn't a transmitter?
    a. acetylcholine
    b. amino acids
    c. monamines
    d. neuropeptides
    e. nucleic acids

71. A resting membrane potential along a neuron is maintained by:
    a. diffusion
    b. ion channels
    c. the sodium-potassium pump
    d. both a and b
    e. both b and c

72. The depolarization of the neuronal membrane is produced by any factor that does one of the following:
    a. increases the membrane's permeability to sodium ions
    b. inhibits the membrane's permeability to sodium ions
    c. stimulates the sodium-potassium pump
    d. increases the membrane's permeability to potassium ions
    e. decreases the membrane's permeability to calcium ion

73. A progressive degeneration of myelin sheaths in neurons of the brain and spinal cord is called:
    a. senility
    b. Huntington's chorea
    c. Parkinson's disease
    d. multiple sclerosis
    e. myasthenia gravis

74. The most common means by which cells communicate with one another is:
    a. gap junctions
    b. direct cell to cell communication
    c. extracellular chemical signals
    d. both a and b
    e. none of the above

75. Which of the following describe the function of graded potentials?
    a. are local changes in membrane potential that occur in varying degrees of magnitude
    b. serve as short distance signals
    c. serve as long-distance signals
    d. both a and b
    e. both a and c

76. The cells of excitable and nonexcitable tissue share which of the following properties?
    a. a threshold potential
    b. a resting membrane potential
    c. an ability to open gated channels
    d. all of the above (a-c)
    e. none of the above (a-c)

77. Which of the following occurs during the rising phase of the nerve action potential?
    a. The potassium potential is greater than the sodium potential.
    b. The sodium potential is greater than the potassium potential.
    c. Both sodium and potassium potentials are the same.
    d. Sodium efflux occurs.
    e. Both a and b are correct.

78. Which of the following occurs at the peak of an action potential?
    a. The electrical gradient for potassium tends to move this ion outward.
    b. The concentration gradient for potassium tends to move this ion outward.
    c. The potassium permeability greatly increases.
    d. Both a and b are correct.
    e. All of the above are correct.

79. Which of the following is responsible for the falling phase of an action potential?
    a. Sodium gates open.
    b. The sodium-potassium ATPase pump restores ions to their original location.
    c. The increased permeability to sodium.
    d. All of the above are correct.
    e. None of the above (a-c) are correct.

80. When is an excitable membrane more permeable to potassium than sodium?
    a. at a resting potential
    b. during the rising phase of the action potential
    c. during the falling phase of the action potential
    d. both a and b
    e. both a and c

81. Where does saltatory conduction occur?
    a. in non myelinated nerve fibers
    b. in myelinated nerve fibers
    c. in all nerve fibers
    d. in all excitable tissues
    e. only in heart tissue

82. Which of the following statements concerning the propagation of an action potential is incorrect?
    a. Saltatory conduction occurs in myelinated nerve fibers.
    b. During conduction by local current flow, there is a flow of current between the active and adjacent inactive area of the cell membrane.
    c. The action potential jumps from one Schwann cell to an adjacent Schwann cell in a myelinated fiber.
    d. Saltatory conduction is faster than conduction by local current flow.
    e. Conduction by local current flow is the method of propagation in unmyelinated fibers.

83. If an action potential were initiated in the center of an axon, in which direction would it travel?
    a. toward the cell body
    b. away from the cell body
    c. in both directions
    d. in neither direction
    e. none of the above

84. Which of the following is not a graded potential?
    a. an action potential
    b. an excitatory postsynaptic potential
    c. an inhibitory postsynaptic potential
    d. a grand postsynaptic potential
    e. a summation potential

85. When does spatial summation occur in a postsynaptic neuron?
    a. when several EPSPs from a single presynaptic input sum to reach threshold
    b. when EPSPs from several presynaptic inputs sum to reach threshold
    c. upon simultaneous interaction of an EPSP and a IPSP
    d. when several IPSPs from a single presynaptic input sum to hyperpolarize the membrane
    e. none of the above

86. At an excitatory synapse:
    a. There is increased permeability of the subsynaptic membrane to both sodium and potassium.
    b. A small hyperpolarization occurs.
    c. An action potential in the presynaptic neuron always causes an action potential in the postsynaptic neuron.
    d. Two of the above are correct.
    e. All of the above (a-c) are correct.

87. An IPSP:
    a. is produced by increased sodium and potassium potentials
    b. is produced by increased potassium or increased chloride potentials
    c. is a small depolarization of the postsynaptic cell
    d. both a and c
    e. both b and c

88. If presynaptic neurons "X" and "Z" are stimulated simultaneously, what changes would you expect to occur in the postsynaptic neuron?
    a. a single EPSP
    b. a single IPSP
    c. temporal summation of EPSPs
    d. spatial summation of EPSPs
    e. no change: an IPSP and EPSP would cancel each other out

89. If presynaptic neuron "Z" is repetitively stimulated very rapidly, what change would you expect to occur in the postsynaptic neuron?
    a. a single EPSP
    b. a single IPSP
    c. spatial summation of EPSPs
    d. none of the above (a-c)
    e. all of the above (a-c)

Modified Multiple Choice

The following questions refer to comparative concentrations, permeabilities, and potentials under various circumstances. Indicate the relationship between the two items listed in each situation by filling in the appropriate letter in the blank using the following answer code:

(a) = A is greater than B
(b) = B is greater than A
(c) = A and B are equal

90. _____ A. permeability of resting nerve cell membrane to potassium
    B. permeability of a resting nerve cell membrane to sodium

91. _____ A. permeability of a nerve cell membrane to sodium during the rising phase of the action potential
    B. permeability of the nerve cell membrane to potassium during the rising phase of the action potential

92. _____ A. permeability of a resting nerve cell membrane to potassium
    B. permeability of a nerve cell membrane to sodium during the rising phase of an action potential

93. _____ A. permeability of a resting nerve cell membrane to potassium
    B. permeability of a nerve cell membrane to potassium during the falling phase of an action potential

94. _____ A. permeability of a nerve cell membrane to sodium during the falling phase of an action potential
B. permeability of a nerve cell membrane to potassium during the falling phase of an action potential

95. _____ A. concentration of sodium in the intracellular fluid of a nerve cell immediately before an action potential
B. concentration of sodium in the intracellular fluid of a nerve cell immediately following an action potential

96. _____ A. concentration of potassium in the intracellular fluid of a nerve cell immediately before an action potential
B concentration of potassium in the intracellular fluid of a nerve cell immediately following an action potential

97. _____ A. concentration of sodium in the extracellular fluid
B. concentration of sodium in the intracellular fluid of a nerve cell immediately following an action potential

98. _____ A. membrane potential at rest
B. the potential of the same membrane when it is hyperpolarized

99. _____ A. membrane potential at rest
B. the potential of the same membrane when it is depolarized

100. _____ A. membrane potential at rest
B. the potential of the same membrane potential when it reaches threshold

Points to Ponder

1. Which ion (sodium or potassium) do you think is more important in neuronal physiology?

2. If yelling and grunting enhance an athlete's performance, what do you think happens when the game official tells them to stop yelling and grunting?

3. How do humans perceive differences in the intensity of external stimuli?

4. What is the physiological importance of inhibition?

5. In common usage how does the definition of a nerve differ from its physiological definition?

6. What are five basic properties of neurons?

7. Why are some basic principles of electricity needed to understand nerve physiology?

8. How many synapses does a neuron have?

9. How do pre- and post-synaptic neurons differ?

10. What is a neuromodulator?

Clinical Perspectives

1. Why are people with multiple sclerosis likely to lose their ability to control skeletal muscles?

2. Suppose that your doctor is performing neurological tests on your arm using tiny needle electrodes. In doing so he places an electrode adjacent to an axon halfway between the cell body and the axon terminal. When the electrical stimulus is applied which way does the signal travel in the axon?

3. Nerve gas exerts its odious effects by inhibiting acetylcholine esterase in skeletal muscles. Since acetylcholine is not degraded, what will the results be in this clinical situation?

4. Muscle weakness in the disease myasthenia gravis is due to the fact the acetylcholine receptors are blocked and destroyed by antibodies secreted by the immune system of the effected person. What clinical results will become apparent in the effected person?

5. Patient A has clinical depression. After eating a meal of shellfish, she falls out of her chair and experiences difficulty breathing. The paramedics find a prescription for a monoamine oxidase inhibitor in her pocket. What may have happened to patient A?

6. Monoamine oxidase (MAO) is an enzyme in the endings of presynaptic axons that breaks down catecholamines and serotonin after they have been taken up from the synaptic cleft. For what clinical situations would you use MAO inhibitors?

Experiment of the Day

Demonstrate the following using a set of dominos.
   a. collateral axon
   b. convergence
   c. divergence
   d. threshold

The Central Nervous System

Chapter Overview

The nervous system is one of the two major regulatory systems of the body, the other being the endocrine system. In order to maintain homeostasis, the body, with its many cells, tissues, organs, and systems, must be completely aware of any and all changes in the internal and external environment. The nervous system provides this information using affectors and afferent neurons. The central nervous system, utilizing various components of the brain, processes this information and initiates a response. Rapid responses are performed by the nervous system through efferent neurons and effectors. The fastest response is the reflex. The endocrine system is used when the response requires duration. To perform these essential functions the brain is divided, and subdivided, and these divisions are interconnected and integrated in such a manner that all parts of the body are controlled. Thus, the brain controls the responses through coordination of the various body systems. In this perspective the nervous system controls homeostasis to preserve the cells of the body.

Chapter Outline

ORGANIZATION OF THE NERVOUS SYSTEM
The nervous system is organized into the central nervous system and the peripheral nervous system.

- The nervous system is organized into the central nervous system (CNS), consisting of the brain and spinal cord, and the peripheral nervous system (PNS), consisting of nerve fibers that carry information between the CNS and other parts of the body.
- The afferent division of the PNS carries information to the CNS.
- Instructions from the CNS are transmitted via the efferent division to effector organs (muscles or glands).
- The efferent division is divided into the somatic nervous system, which consists of motor neurons that supply the skeletal muscles, and the autonomic nervous system fibers, which innervate smooth muscle, cardiac muscle, and glands.
- The latter system is subdivided into the sympathetic nervous system and the parasympathetic nervous system, both of which innervate most of the organs supplied by the autonomic nervous system.
- In addition to the CNS and PNS, the enteric nervous system is an extensive nerve network in the wall of the digestive tract. Digestive activities are controlled by the autonomic nervous system, the enteric nervous system, and hormones.

The three functional classes of neurons are afferent neurons, efferent neurons, and interneurons.

- Three classes of neurons make up the nervous system: afferent neurons, efferent neurons, and interneurons.
- At its peripheral ending, an afferent neuron has a sensory receptor that generates action potentials in response to a particular type of stimulus.
- The afferent neuron cell body is located adjacent to the spinal cord.
- A long peripheral axon extends from the receptor to the cell body, and a short central axon passes from the cell body into the spinal cord.
- The cell bodies of efferent neurons originate in the CNS.
- Efferent axons leave the CNS to course their way to the effector organs they innervate.
- About 99% of all neurons are interneurons which lie entirely within the CNS.
- They serve two main roles: (1) Interneurons lie between the afferent and efferent neurons and are important in the integration of peripheral responses to peripheral information. (2) Interconnections between interneurons themselves are responsible for abstract phenomena associated with the "mind."

Glial cells support the interneurons physically, metabolically, and functionally.

- About 90% of the cells within the CNS are not neurons but glial cells or neuroglia.
- Glial cells do not initiate or conduct nerve impulses.

- The glial cells serve as connective tissue of the CNS and as such help support the neurons both physically and metabolically.
- The four major types of glial cells are astrocytes, oligodendrocytes, microglia. And ependymal cells.

Astrocytes.

- Astrocytes provide a number of critical functions: (1) As the main "glue" they hold the neurons together in proper spatial relationships. (2) Astrocytes serve as a scaffold to guide neurons to their proper final destination during fetal brain development. (3) Astrocytes induce the small blood vessels of the brain to undergo the anatomical and functional changes that are responsible for the establishment of the blood-brain barrier. (4) They help transfer nutrients from blood to neurons. (5) They are important in the repair of brain injuries. (6) Astrocytes play a role in neurotransmitter activity. (7) They take up excess potassium ions from the ECF. (8) Finally, astrocytes have receptors for the same neurotransmitters as do neurons.

Oligodendrocytes.

- Oligodendrocytes form the insulative myelin sheaths around axons in the CNS.

Ependymal cells.

- Ependymal cells line the internal cavities of the CNS.

Microglia.

- The microglia are the phagocytic scavengers of the CNS.

Ependymal cells

- The ependymal cells line the ventricles and help form cerebrospinal fluid.
- They serve as neural stem cells with the potential of forming not only other glial cells but new neurons.

PROTECTION AND NOURISHMENT OF THE BRAIN

The delicate CNS is well protected

- Four major features help protect the CNS: (1) The CNS is enclosed in hard bony structures. (2) Three protective and nourishing membranes, the meninges, lie between the bony covering and the nervous tissue. (3) The brain floats in a special cushioning fluid, the cerebrospinal fluid (CSF). (4) A highly selective blood-brain barrier limits access of blood borne materials into the brain tissue.
- The brain is carefully shielded from harmful changes in the blood by the blood-brain barrier.
- Unlike the rather free exchange across capillaries elsewhere in the body, there are strict limitations on permissible exchanges across brain capillaries.
- The blood-brain barrier consists of both anatomical and physiological factors.
- In brain capillaries the cells are joined by tight junctions that completely seal the capillary wall.
- The only possible exchanges are through the capillary cells themselves.
- Certain areas of the brain are not subject to the blood-brain barrier, most notably a portion of the hypothalamus.

- Functioning of the hypothalamus depends on its "sampling" the blood and adjusting its controlling output accordingly in order to maintain homeostasis.

The brain depends on constant delivery of oxygen and glucose by the blood.

- The brain cannot produce ATP in the absence of oxygen.
- Unlike other tissues, the brain uses only glucose in energy production but does not store any of this nutrient.
- Because of its high rate of demand for ATP, under resting conditions the brain uses 20% of the O_2 and 50% of the glucose consumed in the body. In addition it receives 15% of blood flow.
- Therefore, the brain is absolutely dependent on a continuous, adequate blood supply of oxygen and glucose.
- Brain damage may occur in spite of protective mechanisms.
- Direct brain damage occurs if the brain is violently shaken or jarred by a forceful impact.

OVERVIEW OF THE CENTRAL NERVOUS SYSTEM

- Newer, more sophisticated regions of the brain are piled on top of older, more primitive regions.
- The brain stem, the oldest and smallest region of the brain, is continuous with the spinal cord.
- It controls many of the life-sustaining activities, such as breathing, circulation, and digestion.
- Attached at the top rear portion of the brain stem is the cerebellum, which is concerned with maintaining proper position of the body in space and subconscious coordination of motor activity.
- On top of the brain stem, tucked within the interior of the cerebrum, is the diencephalon.
- It houses the hypothalamus, which controls many homeostasis functions, and the thalamus, which performs some primitive sensory processing.
- On top of this "cone" of lower brain regions is the cerebrum.

- The outer layer of the cerebrum is the cerebral cortex, which caps an inner core that houses the basal nuclei.

CEREBRAL CORTEX
The cerebral cortex is an outer shell of gray matter covering an inner core of white matter.

- The cerebral cortex plays a key role in the most sophisticated neural functions, such as voluntary initiation of movement, final sensory perception, conscious thought, language, personality traits, and other factors we associate with the mind or intellect.
- The cerebrum is divided into two halves, the right and left cerebral hemispheres.
- They are connected to each other by the corpus callosum.
- Each hemisphere is composed of a thin outer shell of gray matter, the cerebral cortex, covering a thick central core of white matter.
- Located deep within the white matter is another region of gray matter, the basal nuclei.
- Gray matter consists predominately of densely packaged cell bodies and their dendrites as well as glial cells.
- Bundles or tracts of myelinated nerve fibers constitute the white matter.
- The fiber tracts in the white matter transmit signals from one part of the cerebral cortex to another, or between the cortex and other regions of the CNS.
- Such communication enables integration between different areas of the cortex and elsewhere.

The four pairs of lobes in the cerebral cortex are specialized for different activities.

- Each half of the cortex is divided into four major lobes: the occipital, temporal, parietal, and frontal lobes.
- The occipital lobes, which are located posteriorly, are responsible for initially processing visual input.
- Sound sensation is initially received by the temporal lobes, located laterally.
- The parietal lobes lie to the rear of the central sulcus on each sideThe frontal lobes are responsible for three main functions: (1) voluntary motor activity, (2) speaking ability, and (3) elaboration of thought

The parietal lobes are responsible for somatosensory processing.

- It is the site for initial cortical processing and perception of both somesthetic and proprioceptive input.
- The parietal lobes are primarily responsible for receiving and processing sensory input such as touch, pressure, heat, cold, and pain from the surface of the body.
- These sensations are collectively known as somesthetic sensations.
- The parietal lobes also perceive awareness of body position, a phenomenon referred to as proprioception.
- The somatosensory cortex, the site for initial cortical processing of this somesthetic and proprioceptive input, is located at the front of each parietal lobe immediately behind the central sulcus.

The primary motor cortex located in the frontal lobes controls the skeletal muscles

- The frontal lobes, lying at the front of the cortex, are responsible for three main functions: (1) voluntary motor activity, (2) speaking ability, and (3) elaboration of thought.
- The area at the rear of the frontal lobe immediately in front of the central sulcus and adjacent to the somatosensory cortex is the primary motor cortex.
- As in sensory processing, the motor cortex on each side of the brain primarily controls muscles on the opposite side of the body.

The higher motor areas are also important in motor control.

- Lower brain regions and the spinal cord control involuntary skeletal muscle activity, such as the maintenance of posture.
- Although fibers originating from the motor cortex can activate motor neurons to bring about the muscle contraction, the motor cortex itself does not initiate voluntary movement.
- The higher motor areas of the brain believed to be involved in the voluntary decision-making period include the supplementary motor area, the premotor cortex and the posterior parietal cortex.
- These higher areas all command the primary motor cortex.
- Furthermore, the cerebellum appears to play an important role in the planning, initiation, and timing of certain kinds of movement by sending input to the motor areas of the cortex.
- The supplementary motor area lies on the medial surface of each hemisphere anterior to the primary motor cortex.
- Stimulation of various regions of this motor area brings about complex patterns of movement, such as the opening or closing of the hand.
- The premotor cortex located on the lateral surface of each hemisphere in front of the primary motor cortex, is believed to be important in orienting the body and arms toward a specific target.

- The premotor cortex is guided by sensory input processed by the posterior parietal cortex, a region that lies posterior to the primary somatosensory cortex.

Different regions of the cortex control different aspects of language..

- The areas of the brain responsible for language ability are found in only one hemisphere: the left hemisphere in the vast majority of the population.
- The primary areas of cortical specialization for language are Broca's and Wernicke's. Broca's area, which is responsible for speaking ability, is located in the left frontal lobe in close association with the motor areas of the cortex, which control the muscles necessary for articulation.
- Wernicke's area, located in the left cortex at the juncture of the parietal, temporal, and occipital lobes, is concerned with language comprehension.

The association areas of the cortex are involved in many higher functions.

- The prefrontal association cortex is the front portion of the frontal lobe just anterior to the premotor cortex.
- The roles attributed to this region are: (1) planning for voluntary activity; (2) decision making (3) creativity; and (4) personality traits
- The parietal-temporal-occipital association cortex is found at the interface of the three lobes for which it is named.
- In this location, this cortical area pools and integrates somatic, auditory, and visual sensations projected from these three lobes for complex perceptual processing.
- This region is also involved in the language pathway connecting Wernicke's area to the visual and auditory cortices.
- The limbic association is located on the bottom and adjoining inner portion of each temporal lobe.
- This area is concerned primarily with motivation and emotion and is extensively involved in memory.

The cerebral hemispheres have some degree of specialization.

- The left cerebral hemisphere excels in performance of logical, analytical, sequential, and verbal tasks, such as math, language forms, and philosophy.
- The right cerebral hemisphere excels in nonlanguage skills, especially spatial perception, and artistic and musical endeavors.

An electroencephalogram is a record of postsynaptic activity in cortical neurons.

The electroencephalogram (EEG) has three major uses: (1) It is often used as a clinical tool in the diagnosis of cerebral dysfunction. (2) The EEG is used in the legal determination of brain death. (3) The EEG is also used to distinguish various stages of sleep.

BASAL NUCLEI, THALAMUS, AND HYPOTHALAMUS

The basal nuclei play an important inhibitory role in motor control.

- The subcortical regions include the basal nuclei, located in the cerebrum, and the thalamus and hypothalamus, located in the diencephalon.
- The basal nuclei play a complex role in the control of movement in addition to having nonmotor functions that are less well understood.
- The basal nuclei are important in: (1) inhibiting muscle tone throughout the body; (2) selecting and maintaining purposeful motor activity while suppressing useless or unwanted patterns of

movement; and (3) helping monitor and coordinate slow, sustained contractions, especially those related to posture and support.
- Loss of dopamine producing neurons in the basal nuclei result in Parkinson's disease.

The thalamus is a sensory relay station and is important in motor control.

- It is speculated that the thalamus positively reinforces voluntary motor behavior initiated by the cortex, whereas the basal nuclei modulate this activity by exerting an inhibitory effect on the thalamus to eliminate antagonistic or unnecessary movements.
- The thalamus is a sensory relay station and is important in motor control.
- The diencephalon consists of two main parts, the thalamus and the hypothalamus.
- The thalamus serves as a "relay station" and synaptic integrating center for preliminary processing of all sensory input on its way to the cortex.
- The thalamus, along with the brain stem and cortical association areas, is important in our ability to direct attention to stimuli of interest.

The hypothalamus regulates many homeostatic functions.

- The hypothalamus is the area of the brain most notably involved in the direct regulation of the internal environment.
- The hypothalamus: (1) controls body temperature; (2) controls thirst and urine output; (3) controls food intake; (4) controls anterior pituitary hormone secretion; (5) produces posterior pituitary hormones; (6) controls uterine contractions and milk ejection; (7) serves as a major autonomic nervous system coordinating center, which in turn affect all smooth muscle, cardiac muscle, and exocrine glands; and (8) plays a role in emotional and behavioral patterns. (9) participates in the sleep-wake cycle.

EMOTIONS, BEHAVIOR, AND MOTIVATION

The limbic system plays a key role in emotion.

- The limbic system includes portions of each of the following: the lobes of the cerebral cortex, the basal nuclei, the thalamus, and the hypothalamus.

The limbic system and higher cortex participate in the control of basic behavior patterns.

- This complex interacting network is associated with emotions, basic survival, and sociosexual behavioral patterns, motivation, and learning.
- Homeostasis drives represent the subjective urges associated with specific bodily needs that motivate appropriate behavior to satisfy those needs.

Motivated behaviors are goal directed.

- Motivation is the ability to direct behavior toward specific goals.
- Homeostatic drives represent the subjective urges associated with specific body needs that motivate appropriate behavior to satisfy those needs.

Norepinephrine, dopamine, and serotonin are neurotransmitters in pathways for emotions and behavior.

- The underlying neurophysiological mechanisms responsible for the psychological observations of motivated behavior and emotions largely remain a mystery, although the neurotransmitters norepinephrine, dopamine, and serotonin all have been implicated.
- Norepinephrine and dopamine, both chemically classified as catecholamines, are known to be transmitters in the regions that elicit the highest rates of self-stimulation in animals equipped with do-it-yourself devices.
- Schizophrenia, a mental disorder characterized by delusions and hallucinations, probably results from excess dopamine transmissions.
- Serotonin and norepinephrine are synaptic messengers in the regions of the brain involved in pleasure and motivation.
- A functional deficiency of serotonin or norepinephrine or both is implicated in depression, a disorder characterized by a pervasive unpleasant mood accompanied by a general loss of interest and ability to experience pleasure.
- Amphetamine stimulates the release of dopamine from dopamine-secreting neurons.
- A functional deficiency of serotonin or norepinephrine or both is implicated in depression.

LEARNING AND MEMORY

Learning is the acquisition of knowledge as a result of experiences.

- Learning is the acquisition of knowledge or skills as a consequence of experience, instruction, or both.
- Learning is a change in behavior that occurs as a result of experiences.

Memory is laid down in stages.

- Memory is the storage of acquired knowledge for later recall.
- Learning and memory form the basis by which individuals adapt their behavior to their particular external circumstances.
- The neural change responsible for retention or storage of knowledge is known as a "memory trace."
- Concepts, not verbatim information, are generally stored.

- Short-term memory lasts for seconds to hours, whereas long-term memory is retained for days to years.
- Occasionally, individuals suffer from a lack of memory that involves whole portions of time rather than isolated bits of information. This condition is called "amnesia."
- Retrograde amnesia is the inability to recall recent past events.
- Anterograde amnesia is the inability to store memory in long-term storage for later retrieval.

Short-term memory and long-term memory involve different molecular mechanisms.

- Short-term memory involves transient modifications in the function of preexisting synapses.
- Long-term memory storage, requires the activation of specific genes that control synthesis of proteins, needed for lasting structural or functional changes at specific synapses

Memory traces are present in multiple regions of the brain.

- The hippocampus, the elongated, medial portion of the temporal lobe that is part of the limbic system, plays a vital role in short-term memory involving the integration of various related stimuli and is also crucial for consolidation into long-term memory. The hippocampus and the surrounding regions play an important role in declarative memories.

- Alzheimer's disease accounts for about two-thirds of the cases of senile dementia, which refers to a generalized age-related diminution of mental abilities. Alzheimer's can only be confirmed at autopsy upon finding the characteristic brain lesions associated with the disease: neuritic plaques and neurofibrillary tangles dispersed throughout the cerebral cortex, and degeneration of cell bodies of certain neurons in the basal brain.
- Three recent discoveries have offered hope for diagnosing Alzheimer's: new brain imaging techniques, a possible skin test for defective potassium channels, and a simple eye-drop test.
- The cerebellum seems to play an essential role in procedural memories involving motor skills gained through repetitive training.

The prefrontal cortex and working memory

- The prefrontal cortex serves as a temporary storage site for holding relevant data online and is largely responsible for the so-called executive functions involving manipulation and integration of this information for planning, juggling competing priorities, problem solving, and organizing activities

CEREBELLUM
The cerebellum is important in balance and in planning and execution of voluntary movement.

- The cerebellum, which is attached to the back of the upper portion of the brain stem, lies underneath the occipital lobe of the cortex.
- The cerebellum consists of three functionally distinct parts, the vestibulocerebellum, the spinocerebellum, and the cerebrocerebellum.
- The vestibulocerebellum is important for the maintenance of balance and controls eye movement.
- The spinocerebellum enhances muscle tone and coordinates skilled, voluntary movements.

- The cerebrocerebellum plays a role in the planning and initiation of voluntary activity by providing input to the cortical motor areas.
- This is also the cerebellar region involved in procedural memories.
- The cerebellum and basal nuclei both monitor and adjust motor activity commanded from the motor cortex.
- The cerebellum does not have any direct influence on the efferent motor neurons.
- The cerebellum helps maintain balance, smoothes out fast, phasic motor activity, and enhances muscle tone.

BRAIN STEM

The brain stem is a vital link between the spinal cord and higher brain regions.

- The brain stem consists of the medulla, pons, and midbrain.
- All incoming and outgoing fibers traversing between the periphery and higher brain centers must pass through the brain stem, with incoming fibers relaying sensory information to the brain and outgoing fibers carrying command signals from the brain.
- The functions of the brain stem include the following:
  1. Except for the vagus, the cranial nerves arise from the brain stem and supply structures in the head and neck with both sensory and motor fibers. Most of the branches of the vagus supply organs in the thoracic and abdominal cavities.
  2. Collected within the brain stem are neuronal clusters that control heart and blood vessel functions, respiration, and many digestive activities.
  3. The brain stem also plays a role in modulating the sense of pain.
  4. The brain stem plays a role in the regulation of muscle reflexes involved with equilibrium and posture.
  5. The reticular formation, found in the brain stem, receives and integrates all synaptic input. The reticular activating system (RAS) controls the overall degree of cortical alertness and is important in the ability to direct attention.
  6. The centers responsible for sleep are also housed within the brain stem.

Sleep is an active process consisting of alternating periods of slow-wave and paradoxical sleep.

- Consciousness refers to subjective awareness of the external world and self.
- Even though the final level of awareness resides in the cerebral cortex and a crude sense of awareness is detected by the thalamus, conscious experience depends on the integrated functioning of many parts of the nervous system.
- The following states of consciousness are listed in decreasing order of level of arousal: maximum alertness, wakefulness, sleep, and coma.
- Sleep is an active process.
- There are two types of sleep, characterized by different EEG patterns and different behaviors: slow-wave sleep and paradoxical, or REM, sleep.
- Slow-wave sleep occurs in four stages, each displaying progressively slower EEG waves of higher amplitude.
- In slow-wave sleep, the person still has considerable muscle tone and frequently shifts body position.
- The muscles are completely relaxed in paradoxical sleep with no movement taking place.
- Paradoxical sleep is further characterized by rapid eye movement (REM).
- The EEG pattern during paradoxical sleep is similar to that of a wide-awake, alert individual.

- A person cyclically alternates between the two types of sleep throughout the night.

The sleep-wake cycle is controlled by interactions among three neural systems.

- The sleep-wake cycle, as well as the various stages of sleep, is believed to be due to the cyclical interplay of three different neural systems in the brain stem: (1) an arousal system, (2) a slow-wave sleep center, and (3) a paradoxical sleep center.
- Some scientists speculate that paradoxical sleep is necessary to allow the brain to "shift gears" to accomplish the long-term structural and chemical adjustments necessary for learning and memory.
- A specific amount of paradoxical sleep appears to be required.

The function of sleep is unclear.

- Even though humans spend about a third of their lives sleeping, the reason sleep is needed largely remains a mystery.
- Sleep is not associated with a reduction in neural activity but rather a change in activity.

SPINAL CORD
The spinal cord extends through the vertebral canal and is connected to the spinal nerves.

- Exiting through the large hole in the base of the skull, the spinal cord is enclosed by the protective vertebral column as it descends through the vertebral canal.
- Paired spinal nerves emerge from the spinal cord through spaces formed between the bony, winglike arches of adjacent vertebrae.
- There are eight pairs of cervical nerves, twelve thoracic nerves, five lumbar nerves, five sacral nerves, and one coccygeal nerve.
- The spinal cord extends only to the level of the first or second lumbar vertebra.
- The remaining nerves are greatly elongated in order to exit the vertebral column at their appropriate space.

The white matter of the spinal cord is organized into tracts.

- Tracts are bundles of nerve fibers with similar functions.
- The gray matter in the spinal cord forms a butterfly-shaped region on the inside and is surrounded by the outer white matter.
- The gray matter of the cord consists primarily of neuronal cell bodies and their dendrites, short interneurons, and glial cells.
- The white matter is organized into tracts.
- Each of these tracts begins or ends within a particular area of the brain.
- Some are ascending tracts that transmit to the brain signals derived from afferent input.
- Others are descending tracts that relay messages from the brain to the efferent neurons.

Spinal nerves carry both afferent and efferent fibers.

- Afferent fibers carrying incoming signals enter the spinal cord through the dorsal root.
- Efferent fibers carrying outgoing signals leave the spinal cord through the ventral root.
- The cell bodies of the efferent neurons are in the gray matter of the cord.
- The dorsal and ventral roots at each level join to form a spinal nerve.
- A nerve is a bundle of peripheral neuronal axons, some afferent and some efferent, enclosed by a connective tissue covering and following the same pathway
- The spinal nerves and the cranial nerves constitute the peripheral nervous system.

The spinal cord is responsible for the integration of many basic reflexes.

- The neural pathway involved in accomplishing reflex activity is known as a "reflex arc," which typically includes five basic components: a receptor, an afferent pathway, an integrating center, an efferent pathway, and an effector.

- There are several types of reflexes involving the reflex arc.

- The simplest reflex is the stretch reflex, in which an afferent neuron originating at a stretch-detecting receptor in a skeletal muscle terminates directly on the efferent neuron supplying the same skeletal muscle to cause it to contract and counteract the stretch.

- Pathways for unconscious responsiveness digress from the typical reflex arc in two general ways: responses mediated at least in part by hormones, and local responses that do not involve either nerves or hormones.

Key Terms

Afferent division
Afferent neuron
Amnesia
Amygdala
Aphasias
Astrocytes
Autonomic nervous system
Blood-brain barrier
Brain stem
Broca's area
Cauda equine
Central nervous system (CNS)
Central sulcus
Cerebellum
Cerebrocerebellum
Cerebrospinal fluid
Cerebrum
Cognition
Consciousness
Consolidation
Corpus callusum
Depression
Diencephalon
Dorsal root
Dorsal root ganglion
Dyslexia
Effector organs
Efferent division
Efferent neuron

Electroencephalogram (EEG)
Emotion
Ependymal cells
Epilepsy
Frontal lobes
Glial cells
Gray matter
Hippocampus
Homeostatic drives
Homunculus
Hypothalamus
Interneuron
Language
Learning
Limbic system
Long-term memory
Medulla
Memory
Meninges
Microglia
Midbrain
Motor homunculus
Motor program
Motivation
Nerve
Neuroendocrinology
Neuroglia
Occipital lobes
Oligodendrocytes
Parasympathetic nervous system
Parietal lobes
Peripheral nervous system (PNS)
Parkinson's disease (PD)
Pons

Proprioception
Psychoactive drugs
Reflex arc
Sensory
homunculus
Sensory receptors
Short-term memory
Sleep
Sleep-wake cycle
Somatic nervous
system

Somatic sensations
Spinal reflex
Spinocerebellum
Sympathetic nervous system
Temporal lobes
Thalamus
Vestibulocerebellum
Wernicke's area
White matter
Withdrawal reflex
Working memory

Review Exercises
Answers are in the appendix

True/False-if an answer is false, underline the incorrect portion of the statement and rewrite it to make the statement true.

_____ 1. The brain stem, the oldest region of the brain, is continuous with the spinal cord.

_____ 2. Astrocytes are the glial cells in the CNS that produce myelin,

_____ 3. Simple awareness of touch, pressure, or temperature is detected by the thalamus.

_____ 4. The premotor motor areas of the frontal lobe are guided by sensory input. .

_____ 5. The cerebellum appears to play an important role in the planning, initiation, and timing of certain kinds of movement by sending input to the motor areas of the cortex.

_____ 6. In most people, the areas of the brain responsible for language ability are found in only one hemisphere—the right hemisphere.

_____ 7. The brain receives an unusually large amount of blood flow relative to its size.

_____ 8. The central nervous system (CNS) consists of the brain and the spinal cord.

_____ 9. The efferent division carries information to the CNS, apprising it of the external environment and providing status reports on internal activities being regulated by the nervous system.

_____ 10. Afferent neurons are shaped differently than efferent neurons and interneurons.

_____ 11. The efferent neuron cell body is located adjacent to the spinal cord.

_____ 12. The cell bodies of afferent neurons originate in the CNS.

_____ 13. Interneurons lie between the afferent and efferent neurons.

_____ 14. Interconnections between interneurons are responsible for the abstract phenomena associated with the mind.

_____ 15. The ependymal cells line the ventricles and help form cerebrospinal fluid

_____ 16. The cerebellum does not have any direct influence on the efferent motor neurons.

_____ 17. The cerebellum plays a role in the planning and initiation of voluntary activity by providing input to the cortical motor areas.

_____ 18. The vestibulocerebellum helps maintain balance, smoothes out fast, phasic motor activity, and enhances muscle tone.

_____ 19. About 90% of the cells within the CNS are neurons.

_____ 20. Glial cells do not initiate or conduct nerve impulses.

_____ 21. Microglial cells form myelin sheaths in CNS.

_____ 22. Ependymal cells line the internal cavities of brain and spinal cord.

_____ 23. The blood brain barrier is not uniform throughout the brain.

_____ 24. Cerebrospinal fluid is associated wit the brain, but not the spinal cord.

_____ 25. The blood-brain barrier protects the delicate brain and spinal cord from chemical fluctuations in the blood and minimizes the possibility that potentially harmful blood borne substances might reach the central neural tissue.

_____ 26. The brain's only energy source is glucose. .

_____ 27. Damaged brain cells release excessive amounts of glutamate.

_____ 28. The spinal cord gives rise to 12 pairs of spinal nerves.

_____ 29. The vagus nerve is the only cranial nerve that does not originate from the brain stem.

_____ 30. Ascending tracts relay messages from the brain to efferent neurons.

_____ 31. Efferent fibers carrying incoming signals enter the spinal cord through the dorsal root.

_____ 32. The cell bodies for the efferent neurons originate in the gray matter and send axons out through the ventral root.

_____ 33. The 31 pairs of spinal nerves, along with the 12 pairs of cranial nerves that arise from the brain, constitute the peripheral nervous system.

_____ 34. The skin has different receptors for warmth, cold, light touch, pressure, and pain.

_____ 35. All incoming and outgoing fibers traversing between the periphery and higher brain centers must pass through the brain stem.

_____ 36. Most of the branches of the vagus nerve supply organs in the thoracic and abdominal cavities.

_____ 37. The following states of consciousness are listed in increasing order of level of arousal: coma, sleep, wakefulness, and maximum alertness.

_____ 38. The brain's overall level of activity is reduced during sleep.

_____ 39. Paradoxical sleep occupies 80% of total sleeping time throughout adolescence and most of adulthood.

_____ 40. Slow-wave sleep is characterized by REM.

_____ 41. The subcortical regions include the basal nuclei, located in the cerebrum, and the thalamus and hypothalamus, located in the diencephalon.

_____ 42. The thalamus exerts an inhibitory effect on motor activity by acting through neurons in the brain.

_____ 43. The hypothalamus is the area of the brain most notably involved in the direct regulation of the internal environment.

_____ 44. Loss of dopamine in the basal nuclei is characteristic of Parkinson's disease.

_____ 45. Short-term memory is believed to involve relatively permanent functional or structural changes between existing neurons in the brain.

_____ 46. Sleep is not associated with a reduction in neural activity but rather a change in activity

_____ 47. Numerous hormones and neuropeptides are known to affect learning and memory processes.

_____ 48. Both the nervous and the endocrine system ultimately alter their target cells by releasing chemical messengers.

_____ 49. Both the nervous and the endocrine systems have an influence on other major control systems.

_____ 50. The nervous system has a very slow speed of response.

_____ 51. The EEG is used in the legal determination of brain death

_____ 52. The endocrine system is responsible for coordinating rapid, precise responses.

_____ 53. In short term memory permanent changes in cellular structure underlie the storage of information.

_____ 54. In the nervous system, neurotransmitters are released into the synaptic knob.

Fill in the Blanks

55. The _____ regulates muscle tone and coordinates skilled, voluntary movements.

56. The _____ is important for the maintenance of balance and control of eye movement.

57. The _____ plays a role in the planning and initiation of voluntary activity by providing input to the cortical motor areas.

58. There are two types of reflexes: _____ (_____) reflexes and _____ (_____) reflexes.

59. The spinal nerves are named according to the region of the vertebral column from which they emerge: There are eight pairs of _____ nerves, twelve _____ nerves, five _____ nerves, five _____ nerves, and one coccygeal nerve.

60. The thick bundle of elongated nerve roots within the lower vertebral canal is known as the _____.

61. _____ fibers carrying incoming signals enter the spinal cord through the dorsal root; _____ fibers carrying outgoing signals leave through the ventral root.

62. The simplest reflex is a _____ reflex where the afferent neuron synapses directly into the efferent neuron.

63. The dorsal and ventral roots at each level join to form a(n) _____.

64. A(n) _____ is a bundle of peripheral neuronal axons, some afferent and some efferent, which are enclosed by a connective-tissue covering.

65. The diencephalon houses two brain components: (1) _____ and (2) _____.

66. The _____ is the largest portion of the human brain.

67. Each hemisphere is composed of a thin outer shell of gray matter called _____.

68. _____ matter consists predominantly of densely packaged cell bodies and their dendrites as well as glial cells. Bundles or tracts of myelinated nerve fibers (axons) constitute the _____ matter.

69. The _____ are primarily responsible for receiving and processing sensory input such as touch, pressure, heat, cold, and pain from the surface of the body. These sensations are collectively known as _____.

70. _____ area, which is responsible for speaking ability, is located in the left _____. _____ area, located in the left _____ at the juncture of the parietal, temporal, and occipital lobes, is concerned with language comprehension.

71. About 90% of the cells within the CNS are _____ or _____.

72. _____ serve as a scaffold to guide neurons to their proper final destination during fetal brain development.

73. _____ form the insulative myelin sheaths around axons in the CNS.

74. _____ phagocytic cells are the scavengers of the CNS.

75. The innermost meningeal layer, the _____, is the most fragile. It is highly vascular and closely adheres to the surfaces of the brain and spinal cord.

76. Cerebrospinal fluid is formed primarily by the _____, found in particular regions of the ventricle cavities of the brain.

77. The brain stem consists of the _____, _____, and _____.

78. Running throughout the entire brain stem and into the thalamus is a widespread network of interconnected neurons called the _____.

79. _____ refers to subjective awareness of the external world and self.

80. _____ refers to the total unresponsiveness of a living person to external stimuli, caused either by brain stem damage or by widespread depression of the cerebral cortex.

81. _____ sleep can be considered either the deepest sleep or the lightest sleep.

82. In _____ sleep a person still has considerable muscle tone and frequently shifts body position.

83. In _____ sleep, muscles are completely relaxed with no movement taking place.

84. The subcortical regions include the _____, located in the cerebrum, and the _____ and _____, located in the diencephalon.

85. The _____ serves as a "relay station" and synaptic integrating center for preliminary processing of all sensory input on its way to the cortex.

86. The _____ controls body temperature, food intake, and thirst.

87. _____ is the ability to direct behavior toward specific goals.

88. _____ is the acquisition of knowledge or skills as a consequence of experience, instruction, or both, and is a change in behavior that occurs as a result of experiences.

89. The process of transferring and fixing short-term memory traces into long-term memory stores is known as _____.

90. A _____ is a bundle of peripheral neuronal axons, some afferent and some efferent, enclosed by a connective tissue covering and following the same pathway

91. The nervous system is organized into the central nervous system, consisting of the _____ and _____, and the _____.

92. The _____ carries information to the CNS.

93. Three classes of neurons make up the nervous system: (1) _____, (2) _____, and (3) _____.

94. Interneurons lie entirely within the _____.

Matching
Match function to type of glial cell.

a. Astrocytes
b. Ependymal cells
c. Microglia
d. Oligodendrocytes

_____ 95.   Contribute to formation of cerebrospinal fluid
_____ 96.   Form myelin sheaths in CNS
_____ 97.   Induce formation of blood-brain barrier
_____ 98.   Serve as scaffold during fetal brain development
_____ 99.   Take up excess potassium to help maintain proper brain ECF ion concentration and normal neural excitability
_____ 100. Physically support neurons in proper spatial relationship
_____ 101. Play a role in defense of brain as phagocytic scavengers
_____ 102. Form neural scar tissue
_____ 103. Line internal cavities of brain and spinal cord
_____ 104. Take up and degrade released neurotransmitters into raw materials for synthesis of more neurotransmitters by neurons
_____ 105. Possess receptors for neurotransmitters, which may be important in a chemical signaling system

Multiple Choice

106. Which of the nerve cells loses its ability to undergo cell division?
    a. astrocytes
    b. ependymal cells
    c. interneurons
    d. oligodendrocytes
    e. microglia

107. Which is the outermost covering of the brain?
    a. pia mater
    b. arachnoid mater
    c. dura mater
    d. perineural sac
    e. epineural sac

108. Which type of sleep is characterized by having progressively slower EEG waves of higher amplitude?
    a. REM sleep
    b. slow-wave sleep
    c. paradoxical sleep
    d. rhythmic sleep
    e. parenthetical sleep

109. This disease is characterized neurofibrillary plaques and tangles.
    a. cerebrovascular accident
    b. dyslexia
    c. aphasias
    d. Alzheimer's
    e. Parkinson's

110. In which form of short-term memory is calcium entry into the presynaptic terminal enhanced?
    a. paradoxical
    b. sensitization
    c. aphasias
    d. habituation
    e. amnesia

111. Which structure controls eye movement?
    a. spino cerebellum
    b. cerebrocerebellum
    c. vestibulocerebellum
    d. medulla oblongata
    e. retinal plexus

112. Deficiency of serotonin and which other neurotransmitter is implicated in depression?
    a. norepinephrine
    b. dopamine
    c. GABA
    d. Acetylcholine
    e. CCK

113. Which of the following is a language disorder caused by damage to specific cortical areas?
    a. amblyopia
    b. hydrocephalus
    c. Alzheimer's disease
    d. Schizophrenia
    e. aphasia

114. This disorder is diagnosed when a large collection of neurons abnormally undergo synchronous action potentials that produce stereotypical involuntary spasms and alterations in behavior.
    a. Alzheimer's disease
    b. Schizophrenia
    c. Parkinson's disease
    d. epilepsy
    e. dyslexia

115. Which area of the brain is associated with procedural memories?
    a. hippocampus
    b. basal nuclei
    c. cerebellum
    d. corpus callosum
    e. thalamus

Modified Multiple Choice

Indicate which brain structure is associated with each function by writing the appropriate letter in the blank using the following answer code:

(a) = cerebellum
(b) = hypothalamus
(c) = brain stem
(d) = basal nuclei
(e) = cerebral cortex

116. _____ Involved in food intake.
117. _____ Structure that initiates all voluntary movement.
118. _____ Inhibits muscle tone throughout the body.
119. _____ Damage to this structure is associated with a resting tremor.
120. _____ Controls thirst and urine output, .
121. _____ Contains the centers for respiration and heart and blood vessel function.
122. _____ Helps monitor and coordinate slow, sustained contractions with respect to balance.
123. _____ Contains the autonomic nervous system coordinating center.
124. _____ Disorders of this structure produce an intention tremor.
125. _____ Plays an important role in emotion and behavior patterns.
126. _____ Coordinates motor activity.
127. _____ The structure that accomplishes final sensory perception.
128. _____ Consists of the medulla, pons, and medulla.

Points to Ponder

1. Suppose you had a dream last night in which you, David Crockett, and Columbus were building a space laboratory on the moon. How did your brain come up with such a story?

2. After playing tennis yesterday you stopped at a fast food store for something to drink and then decided to have a snack. What were the cues your brain received and how were they processed?

3. It has been demonstrated by experimentation that learning can occur by rewarding the animal if it responds in the desirable manner. If the desirable response is for the rat to run through a maze, and we reward the animal with food upon completion of the task, would it be correct to assume that learning occurs only if the rat is stressed? What happens if you feed the rat first?

4. How do the meninges support and/or protect the brain?

5. What is the functional significance of the gyri and sulci of the cerebral cortex?

6. In what ways are the physiological functions of stimulants and antidepressants similar? How do they differ physiologically?

7. Why do most people dream every night?

8. What happens when the corpus callosum is severed?

9. How are nuclei and ganglia similar? How do they differ?

10. What is the role of the processing center in the spinal cord?

Clinical Perspectives

1. What would you suggest to a person complaining of frequent severe headaches of long duration?

2. What loss in body function would you expect in a car crash victim with extensive injury to the cerebellum, the left temporal lobe, and the left occipital lobe?

3. Based on what you know about the brain, why does a person with Parkinson's disease suffer from muscle rigidity, resting tremors, and difficulty in initiating voluntary movements?

4. How do general anesthetics produce unconsciousness by depressing the RAS?

5. Explain how "split-brain" patients have contributed to research on the function of the cerebral hemispheres.

6. Fetal alcohol syndrome, produced by consuming too much alcohol during pregnancy, damages the corpus callosum and basal nuclei. What effects would damage to these areas have after the child is born?

7. If a patient is taking a drug that effects protein synthesis, how would the drug effect short and long term memory?

8. What would you tell a Parkinson's patient about his/her illness?

9. Based on your knowledge of the brain, what would you tell a patient about the cause of his or her headache?

10. A young college student begins to have lower-back pain. The pain radiates down the student's left leg and is intense when the student coughs and sneezes. In addition, there is some loss of motor function in the left leg. What neurological problem exists? What are the neurological causes of the symptoms? Which spinal nerve is probably involved?

6

The Peripheral Nervous System: Afferent Division and Special Senses

Chapter Overview

The nervous system, one of the two major regulatory systems of the body, consists of the central nervous system (CNS), composed of the brain and spinal cord, and the peripheral nervous system, composed of the afferent and efferent fibers that relay signals between the CNS and the periphery. The afferent division of the peripheral nervous system detects, encodes, and transmits peripheral signals to the CNS for processing. It is the communication link by which the CNS is informed about the internal and external environment. Perception plays a major role in maintaining homeostasis. Any change in the environment must be dealt with by the organism if homeostasis is to be maintained. The necessary internal and external environmental information is picked up at the peripheral endings of afferent neurons. These specialized endings are classified as follows: photoreceptors, mechanoreceptors, thermoreceptors, osmoreceptors, chemoreceptors, and nociceptors. All of these receptors depolarize or hyper-polarize to produce a receptor or generator potential that is a graded potential. Graded potentials initiate action potentials, thus the peripheral nervous system delivers the signal to the CNS for interpretation. Each type of receptor is part of a designed pathway going to a specific part of the CNS. With the information reaching the CNS, the nervous system performs its all important role of maintaining homeostasis. In this chapter special emphasis is placed on: pain (nociceptors), vision (photoreceptor), hearing and equilibrium (mechanoreceptors), and taste and smell (chemoreceptors).

Chapter Outline

INTRODUCTION
Receptors have differential sensitivities to various stimuli.

- At their peripheral endings afferent neurons have receptors that apprise the CNS of detectable changes in both the external world and the internal environment by generating action potentials in response to the stimuli.
- Stimuli bring about graded potentials known as receptor potentials. The conversion of stimulus energy into a receptor potential is known as sensory transduction.
- Photoreceptors are responsive to light.
- Mechanoreceptors are sensitive to mechanical energy.
- Thermoreceptors are sensitive to heat and cold.
- Osmoreceptors detect changes in the concentration of solutes in the body fluids and the resultant changes in osmotic activity.
- Chemoreceptors are sensitive to specific chemicals.
- Nociceptors, or pain receptors, are sensitive to tissue damage such as cutting or burning.
- Intense stimulation of any receptor is also perceived as painful.

Uses for information detected by receptors

- The information detected by receptors is conveyed via afferent neurons to the CNS, where it is used for a variety of purposes: (1) it us essential for control of efferent output; (2) processing of sensory input by the reticular activating system in the brain stem is critical for cortical arousal and consciousness; (3) central processing of sensory information to give rise to perception; (4) information can be stored for future reference; (5) it can impact our emotions.

A stimulus alters the receptor's permeability, leading to a graded receptor potential.

- A receptor may be either a specialized ending of the afferent neuron or a separate cell closely associated with the peripheral ending of the neuron.
- The receptor potential is a graded potential whose amplitude and duration can vary, depending on the strength and the rate of application or removal of the stimulus.
- In the case of a separate receptor, a receptor potential triggers the release of a chemical messenger that diffuses across the small space separating the receptor from the ending of the afferent neuron.
- Binding of the chemical messenger with specific protein receptor sites on the afferent neuron opens chemical messenger-gated sodium channels.
- In the case of a specialized afferent ending, local current flow between the activated receptor ending undergoing a generator potential and the cell membrane adjacent to the receptor brings about the opening of voltage-gated sodium channels in this adjacent region.

Receptor potentials may initiate action potentials in the afferent neuron.

- A larger receptor potential cannot bring about a larger action potential (all-or-none law), but it can induce more rapid firing of action potentials.
- The more rapidly an afferent fiber fires, the more neurotransmitter it releases. This neurotransmitter influences the next cell in the neural pathway, passing on information about stimulus strength
- Stimulus intensity is distinguished both by the frequency of action potentials generated in the afferent neuron (frequency code) and by the number of receptors activated within the area (population code).

Receptors may adapt slowly or rapidly to sustained stimulation.

- There are two types of receptors—tonic receptors and phasic receptors—based on their speed of adaptation.
- Tonic receptors do not adapt at all or adapt slowly.
- Phasic receptors are rapidly adapting receptors.
- Phasic receptors are useful in situations where it is important to signal a change in stimulus intensity rather than to relay status quo information.

Visceral afferents carry subconscious input whereas sensory afferents carry conscious input.

- The afferent division of the peripheral nervous system sends information about the internal and external environment to the CNS.
- The incoming pathway for subconscious information derived from the internal viscera is called a "visceral afferent."
- Afferent input that reaches the level of conscious awareness is known as "sensory information" and the incoming pathway is a "sensory afferent."

Each somatosensory pathway is "labeled" according to modality and location.

- On reaching the spinal cord, afferent information has two possible destinies: (1) it may become part of a reflex arc, bringing about an appropriate effector response, or (2) it may be relayed upward to the brain via ascending pathways for further processing and possible conscious awareness.

- Somatosensory pathways consist of discrete chains of neurons synaptically interconnected in a particular sequence to accomplish progressively more sophisticated processing of the sensory informationDifferent types of incoming information are kept separated within specific labeled lines as the information passes between the periphery and the cortex.

Acuity is influenced by receptive field size and lateral inhibition.

- Each sensory neuron responds to stimulus information only within a circumscribed region of the skin surface surrounding it; this region is known as its receptive field.
- The smaller the receptive field in a region, the greater its acuity or discriminative ability.
- More cortical space is allotted for sensory reception from areas with smaller receptive fields and, accordingly, greater tactile discriminative ability.
- The most strongly activated signal pathway originating from the center of the stimulus area inhibits the less excited pathways from the fringe areas.This occurs via inhibitory interneurons that pass laterally between ascending fibers serving neighboring receptive fields.
- Perception is the conscious awareness of surroundings derived from interpretation of sensory input.Perception is our conscious interpretation of the external world as created by the brain from a pattern of nerve impulses delivered to it from sensory receptors

PAIN
Stimulation of nociceptors elicits the perception of pain plus motivational and emotional responses.

- Pain is primarily a protective mechanism meant to bring to conscious awareness the fact that tissue damage is occurring or is about to occur.
- Because of their value to survival, nociceptors do not adapt to sustained or repetitive stimulation. Storage of painful experiences in memory helps us avoid potentially harmful events in the future

- None of the nociceptors have specialized receptor structures; they are all naked nerve endings.
- All nociceptors can be sensitized by the presence of prostaglandins.
- Signals arising from mechanical and thermal nociceptors are transmitted over large myelinated fast pain pathways
- Impulses from polymodal nociceptors are carried by small unmyelinated slow pain pathways.

The brain has a built-in analgesic system.

- In addition to the chain of neurons connecting peripheral nociceptors with higher CNS structures for pain perception, the CNS also contains a neuronal system that suppresses pain.
- These two regions are thought to be part of a descending analgesic pathway that blocks the release of substance P from afferent pain fiber terminals.
- Endogenous opiates serve as analgesic neurotransmitters; they are released from a descending analgesic pathway and bind with opiate receptors on the afferent pain fiber terminal.
- This binding suppresses the release of substance P.

EYE: VISION
The eye is a fluid-filled sphere enclosed by three specialized tissue layers.

- Most of the eyeball is covered by a tough outer layer of connective tissue, the sclera.
- Anteriorly, the outer layer consists of the transparent cornea through which light rays pass into the interior of the eye.
- The middle layer underneath the sclera is the highly pigmented choroid.
- The choroid layer becomes specialized anteriorly to form the ciliary body and the iris.
- The innermost coat under the choroid is the retina, which consists of an outer pigmented layer and an inner nervous tissue layer.
- The retina contains the rods and cones.

- The anterior cavity between the cornea and lens contains a clear watery fluid, the aqueous humor, and the larger posterior cavity between the lens and the retina contains a semifluid, jellylike substance, the vitreous humor.
- The aqueous humor carries nutrients for the cornea and lens.
- Aqueous humor is produced by a capillary network within the ciliary body.
- Excess aqueous humor will accumulate in the anterior cavity causing the intraocular pressure to rise.
- This condition is known as "glaucoma."

The amount of light entering the eye is controlled by the iris.

- The iris contains two sets of smooth muscle networks, one circular and the other radial.
- The pupil gets smaller when the circular muscle contracts.
- When the radial muscle shortens, the size of the pupil increases.
- Iris muscles are controlled by the autonomic nervous system.
- Parasympathetic stimulation causes papillary constriction and sympathetic stimulation causes papillary dilation.

The eye refracts the entering light to focus the image on the retina.

- The bending of a light ray (refraction) occurs when the ray passes from a medium of one density into a medium of a different density.
- A lens with a convex surface converges light rays, bringing them closer together, a requirement for bringing an image to a focal point.
- Refractive surfaces of the eye are convex.
- The two structures most important in the eye's refractive ability are the cornea and the lens.
- The refractive power of the cornea remains constant because its shape never changes whereas the lens can change its shape to changes its refractive capabilities.

Accommodation increases the strength of the lens for near vision.

- The ability to adjust the strength of the lens so that both near and far sources can be focused on the retina is known as "accommodation."
- The strength of the lens depends on its shape, which in turn is regulated by the ciliary muscle.
- The ciliary body has two major components: the ciliary muscle and the capillary network, which produces the aqueous humor.
- The ciliary muscle is attached to the lens by suspensory ligaments.
- When the ciliary muscle contracts, the lens assumes a more spherical shape.
- The greater the curvature of the lens increases its strength, causing greater bending of light rays.
- With loss of elasticity, the lens is no longer able to assume the spherical shape required to accommodate for near vision.
- This age-related reduction in accommodative ability, presbyopia, affects most people by middle age.
- Other common vision disorders are myopia and hyperopia.

Light must pass through several retinal layers before reaching the photoreceptors.

- The major function of the eye is to focus light rays from the environment on the rods and cones, the photoreceptor cells of the retina.
- The photoreceptors then transform the light energy into electrical signals for transmission to the CNS.
- The neural portion of the retina consists of three layers: (1) the outermost layer containing the rods and cones, whose light-sensitive ends face the choroid; (2) a middle layer of bipolar neurons; and (3) an inner layer of ganglion cells. Axons of the ganglion cells join together to form the optic nerve.
- The point on the retina at which the optic nerve leaves and through which blood vessels pass is the optic disc.

- In the fovea, which is a pinhead-sized depression located in the exact center of the retina, the bipolar and ganglion cell layers are pulled aside so that light directly strikes the photoreceptors.
- This feature, coupled with the fact that only cones are found here, makes the fovea the point of most distinct vision.

Phototransduction by retina cells converts light stimuli into neural signals.
- Photoreceptors consists of three parts: (1) an outer segment, which lies closest to the eye's exterior, facing the choroid, and detects the light stimulus; (2) an inner segment, which contains the metabolic machinery; and (3) a synaptic terminal, which lies closest to the eye's interior, facing the bipolar neurons, and transmits the signal generated in the photoreceptor upon light stimulation to the next cells in the visual pathway.
- The outer segment is rod-shaped in rods and cone-shaped in cones.
- The photopigment, which is found in the outer segment undergoes a series of changes when exposed to light, These changes ultimately lead to the generation of action potentials in ganglion cells, which transmit this information to the brain for visual processing
- A phtotpigment consists of an enzymatic protein called opsin combined with retineal, a derivative of vitamin A.
- There are four different photopigments, one in rods and one in each of three types of cones.
- Rhodopsin, the rod photopigment, cannot discriminate between various wavelengths.
- Rods provide vision only in shades of gray by detecting different intensities.
- The photopigments in the three types of cones—red, green, and blue—respond selectively to various wavelengths, making color vision possible.
- Phototransduction, the mechanism of excitation, is basically the same for all photoreceptors.

- When a photopigment molecule absorbs light it dissociates into its retinene and opsin components.
- This light-induced breakdown and subsequent activation of the photopigment bring about a hyperpolarizing receptor potential that influences transmitter release from synaptic terminal of the photoreceptor.
- Photoreceptors are inhibited by the stimulus (light) and excited in the absence of the stimulus (darkness).
- The light-induced biochemical change in the photopigment leads to closing of chemical messenger-gated sodium channels in the outer segment's membrane.
- During the light excitation processes, sodium channels close and causes membrane hyperpolarization.
- This hyperpolarization, which is the receptor potential, passively spreads from the outer segment to the synaptic terminal of the photoreceptor.
- The photoreceptors synapse with bipolar cells.
- These cells terminate on the ganglion cells, whose axons form the optic nerve for transmission of signals to the brain.
- The transmitter released from the photoreceptors synaptic terminal has an inhibitory action on the bipolar cells.
- The reduction in transmitter release that accompanies light-induced receptor hyperpolarization decreases this inhibitory action on the bipolar cells.
- Bipolar cells display graded potentials similar to the photoreceptors.
- Action potentials do not originate until the ganglion cells, the first neurons in the chain that must propagate the visual message to the brain.

Rods provide indistinct gray vision at night whereas cones provide sharp color vision during the day.
- Cones are abundant in the center of the retina.
- Rods are most abundant in the periphery.

- Cones provide sharp vision with high resolution for fine detail.
- Rods have low acuity but high sensitivity, so they respond to the dim light of night.
- There is little convergence of neurons in the retinal pathway for cone output.
- In contrast, there is much convergence of neurons in rod pathways.

Color vision depends on the ratios of stimulation of the three cone types.
- The photopigments in cones are one of three types, blue, green or red.
- Each pigment absorbs light of different wave lengths.
- The pigment in various objects selectively absorbs particular wavelengths of light transmitted to them from light-emitting sources, and the unabsorbed wavelengths are reflected from the objects' surfaces.
- These reflected light rays enable us to see the objects.
- Our perception of the many colors of the world depends on the three cone types' various ratios of stimulation in response to different wavelengths.
- DNA studies have shown that men with normal color vision have a variable number of genes coding for cone pigments.

The sensitivity of the eyes can vary markedly through dark and light adaptation.
- When you go from bright sunlight into darkened surroundings, you cannot see anything at first, but gradually you begin to distinguish objects as a result of the process of dark adaptation.
- When you move from the dark to the light, your eyes are very sensitive to the light at first.
- The sensitivity of the eyes decreases and normal contrasts can once again be detected, a process known as "light adaptation."
- Because retinal is a derivative of vitamin A, adequate amounts of this nutrient must be available for synthesis of photopigments. Night blindness occurs as a result of dietary deficiencies of vitamin A.

Visual information is separated before reaching the visual cortex.
- The field of view that can be seen without moving the head is known as the visual field.
- Because of the pattern of wiring between the eyes and the visual cortex, the left half of the cortex receives information only from the right half of the visual field as detected by both eyes, and the right half receives input only from the left half of the visual field of both eyes

The thalamus and visual cortices elaborate the visual message.
- The first stop in the brain for information in the visual pathway is the thalamus.
- It separates information received from the eyes and relays it to different zones in the cortex, each of which processes different aspects of the visual stimulus (color, form, depth, and movement).

Hierarchy of visual cortical processing
- Within the cortex, visual information is first processed in the primary visual cortex, then is sent to higher-level visual areas for even more complex processing and abstraction

Visual input goes to other areas of the brain not involved in visual perception.
- Not all fibers in the visual pathway terminate in the visual cortices.
- Following are examples of nonsight activities dependent on input from the retina: (1) contribution to cortical alertness and attention (2) control of pupil size, , (3) control of eye movements.
- Three percent of the ganglion cells are not involved in visual processing. They instead make melanopsin which plays a key role in setting the body's "biological clock."

EAR: HEARING AND EQUILIBRIUM
Sound waves consist of alternate regions of compression and rarefaction of air molecules.

- Hearing is the neural perception of sound energy.
- Sound waves are traveling vibrations of air that consist of regions of high pressure caused by compression of air molecules alternating with regions of low pressure caused by rarefaction of the molecules.
- Sound waves can also travel through media other than air, such as water.
- Sound is characterized by its pitch, intensity, and timbre.

The external ear plays a role in sound localization.

- The pinna collects sound waves and channels them down the external ear canal.

The tympanic membrane vibrates in unison with sound waves in the external ear.

- The tympanic membrane, which is stretched across the entrance to the middle ear, vibrates when struck by sound waves.
- The middle ear transfers the vibratory movements of the tympanic membrane to the fluid of the inner ear.

The middle ear bones convert tympanic-membrane vibrations into fluid movements in the inner ear.

- This transfer is facilitated by a moveable chain of three small bones (the malleus, incus, and stapes) that extend across the middle ear.

The cochlea contains the organ of Corti, the sense organ for hearing.

- The stapes is attached to the oval window, the entrance into the fluid-filled cochlea.
- Important structures associated with the cochlea include: the cochlear duct, round window, oval window and basilar membrane.

Hair cells in the organ of Corti transduce fluid movements into neural signals.

- The snail-shaped cochlear portion of the inner ear is a coiled tubular system lying deep within the temporal bone.

- The cochlea is divided throughout most of its length into three fluid-filled longitudinal compartments (the cochlear duct, scala tympani, and scala vestibuli).
- The basilar membrane forms the floor of the cochlear duct and bears the organ of Corti, the sense organ for hearing.
- The organ of Corti contains hair cells that are the receptors of sound.
- Pressure waves through the basilar membrane cause this membrane to vibrate in synchrony with the pressure wave.
- Since the organ of Corti rides on the basilar membrane, the hair cells also move up and down as the basilar membrane oscillates.
- This back and forth mechanical deformation of the hairs alternately opens and closes mechanically gated ion channels in the hair cell, resulting in alternating depolarizing and hyperpolarizing potential changes—the receptor potential—at the same frequency as the original sound stimulus.
- The inner hair cells are specialized receptor cells that communicate via a chemical synapse with the terminals of afferent nerve fibers making up the auditory nerve.
- The outer hair cells enhance the response of the inner hair cells, the real auditory sensory receptors.

Pitch discrimination depends on the region of the basilar membrane that vibrates.

- Different regions of the basilar membranes naturally vibrate maximally at different frequencies; that is, each frequency displays peak vibration at a different position along the membrane.
- Intensity discrimination depends on the amplitude of the vibration.
- The greater tympanic membrane deflection is converted into greater amplitude of basilar membrane movement in the region of peak responsiveness.

Loudness discrimination depends on the amplitude of the vibration.

- The CNS interprets greater basilar membrane oscillation as louder sound.

- Pitch discrimination depends on *where* the basilar membrane maximally vibrates and loudness discrimination depends on *how much* this place vibrates.

Deafness is caused by defects either in conduction or neural processing of sound waves.

- Hearing loss, which affects about 10% of all Americans, is the second most common physical disability in the United States

- Conductive deafness occurs when sound waves are not adequately conducted through the external and middle portions of the ear to set the fluids in the inner ear in motion.

- In sensorineural deafness, the sound waves are transmitted to the inner ear, but they are not translated into nerve signals that are interpreted by the brain as sound sensations.

- Neural presbycusis is a degenerative, age-related process that occurs as hair cells "wear out" with use.

- On average. We lose more than 40% of our cochlear hair cells by age 65

The vestibular apparatus is important for equilibrium by detecting position and motion of the head.

- The vestibular apparatus consists of two sets of structures within the temporal bone near the cochlea; the semicircular canals and the otolith organs.

- The semicircular canals detect rotational or angular acceleration or deceleration of the head.

- Each ear contains three semicircular canals arranged three-dimensionally in planes that lie at right angles to each other.

- The receptive hair cells of each semicircular canal are situated on top of a ridge located in the ampulla.

- The hairs are embedded in the cupula.

- As the head starts to move, the fluid within the semicircular canal does not move in the direction of the rotation but lags behind because of its inertia.

- This nerve unites with the auditory nerve from the cochlea to form the vestibulocochlear nerve.

- The otolith organs provide information about the position of the head relative to gravity and detect changes in rate of linear motion.

- The utricle and saccule are saclike structures housed within a bony chamber situated between the semicircular canals and the cochlea.

- The hairs of the receptive hair cells in these sense organs also protrude into an overlying gelatinous sheet, whose movement displaces the hairs and results in changes in hair cell potential.

- When a person is in an upright position, the hairs within the utricles are oriented vertically and the saccule hairs are lined up horizontally.

- Signals are carried through the vestibulocochlear nerve to the vestibular nuclei in the brain stem and to the cerebellum.

- Here the vestibular information is integrated with input from the skin surface, eyes, joints and muscles for: (1) maintaining balance and desired posture, (2) controlling the external eye muscles so that the eyes remain fixed on the same point, and (3) perceiving motion and orientation.

CHEMICAL SENSES: TASTE AND SMELL
Taste receptor cells are located primarily within tongue taste buds.

- The chemoreceptors for taste sensation are packaged in taste buds.

- Each taste bud has a small opening, the taste pore.

- Taste receptor cells are modified epithelial cells with microvilli.

- The plasma membrane of the microvilli contains receptor sites that bind selectively with chemical molecules in the environment.

Taste discrimination is coded by patterns of activity in various taste bud receptors.

- Binding of a taste-provoking chemical, a tastant, with a receptor cell alters its ionic channels to produce a depolarizing receptor potential.
- The receptor potential initiates an action potential within the terminal endings of afferent nerve fibers with which the receptor cell synapses.
- Signals in these sensory inputs are conveyed via synaptic stops in the brain stem and thalamus to the cortical gustatory area.
- All tastes are varying combinations of five primary tastes: salty, sour, sweet, bitter, and umani
- Taste perception is also influenced by information from other receptors, especially odor.

The olfactory receptors in the nose are specialized endings of renewable afferent neurons.

- The olfactory mucosa contains olfactory receptors, supporting cells, and basal cells.

Various parts of an odor are detected by different olfactory receptors and sorted into "smell files."

- To be smelled, a substance must be: (1) sufficiently volatile that some of its molecules can enter the nose and (2) sufficiently water-soluble that it can dissolve in the mucus layer coating the olfactory mucosa.

- Binding of an appropriate scent signal to an olfactory receptor activates a G protein, triggering intracellular reactions that lead to opening of sodium channels.
- The resultant ion movement brings about a receptor potential.
- The receptor potential generates the action potential in the afferent fiber.
- The afferent fibers pass through the bone and immediately synapse in the olfactory bulb.
- Each olfactory bulb is lined by small ball-like neural junctions known as "glomeruli."
- Because each glomerulus receives signals only from receptors that detect a particular odor component, the glomeruli serve as "smell files."
- Fibers leaving the olfactory bulb travel in two different routes: (1) a subcortical route going to regions of the limbic system, and (2) a thalamic-cortical route.

Odor discrimination is coded by patterns of activity in the olfactory bulb glomeruli.

- Odor discrimination is based on different patterns of glomeruli activity by various scents.

The olfactory system adapts quickly.

- The olfactory system adapts quickly.
- Adaptation is specific for a particular odor and responsiveness to other odors remains unchanged.

Key Terms

Accommodation
Acuity
Afferent division
Analgesic system
Astigmatism
Basilar membrane
Cataract
Chemoreceptors
Cochlea
Cochlear implants
Cones
Deafness
Fast pain pathway
Generator potentials
Glaucoma
Gustation
Intensity discrimination
Lateral inhibition
Macular degeneration
Mechanoreceptors
Morphine
Motion sickness
Myopia
Nociceptors
Off response
Opiate receptors
Opsin
Optic disc
Organ of Corti
Osmoreceptors
Ossicles
Otolith organs
Oval window

Pacinian corpuscle
Pain
Perception
Peripheral nervous system
Phasic receptors
Pheromones
Photoreceptors
Pitch discrimination
Receptive field
Receptor potential
Receptors
Retinal
Rhodopsin
Rods
Saccule
Scala media
Scala vestibuli
Sclera
Semicircular canals
Sensory afferent
Slow pain pathway
Somatic sensation
Substance P
Tastant
Taste bud
Thermoreceptors
Tonic receptors
Transduction
Tympanic membrane
Umami
Utricle
Vestibular nucleus
Visceral afferent
Vitreous humor

Review Exercises

Answers are in the appendix.

True/False-if an answer is false, underline the incorrect portion of the statement and rewrite it to make the statement true.

_____ 1. Nociceptors are not able to adapt to sustained pain.

_____ 2. All nociceptors can be sensitized by the presence of prostaglandins.

_____ 3. Impulses from polymodal nociceptors are carried by slow pain pathways.

_____ 4. Visceral afferents would carry conscious information regarding pain from an organ.

_____ 5. Endogenous opiates serve as analgesic neurotransmitters; they are released from a descending analgesic pathway and bind with opiate receptors on the afferent presynaptic terminal; this suppresses the release of substance Q.

_____ 6. Pain impulses originating at nociceptors are transmitted to the CNS via one of two types of efferent fibers.

_____ 7. Taste receptor cells are modified epithelial cells with microvilli.

_____ 8. Binding of a taste-provoking chemical with a receptor cell alters its ionic channels to produce a depolarizing receptor potential.

_____ 9. Both taste and smell sensations play a major role in our ability to taste food.

_____ 10. Olfactory receptors are specialized endings of afferent neurons.

_____ 11. Night blindness can be caused by excess levels of Vitamin A.

_____ 12. Our sensitivity to a new odor rapidly diminishes after a short period of exposure to it, even though the odor source continues to be present.

_____ 13. Intense stimulation of any receptor is perceived as painful.

_____ 14. A receptor may be either a specialized ending of the afferent neuron or a separate cell closely associated with the peripheral ending of the neuron.

_____ 15. Binding of a chemical messenger with specific protein receptor sites on the afferent neuron closes chemical messenger-gated sodium channels.

_____ 16. A larger receptor potential cannot bring about a larger action potential (all-or-none law), but it can induce more rapid firing of action potentials.

_____ 17. Tonic receptors are useful in situations where it is important to signal a change in stimulus intensity rather than to relay status quo information.

_____ 18. Stimulation of the iris by parasympathetic nervous causes papillary constriction.

_____ 19. Each sensory neuron responds to stimulus information only within a circumscribed region of the skin surface surrounding it.

_____ 20. Rods contain pigments that can detect red, blue and green colors.

_____ 21. The vitreous humor is important in maintaining the spherical shape of the eyeball.

_____ 22. Iris muscles are controlled by the autonomic nervous system.

_____ 23. The strength of the lens depends on its shape, which in turn is regulated by the ciliary muscle.

_____ 24. Retinene is identical in all four photopigments, but the photoreceptor's opsin vary slightly.

_____ 25. Photoreceptor cells are hyperpolarized when exposed to their light stimulus.

_____ 26. The sodium channels of a photoreceptor are closed in the absence of stimulation, that is, in the dark.

_____ 27. Sound waves can only travel through air and other media like water.

_____ 28. Sound is characterized by its pitch (tone), intensity (loudness), and timbre (quality).

_____ 29. The tympanic membrane collects sound waves and channels them down the external ear canal.

_____ 30. The middle ear transfers the vibratory movements of the tympanic membrane to the fluid of the inner ear.

_____ 31. The cochlea is divided throughout most of its length into four fluid-filled longitudinal compartments.

_____ 32. The pitch of a sound is determined by the frequency of the sound waves.

_____ 33. The vestibular membrane forms the floor of the cochlear duct, separating it from the scala tympani.

_____ 34. Hair cells are specialized receptor cells that communicate via a chemical synapse with the terminals of afferent nerve fibers making up the auditory nerve.

_____ 35. Photoreceptors synapse with bipolar cells.

_____ 36. Rods are most abundant in the center of the retina.

Fill in the Blank

37. Most of the eyeball is covered by a tough outer layer of connective tissue called _____.

38. LASIK surgery changes the shape of the _____ to correct visions.

39. The ability to adjust the strength of the lens so that both near and far sources can be focused on the retina is known as _____.

40. The lens is an elastic structure consisting of transparent fibers. Occasionally these fibers become opaque so that light rays cannot pass through, a condition known as a _____.

41. The point on the retina at which the optic nerve leaves and through which blood vessels pass is the _____.

42. A photopigment consist of an enzymatic protein called _____ combined with retinal, a derivative of vitamin A.

43. The right visual field is processed by the _____ occipital lobe

44. Each taste bud has a small opening called the _____.

45. Signals from the olfactory receptor cells are first sent to the _____ before moving to higher olfactory centers.

46. The axons of the olfactory receptor cells collectively form the _____.

47. The chemoreceptors for taste sensations are packaged in _____.

48. Fibers leaving the olfactory bulb travel in two different routes:
    (1)_____ and (2)_____.

49. The _____ cells are found in the olfactory bulb and make synaptic connections with the olfactory nerves.

50. Afferent input that does reach the level of conscious awareness is known as _____ and includes somatic sensation and special senses.

51. _____ receptors are rapidly adapting receptors.

52. The _____ is the structure of the eye that bends light without itself changing shape.

53. The incoming pathway for subconscious information derived from the internal viscera is called a _____.

54. If the information detected by the receptors is propagated to the conscious level of the brain, it is called _____.

55. The afferent neuron with its peripheral receptor that first detects the stimulus is known as a(n) _____.

56. Pain perceived as originating in the foot by a person whose leg has been amputated at the knee is known as _____.

57. The _____ houses sensory systems for equilibrium and provides input essential for maintenance of posture and balance.

58. _____ is a chemical made by ganglion cells and is thought to be associated with setting the body's clock..

59. The fluid within the cochlear duct is called _____.

60. The thin _____ membrane separates the cochlear duct from the scala vestibuli.

61. Hair cells are specialized receptor cells that communicate via a chemical synapse with the terminals of afferent nerve fibers making up the _____.

62. Pitch discrimination depends on the shape and properties of the _____ membrane.

63. The _____ provide information about the position of the head relative to gravity and also detect changes in rate of linear motion.

64. Hearing loss affects _____ percent of all US citizens.

65. The conversion of a sensory stimulus into an electrical signal is called _____.

66. One of the neurotransmitters released from these afferent pain terminals is _____, which is believed to be unique to pain fibers.

67. All nociceptors can be sensitized by the presence of _____.

68. Ascending pain pathways have poorly understood destinations in the _____ _____, the _____, and the _____.

69. The built-in analgesic system is dependent on the presence of _____.

70. Axons from _____ cells give rise to the optic nerve.

Matching
Match senses to characteristics. .

a. visionb. hearing
c. touch
d. taste
e. smell

_____ 71. Humans can detect 1,000 different molecules associated with this sense._____
　　　　 72. 　　Timbre is a part of this sense.
_____ 73. Perception of this sense is associated with the parietal lobe.
_____ 74. Detection of this sense requires the function of hair cells.
_____ 75. All of the senses but this one first pass through the thalamus before they are sent to their cortical areas for processing.
_____ 76. By the age of 65 we lose about 40% of this sense.
_____ 77. Receptor cells for this sense are depolarized in the absence of the stimulus.
_____ 78. Associated with the stimulus umani.
_____ 79. Its receptor cells are similar in structure to receptor cells associated with balance.
_____ 80. Receptors for this sense can be tonic or phasic.

Multiple Choice

81. Which receptor has a life span of about ten days?
    a. ear
    b. eye
    c. taste
    d. pain
    e. none of the above

82. Which receptors relay status quo information?
    a. photoreceptors
    b. phasic receptors
    c. nociceptors
    d. transduction receptors
    e. tonic receptors

83. Which eye disorder is characterized by the surface of the cornea being uneven?
    a. astigmatism
    b. cataract
    c. diplopia
    d. glaucoma
    e. hyperopia

84. Which part of the CNS is involved in taste?
    a. cerebellum
    b. lateral geniculate nucleus
    c. gustatory nucleus
    d. cortical gustatory area
    e. periaqueductal gray matter

85. Sensations are used for all of the following except:
    a. stored for reference for latter use
    b. perception of the world in which we live
    c. arousal
    d. control efferent activity. primary analgesic cortex sensations are used for all of the above

86. Which type receptor is associated with pain?
    a. nociceptors
    b. mechanoreceptors
    c. photoreceptors
    d. opiate receptors
    e. vitreous receptors

87. Receptors convert various forms of energy into electrical energy. What is this conversion process called?
    a. depolarization
    b. hyperpolarization
    c. frequency modulation
    d. somesthetic propagation
    e. transduction

88. Which type of stimuli is detected by the cells associated with the neurotransmitter substance P?
    a. specific chemicals
    b. pain
    c. temperature changes
    d. odors
    e. solute concentration changes

89. Which part of the CNS is involved with sight?
    a. periaqueductal gray matter
    b. occipital lobes
    c. reticular function
    d. olfactory bulb
    e. third ventricle

90. Which receptors do not adapt at all or adapt slowly?
    a. tactile
    b. nociceptors
    c. phasic
    d. tonic
    e. taste

91. Which of the following structures when damaged causes "donut hole" vision?
    a. olfactory bulb
    b. vomeronasal organ
    c. macula lutea
    d. cupula
    e. mitral cells

92. Which of the following disorders is not related to age?
    a. neural presbycusis
    b. macular degeneration
    c. presbyopia
    d. hearing loss
    e. none of the above

93. Which of the primary tastes listed below is suggested to include the taste of amino acids and peptides?
    a. sweet
    b. umami
    c. salty
    d. sour
    e. bitter

Points to Ponder

Since perception is not always reality, consider these three questions:

1. What color is a bluebird? Be careful.

2. As you listen to a recording of a train on your stereo you hear the approach of the train in one speaker and it seems to travel across in front of you to exit in the other speaker. How do orchestras accomplish this phenomenon in live concerts?

3. On a very hot day in July, you buy a nice cold canned diet drink containing the artificial sweetener aspartame. Upon drinking the beverage you experience a bitter taste instead of the normal sweet taste. How would you explain this?

Clinical Perspectives

The patient complains of severe dizziness when riding the elevator to her job on the fifteenth floor. Explain how she could solve this problem without walking up the stairs.

Experiment of the Day

1. Place your index finger against the corner of your closed eye. Apply short bursts of gentle pressure on the eyeball. Describe and explain this phenomenon.

2. During your travels about campus, locate some people wearing excessive amounts of perfume or cologne. Explain why they do this. (Hint: Check their other habits.)

7

The Peripheral Nervous System: Efferent Division

Chapter Overview

The nervous system, one of the two major regulatory systems of the body, consists of the central nervous system, composed of the brain and spinal cord, and the peripheral nervous system, composed of the afferent and efferent fibers that relay signals between the CNS and the periphery. Thus, as one of the major control systems of the body, the CNS transmits impulses through the efferent portion of the peripheral nervous system to the effector organs. The transmitted signal is usually the result of a CNS interpretation of conditions that warrant changes in order to maintain homeostasis. Voluntary and involuntary effector organs are controlled using the somatic and autonomic nervous system respectively. The key to this very precise autonomic control is in the neurotransmitters, acetylcholine and norepinephrine, and the dual innervation by the sympathetic and parasympathetic systems. The somatic nervous system acts through the release of a neurotransmitter at the neuromuscular junction to cause skeletal muscle contractions. The efferent division of the peripheral nervous system is the final link in the CNS control of the body. The actions elicited by the efferent neurons are essential for homeostasis.

Chapter Outline

INTRODUCTION
- The efferent division of the peripheral nervous system is the communication link by which the CNS controls the activities of muscles and glands.
- Cardiac muscle, smooth muscle, most exocrine glands, and some endocrine glands are innervated by the autonomic nervous system.
- Skeletal muscle is innervated by the somatic nervous system.
- Two neurotransmitters—acetylcholine and norepinephrine—are released from efferent neuronal terminals to elicit essentially all the neurally controlled effector organ responses.

AUTONOMIC NERVOUS SYSTEM
An autonomic nerve pathway consists of a two-neuron chain.
- Each autonomic nerve pathway consists of a two-neuron chain.
- The cell body of the first neuron is located in the CNS.
- Its axon, the preganglionic fiber, synapses with the cell body of the second neuron, which lies within a ganglion outside the CNS.
- The axon of the second neuron, the postganglionic fiber, innervates the effector organ.
- The autonomic nervous system consists of two divisions—the sympathetic and the parasympathetic nervous system.
- Sympathetic nerve fibers originate in the thoracic and lumbar regions of the spinal cord.

- Most sympathetic preganglionic fibers are very short, synapsing within a sympathetic ganglion chain located along either side of the spinal cord.
- Long postganglionic fibers terminate on the effector organs.
- Parasympathetic preganglionic fibers arise from the cranial and sacral areas of the CNS.
- These fibers are long because they do not end until they reach terminal ganglia, which lie in or near the effector organs.
- Parasympathetic postganglionic fibers are very short.

Parasympathetic postganglionic fibers release acetylcholine; sympathetic ones release norepinephrine.
- Sympathetic and parasympathetic preganglionic fibers release the same neurotransmitter, acetylcholine (ACh).
- Parasympathetic postganglionic fibers release acetylcholine and are called "cholinergic fibers."
- Most sympathetic postganglionic fibers are called "adrenergic fibers" because they release norepinephrine.
- The terminal branches of autonomic fibers contain numerous swellings, or varicosites, which simultaneously release neurotransmitter over a large area of the innervated organ.
- This diffuse release means whole organs instead of discrete cells are typically influenced by autonomic activity.

The sympathetic and parasympathetic nervous systems dually innervate most visceral organs.
- Most visceral organs are innervated by both sympathetic and parasympathetic nerve fibers.

- Visceral afferent information is used to direct activity of the autonomic system.
- Both systems increase the activity of some organs and reduce the activity of others. Innervation of a single target by both branches of the autonomic nervous system is called dual innervation.
- Normally some level of action potential activity exists in both, the sympathetic and the parasympathetic fibers supplying a particular organ.
- This ongoing activity is called "sympathetic" or "parasympathetic" tone or "tonic activity."
- The sympathetic system promotes responses that prepare the body for strenuous physical activity in the face of emergency or stressful situations.
- This response is typically referred to as a "fight-or-flight response."
- The parasympathetic system dominates in quiet, relaxed situations or "rest and digest."
- Dual innervation enables precise control over an organ's activity.

The adrenal medulla is a modified part of the sympathetic nervous system.
- The adrenal medulla is considered to be a modified sympathetic ganglion that does not give rise to postganglionic fibers.
- Upon stimulation by the preganglionic fiber, the adrenal medulla secretes hormones into the blood.
- The hormones are identical or similar to postganglionic sympathetic neurotransmitters: norepinephrine (20%) and epinephrine (80%).

Several different receptor types are available for each autonomic neurotransmitter.
- Responsive tissue cells possess one or more of several different types of plasma membrane receptor proteins for these chemical messengers.
- Two types of acetylcholine receptors have been identified.
- Nicotinic receptors, found in all autonomic ganglia, respond to acetylcholine released from both sympathetic and parasympathetic preganglionic fibers.

- Muscarinic receptors, found on effector cell membranes, bind with acetylcholine released from parasympathetic postganglionic fibers.
- There are two major classes of adrenergic receptors for norepinephrine and epinephrine designated as alpha and beta receptors.
- Both alpha and beta receptors have several subtypes denoted with subscripts 1 and 2.
- Agonist drugs bind with receptors and cause the same response as the neurotransmitter system it is associated with. Antagonist drugs prevent the action of the neurotransmitter.

Many regions of the central nervous system are involved in the control of autonomic activities.

- Some autonomic reflexes, such as urination, defecation, and erection, are integrated at the spinal cord level, but all of these spinal reflexes are subject to control by higher levels of consciousness.
- The medulla, within the brain stem, is the region most directly responsible for autonomic output.
- Centers for controlling cardiovascular, respiratory and digestive activity via the autonomic system are located in the medulla.
- The hypothalamus plays an important role in integrating the autonomic, somatic, and endocrine responses that autonomically accompany various emotional and behavioral states.
- Autonomic activity can also be influenced by the frontal association cortex through its involvement with emotional expression characteristic of the individual's personality.

SOMATIC NERVOUS SYSTEM
Motor neurons supply skeletal muscle.

- Skeletal muscle is innervated by motor neurons, the axons of which constitute the somatic nervous system.
- The cell bodies of these motor neurons are located within the ventral horn of the spinal cord.
- The axon of a motor neuron is continuous from its origin in the spinal cord to its termination on skeletal muscle.
- Motor neuron axon terminals release acetylcholine, which brings about excitation and contraction of the innervated muscle fibers.
- Motor neurons can only stimulate skeletal muscles.

Motor neurons are influenced by many areas of the CNS involved in motor control.

- The only way any other part of the nervous system can influence skeletal muscle activity is by acting on motor neurons
- The somatic system is considered to be under voluntary control, but much of the skeletal muscle activity involving posture, balance, and stereotypical movements is subconsciously controlled.
- The cell bodies of the crucial motor neurons may be selectively destroyed by polio virus.
- Amyotropic lateral sclerosis, Lou Gehrig's disease, is characterized by progressive degeneration and death of motor neurons.

Motor neurons and skeletal muscle fibers are chemically linked at neuromuscular junctions.

- Each axon terminal forms a special junction called a "neuromuscular junction."
- A single muscle fiber is long and cylindrical in shape.
- The axon terminal is enlarged into a terminal button.
- The motor end plate is the specialized portion of the muscle cell membrane immediately under the terminal button.
- Each motor neuron innervates only one muscle cell.

Acetylcholine is the neuromuscular junction neurotransmitter.

- Nerve and muscle cells do not actually come into direct contact at the neuromuscular junction.
- A chemical messenger is used to carry the signal between the neuron terminal and the muscle fiber.
- Propagation of an action potential to an axon terminal triggers the opening of voltage-gated calcium channels in the terminal button.
- Opening the calcium channels permits calcium to diffuse into the terminal button, which in turn causes the release of acetylcholine from several hundred of the vesicles into the cleft.
- The released ACh diffuses across the cleft and binds with specific receptor sites.
- These cholinergic receptors are of the nicotinic type.
- Binding of ACh with these receptor sites induces the opening of chemical messenger-gated channels in the motor end plate.
- When ACh triggers the opening of these channels, considerably more sodium moves inward than potassium outward, bringing about a depolarization of the motor end plate.
- This potential change is known as the "end-plate potential" (EPP).
- It is similar to an EPSP except that the magnitude of an EPP is much larger because: (1) more transmitter is released from a terminal button than from a presynaptic knob in response to the action potential; (2) the motor end plate has a larger surface area and, accordingly, more transmitter receptor sites than a subsynaptic membrane; and (3) many more ion channels are

opened in response to the transmitter receptor complex in the motor end plate.
- An EPP is a graded potential.
- The subsequent action potential triggers contraction of the muscle fiber.

Acetylcholinesterase terminates acetylcholine activity at the neuromuscular junction.

- The muscle's electrical response is turned off by an enzyme in the motor end plate membrane, acetylcholinesterase (AChE), which inactivates ACh.
- Removal of Ach terminates the EPP so that no more action potentials are initiated.

The neuromuscular junction is vulnerable to several chemical agents and diseases.

- The venom of black widow spiders exerts its deadly effect by causing an explosive release of ACh from storage vesicles, the most detrimental consequence of which is respiratory failure.
- Botulinum toxin exerts its lethal blow by blocking the release of ACh from the terminal in the motor neuron.
- Death is due to respiratory failure.
- Curare reversibly binds to the ACh receptor sites on the motor end plate.
- Unlike ACh, however, curare does not alter membrane permeability, nor is it inactivated by AChE.
- When sufficient curare is present the person dies of respiratory paralysis.
- Organophosphates irreversibly inhibit AChE.
- Death from organophosphates is also due to respiratory paralysis.
- In the disease myasthenia gravis, a condition characterized by extreme muscular weakness, the body erroneously produces antibodies against its own motor end plate ACh receptors.

Key Terms

Acetylcholine (Ach)
Acetylcholinesterase
Adrenal medulla
Adrenergic fibers
Agonist
Alpha receptors
Amyotrophic lateral sclerosis
Antagonist
Autonomic nervous system
Beta receptors
Botulism
Cholinergic fibers
Collateral ganglia
Curare
Efferent division
End-plate potential (EPP)
Epinephrine (Adrenaline)
Fight-or-flight response
Lou Gehrig's disease

Motor end plate
Motor neurons
Muscarinic receptors
Myasthenia gravis
Neuromuscular junction
Nicotinic receptors
Noradrenaline
Norepinephrine
Organophosphates
Parasympathetic nervous system
Parasympathetic tone
Peripheral nervous system
Postganglionic fiber
Preganglionic fiber
Somatic nervous system
Sympathetic ganglion chain
Sympathetic nervous system
Sympathetic tone
Terminal button
Visceral afferent

Review Exercises

Answers are in the appendix.

True/False-if an answer is false, underline the incorrect portion of the statement and rewrite it to make the statement true.

_____ 1. Visceral afferent input is used to direct autonomic output.

_____ 2. Nerve and muscle cells occasionally come into direct contact at a neuromuscular junction.

_____ 3. A chemical messenger is used to carry the signal between the neuron terminal and the muscle fiber.

_____ 4. A motor neuron interacts with many muscle cells.

_____ 5. The released ACh diffuses across the cleft and binds with specific receptor sites.

_____ 6. The magnitude of an EPSP is much larger than an EPP.

_____ 7. The muscle's electrical response is turned on by an enzyme present in the motor end plate membrane, which inactivates ACh.

_____ 8. Motor neurons can only stimulate skeletal muscles.

_____ 9. The cells' bodies of the somatic nervous system are located within the lateral horn of the spinal cord.

_____ 10. The somatic system is considered to be under voluntary control.

_____ 11. In the autonomic nervous system there is a single neuron from the origin in the CNS to the effector organ.

_____ 12. Sympathetic nerve fibers originate in the thoracic and lumbar regions of the spinal cord.

_____ 13. The adrenal medulla releases more norepinephrine than epinephrine

_____ 14. Sympathetic preganglionic fibers arise from the cranial and sacral areas of the CNS.

_____ 15. Parasympathetic postganglionic fibers release acetylcholine.

_____ 16. Most visceral organs are innervated by both sympathetic and parasympathetic nerve fibers.

_____ 17. Both systems increase the activity of some organs and reduce the activity of others.

_____ 18. The parasympathetic system promotes responses that prepare the body for strenuous physical activity in the face of emergency or stressful situations.

Fill in the Blank

19. The skeletal muscle is innervated by _____.

20. Motor neurons are considered to be the _____ since the only way any other part of the nervous system can influence skeletal muscle activity is by acting on these motor neurons.

21. As the axon approaches a muscle, it divides into many terminal branches and loses its myelin sheath. Each of these axon terminals forms a special junction called _____.

22. A single muscle cell is called a(n) _____.

23. The axon terminal is enlarged into a knoblike structure called a(n) _____.

24. The specialized portion of the muscle cell membrane immediately under the terminal button is known as the _____.

25. The venom of black widow spiders exerts its deadly effect by causing an explosive release of _____ from the storage vesicles.

26. _____ exerts its lethal blow by blocking the release of ACh from the terminal button in response to an action potential in the motor neuron.

27. _____ are a group of chemicals that modify neuromuscular junction activity.

28. _____ is a condition characterized by loss of ACh receptors in the neuromuscular junction.

29. The _____ innervates the effector organ. The _____ synapses with the cell body of the second neuron.

30. Most sympathetic postganglionic fibers are called _____ because they release noradrenalin.

31. The terminal branches of autonomic fibers contain numerous swellings called _____ _____.

32. About 20 % of the adrenal medullary hormone output is norepinephrine, and the remaining 80 % is the closely related substance _____.

33. _____ respond to acetylcholine released from both sympathetic and parasympathetic preganglionic fibers.

34. _____ are found on effector cell membranes and bind ACh.

Matching
Match the distinguishing feature to the proper division of the autonomic nervous system.

a. sympathetic system
b. parasympathetic system

_____ 35. Activity would increase heart rate
_____ 36. Origin of preganglionic fibers is the brain and sacral region of spinal cord
_____ 37. Short cholinergic postganglionic fibers
_____ 38. Long adrenergic postganglionic fibers

_____ 39. Long cholinergic postganglionic fibers
_____ 40. Activity would stimulate overall digestive function
_____ 41. Origin of postganglionic fibers is the terminal ganglia
_____ 42. Nicotinic receptors for neurotransmitters
_____ 43. Agonist drugs bind to alpha and beta receptors
_____ 44. Muscarinic receptors for neurotransmitters

Multiple Choice

45. Atropine blocks the effect of:
    a. acetylcholine at muscarinic receptors
    b. acetylcholine at nicotinic receptors
    c. acetylcholine at beta receptors
    d. norepinephrine at alpha receptors
    e. norepinephrine at beta receptors

46. Curare:
    a. causes an explosive release of acetylcholine at the neuromuscular junction
    b. blocks the effects of acetylcholine at muscarinic receptors
    c. reversibly binds with acetylcholine receptor sites
    d. irreversibly inhibits acetylcholinesterase
    e. blocks the effect of norepinephrine at both beta receptors

47. Neostigmine:
    a. causes an explosive release of norepinephrine at the neuromuscular junction
    b. blocks the effect of acetylcholine at beta receptors
    c. irreversibly inhibits acetylcholinesterase
    d. prolongs the action of acetylcholine at the neuromuscular junction
    e. blocks the release of acetylcholine at the neuromuscular junction

48. Botulinum toxin:
    a. blocks the release of acetylcholine at the neuromuscular junction
    b. prolongs the action of acetylcholine at the neuromuscular junction
    c. blocks the effect of norepinephrine at the alpha receptors
    d. reversibly binds with acetylcholine receptor sites
    e. irreversibly inhibits acetylcholinesterase

49. Salbutamol:
    a. activates beta adrenergic receptor sites
    b. blocks the effect of norepinephrine at beta receptors
    c. blocks the effect of acetylcholine at nicotinic receptors
    d. activates the cholinergic receptors
    e. irreversibly inhibits acetylcholinesterase

50. The action of this drug would cause excess skeletal muscle twitches:
    a. blocks the effects of norepinephrine at muscarinic receptors
    b. activates muscarinic receptors
    c. activates nicotinic receptors
    d. blocks the effect of norepinephrine at beta receptors
    e. causes an explosive release of norepinephrine at the neuromuscular junction

51. Black widow spider venom:
    a. blocks the effect of norepinephrine at beta receptors
    b. activates muscarinic receptors
    c. activates nicotinic receptors
    d. causes an explosive release of acetylcholine at the neuromuscular junction
    e. irreversibly inhibits acetylcholinesterase

52. Organophosphates:
    a. blocks the effect of acetylcholine at alpha receptors
    b. activates muscarinic receptors
    c. activates nicotinic receptors
    d. causes an explosive release of norepinephrine
    e. irreversibly inhibits acetylcholinesterase

53. Drugs that do this would decrease sympathetic function:
    a. irreversibly inhibits acetylcholinesterase
    b. causes an explosive release of acetylcholine
    c. blocks the release of acetylcholine at the neuromuscular junction
    d. reversibly binds with acetylcholine receptor sites
    e. blocks the release norepinephrine

54. Drugs that do this will increase heart rate:
    a. block the parasympathetic innervation of the heart
    b. block the sympathetic innervation of the heart
    c. inhibit acetylcholinesterase
    d. bind with muscarine
    e. activate cholinergic acid

55. The autonomic nervous system is:
    a. part of the somatic nervous system
    b. considered to be the involuntary branch of the efferent nervous system
    c. part of the efferent division of the peripheral nervous system
    d. two of the above
    e. all of the above

56. Parasympathetic postganglionic fibers:
    a. arise from the ganglion chain located along either side of
    the spinal cord
    b. are cholinergic
    c. secrete norepinephrine
    d. both a and b
    e. both a and c

57. All of the following release acetylcholine except:
    a. sympathetic preganglionic fibers
    b. parasympathetic preganglionic fibers
    c. sympathetic postganglionic fibers
    d. parasympathetic postganglionic fibers
    e. alpha motor neurons

58. The sympathetic nervous system:
    a. is always excitatory
    b. innervates only tissues concerned with protecting the
       body against challenges from outside
    c. dominates in fight or flight situations
    d. is part of the somatic nervous system
    e. is part of the afferent nervous system

59. The sympathetic nervous system:
    a. is part of the somatic nervous system
    b. has cholinergic preganglionic and adrenergic
    postganglionic fibers
    c. originates in the thoracic and lumbar regions of the spinal
    cord
    d. both b and c
    e. a, b and c

60. Which of the following does not characterize the
    sympathetic nervous system?
    a. It promotes responses that prepare the body for strenuous
    physical activity.
    b. It is part of the autonomic nervous system.
    c. It has norepinephrine as its postganglionic
    neurotransmitter.
    d. It is always excitatory (that is, it increases the activity in
    every tissue it innervates).
    e. It is part of the efferent division of the peripheral nervous
    system.

61. The parasympathetic nervous system:
    a. has long preganglionic fibers that end on terminal
       ganglia
    b. dominates in quiet, relaxed situations
    c. releases postganglionic neurotransmitters that bind with
       muscarinic receptors
    d. bind with norepinephrine released from sympathetic
       postganglionic fibers
    e. more than one of the above

62. Nicotinic receptors:
    a. bind with acetylcholine released from parasympathetic postganglionic fibers
    b. respond to acetylcholine released from both sympathetic and parasympathetic fibers
    c. are found primarily in the heart
    d. bind with norepinephrine released from sympathetic postganglionic fibers
    e. more than one of the above

63. The chemical transmitter substance at the neuromuscular junction is:
    a. acetylcholine
    b. the same as the transmitter substance at parasympathetic postganglionic nerve endings
    c. inactivated by organophosphates
    d. more than one of the above
    e. none of the above

64. Acetylcholinesterase:
    a. is stored in vesicles in the terminal button
    b. when combined with receptor sites on the motor-end plate, brings about an end-plate potential
    c. is inhibited by organophosphates
    d. more than one of the above
    e. none of the above

65. The motor end-plate:
    a. contains receptor sites that are capable of binding curare
    b. contains acetylcholinesterase
    c. experiences an increase in permeability to cations when combined with acetylcholine
    d. both b and c
    e. a, b and c

66. The neuromuscular junction:
    a. is the junction between a motor neuron and a skeletal muscle fiber
    b. transmits an action potential between a nerve cell
    c. may produce either an EPSP or an IPSP on the motor end-plate
    d. both a and b
    e. a, b and c

67. Acetylcholinesterase:
    a. is released from the terminal button
    b. destroys acetylcholine
    c. is blocked by curare
    d. both a and b
    e. a, b and c

68. Acetylcholine:
    a. is released from the vesicles when an action potential is propagated to the terminal button of motor neuron
    b. increases the permeability of the motor end-plate to sodium and potassium when combined with receptor sites on the motor end-plate
    c. is the chemical transmitter substance at the neuromuscular junction
    d. two of the above
    e. all of the above

69. Which of the following chemicals paralyzes skeletal muscle by binding to the acetylcholine receptor?
    a. black widow spider venom
    b. curare
    c. organophosphates
    d. DDT
    e. local anesthetics

70. Curare:
    a. strongly binds to acetylcholine receptor sites
    b. inhibits acetylcholinesterase
    c. is found in pesticides and military nerve gases
    d. two of the above
    e. all of the above

Modified Multiple Choice

Indicate which part of the autonomic nervous system is being described by writing the letter in the blank using the following code:

A = sympathetic nervous system
B = parasympathetic nervous system
C = both sympathetic and parasympathetic nervous systems
D = neither sympathetic nor parasympathetic nervous systems

71. _____ preganglionic fibers secrete acetylcholine
72. _____ preganglionic fibers secrete norepinephrine
73. _____ postganglionic fibers secrete acetylcholine
74. _____ postganglionic fibers secrete norepinephrine
75. _____ dominates in fight or flight situations
76. _____ dominates in relaxed situations
77. _____ has a long preganglionic fiber and a short postganglionic fiber
78. _____ has a short preganglionic fiber and a long postganglionic fiber
79. _____ originates in the cranial and sacral regions of the CNS
80. _____ originates in the thoracic and lumbar regions of the CNS
81. _____ innervates smooth muscle, cardiac muscle, and exocrine glands
82. _____ innervates skeletal muscle

Multiple Choice

83. All of the following activities are correctly paired with the division of the autonomic nervous system except:
    a. increased heart rate—sympathetic
    b. decreased salivation—sympathetic
    c. constricts skeletal muscles—parasympathetic
    d. stimulates contraction of the bladder wall—parasympathetic
    e. arises rostrally and sacrally—parasympathetic

84. The cell body of the postganglionic fiber of a parasympathetic pathway leading to a seat gland is located in the:
    a. sweat gland itself
    b. vertebral ganglion
    c. spinal cord
    d. brainstem
    e. dorsal root ganglion

85. A comparison of the components of the autonomic visceral reflex arc with the somatic reflex arc reveals that they differ in the number of:
    a. receptors
    b. afferent neurons
    c. interneurons
    d. efferent neurons
    e. effectors

86. Reflexes help to control:

I.	Heart rate
II.	Blood pressure
III.	Digestive activities

    a. I only
    b. II only
    c. III only
    d. I and II only
    e. I, II, and III

87. If you were to electrically stimulate the parasympathetic nervous system, which of the following would occur?
    a. an erection of the penis
    b. dilation of the pupils
    c. an increased release of glucose by the liver
    d. constriction of the abdominal blood vessels
    e. constriction of the peripheral blood vessels

88. Which of the following does not occur following adrenergic
    stimulation?
    a. dilation of the pulmonary bronchi
    b. dilation of the pupils
    c. increased heart rate
    d. increased basal metabolism
    e. peristalsis of the GI tract

89. The parasympathetic and sympathetic nervous systems
    oppose each other in their effects on the:
    a. liver
    b. heart
    c. skeletal muscles
    d. sweat glands
    e. adrenal glands

90. Stimulation of skeletal muscles by motor neurons is
    essential for muscles to:
    a. maintain their strength
    b. induce contraction
    c. maintain their size
    d. both a and b
    e. a, b, and c

91. One disease known to involve the neuromuscular junction
    is:
    a. myasthenia gravis
    b. polio
    c. Lou Gehrig's disease
    d. both a and b
    e. a, b, and c

Points to Ponder

1. Why do you think veterinarians find it difficult to diagnose toxic levels of organophosphate pesticides in animals?

2. When the opossum "plays possum," what is happening physiologically?

3. What is the relationship between biofeedback and the autonomic nervous system?

4. How does the central nervous system cooperate with the autonomic nervous system to maintain your body temperature?

5. Which divisions of the autonomic nervous system prepare a physiology student, such as yourself, for intense muscular activity? Explain.

6. What is meant by the term "dual intervention"?

7. After being in space for long periods of time, why do space travelers have problems walking?

8. If you get bitten by a Black Widow spider, what problems might occur in your body?

9. Why is the "fight-or-flight" response necessary for the survival of a physiology student such as yourself?

10. What happens in your body if you take the illegal drug cocaine?

Clinical Perspectives

1. How does epinephrine help a person suffering from an asthma attack?

2. How does epinephrine help a person who has just been stung by a wasp and is hypersensitive to bee stings?

3. A patient is accidentally given curare. How would you keep this person alive? Would the same treatment work for an accidental dose of organophosphates?

4. An old antihypertensive drug known as propranolol was widely used because it blocked beta 1 receptors. However, it was discontinued because of its side effects on the respiratory system. What would these effects be?

5. When you get an eye examination, the optometrist or ophthalmologist puts drops of atropine in your eyes. Why? Explain the physiologically and pharmacology behind the action of the eye drops.

Muscle Physiology

Chapter Overview

Muscle cells, subsequently muscle fibers, are specialized to contract. Again the specialization is the extensive development of a basic fundamental function of most cells. Movement of the body, manipulation of objects, and movement of the contents of the body are all important in maintaining homeostasis. Muscles are classified by their appearance, their function, and the nerves that innervate them. The contractile elements are actin and myosin. Individual muscle fibers either contract or they do not contract. Muscles are under the control and coordination of the CNS. These contraction specialists respond to an electrical signal and convert chemical energy into mechanical energy. Muscles perform a wide variety of work in a broad array of responses. Again, we are looking at a body system working to maintain homeostasis, which is essential for the survival of all the cells. Considering all three types, muscles play a very large and important role in maintaining homeostasis.

Chapter Outline

INTRODUCTION
- Almost all living cells possess rudimentary intracellular machinery for producing such movement as redistributing various components of the cell during cell division.
- Muscle cells are the contraction specialists of the body.
- Controlled contraction of muscles allows: (1) purposeful movement of the whole body, or parts of the body, in relation to the environment; (2) manipulation of external objects; (3) propulsion of contents through various hollow internal organs; and (4) emptying the contents of certain organs to the external environment.
- Muscles can be classified in several different ways according to the common characteristics: striated or unstriated, and voluntary (innervated by somatic nervous system) or involuntary (innervated by autonomic nervous system).

STRUCTURE OF SKELETAL MUSCLE
Skeletal muscle fibers are striated by highly organized internal arrangement.
- Skeletal muscles are stimulated to contract via release of acetylcholine (ACh) at neuromuscular junctions between motor neuron terminals and muscle cells.
- A single skeletal muscle cell is known as a "muscle fiber."
- A skeletal muscle consists of a number of muscle fibers lying parallel to each other.
- During embryonic development, the huge skeletal muscle fibers are formed by the fusion of many smaller cells; thus one striking feature is the presence of multiple nuclei in a single muscle cell.
- Another feature is the abundance of mitochondria.
- The most predominant structural feature of a skeletal muscle fiber is the presence of numerous myofibrils.

- Each myofibril consists of a regular arrangement of the proteins myosin (thick filaments) and actin (thin filaments).
- The bands of all the myofibrils lined up parallel to each other collectively lead to the striated appearance of a skeletal muscle fiber.
- An A band consists of a stacked set of thick filaments along with the portions of the thin filaments that overlap on both ends of the thick filaments.
- The thick filaments are found only in the A band and extend its entire width.
- The lighter area within the middle of the A band, where the thin filaments do not reach, is known as the "H zone."
- Only the central portions of the thick filaments are found in this region.
- The I band consists of the remaining portion of the thin filaments that do not project into the A band.
- The I band contains only thin filaments but not the entire length of these filaments.
- In the middle of each I band is a dense, vertical Z line.
- The area between two Z lines is called a "sarcomere," which is the functional unit of skeletal muscle.
- Each relaxed sarcomere consists of one whole A band and half of each of the two I bands located on either side
- A highly elastic protein known as titin extend in both directions from the M line along the length of the thick filament to the Z lines. It contributes to elasticity.

Myosin forms the thick filaments.
- Each thick filament is composed of several hundred myosin molecules packed together in a specific arrangement.
- A myosin molecule is a protein consisting of two identical subunits shaped somewhat like a golf club.
- The globular heads form the cross bridges between thick and thin filaments.

- Each cross bridge has two important sites crucial to the contractile process: an actin binding site and a myosin ATPase site.

Actin is the main structural component of the thin filaments.
- Thin filaments are composed of three proteins: actin, tropomyosin, and troponin.
- Actin molecules are spherical in shape.
- The backbone of a thin filament is formed by actin molecules joined into two strands and twisted together.
- Tropomyosin molecules are threadlike proteins that lie end to end alongside the groove of the actin spiral.
- Tropomyosin covers the actin sites that bind with the cross bridges, thus blocking the interaction that leads to muscle contraction.
- Tropomyosin is stabilized in this blocking position by troponin molecules, which fasten down the ends of each tropomyosin molecule.
- When calcium binds to troponin, the shape of this protein is changed in such a way that tropomyosin is allowed to slide away from its blocking position.
- With tropomyosin out of the way, actin and myosin can bind and interact at the cross bridges, resulting in muscle contraction.

MOLECULAR BASIS OF SKELETAL MUSCLE CONTRACTION
During contraction, cycles of cross-bridge binding and bending pull the thin filaments inward.
- The thin filaments on each side of a sarcomere slide inward toward the A band's center during contraction.
- As they slide inward, the thin filaments pull the Z lines to which they are attached closer together, so the sarcomere shortens.
- This is known as the "sliding-filament mechanism" or "muscle contraction."

- When myosin and actin make contact at a cross bridge, the conformation of the bridge is altered so that it bends inward.
- This so-called power stroke of a cross bridge pulls the thin filament to which it is attached inward.
- Complete shortening is accomplished by repeated cycles of cross-bridge binding and bending.
- Some cross bridges are "holding on" to the thin filaments, while others "let go" to bind with new actin.
- The term excitation–contraction coupling refers to the series of events linking muscle excitation and subsequent contraction

Calcium is the link between excitation and contraction.

- Skeletal muscles are stimulated to contract by release of acetylcholine at neuromuscular junctions.
- At each junction of an A band and an I band, the surface membrane dips into the muscle fiber to form a transverse tubule (T tubule), which runs perpendicularly from the surface of the muscle cell membrane into the central portions of the muscle fiber.
- The T tubule provides a means of rapidly transmitting the surface electrical activity into the central portions of the fiber.
- The sarcoplasmic reticulum is a modified endoplasmic reticulum that consists of a fine network of interconnected tubules surrounding each myofibril.
- Separate segments of sarcoplasmic reticulum are wrapped around each A and I band.
- The ends of each segment expand to form saclike regions, the lateral sacs, which are separated from the adjacent T tubules by a slight gap.
- The sarcoplasmic reticulum lateral sacs store calcium.
- Spread of an action potential down a T tubule triggers the release of calcium from the sarcoplasmic reticulum into the cytosol.
- Consequently, calcium is released from the lateral sacs.
- This released calcium, by slightly repositioning the troponin and tropomyosin molecules, exposes the binding sites on the actin molecules so that they can link with the myosin cross bridges at their complementary binding sites.
- The ATPase site is an enzymatic site that can bind the energy carrier ATP, and split it into ADP and inorganic phosphate, yielding energy in the process.
- The generated energy is stored within the cross bridge to produce a high-energy form of myosin.
- When the muscle fiber is excited, calcium pulls the troponin-tropomyosin complex out of its blocking position so that the energized myosin cross bridge can bind with an actin molecule.
- The energy stored within the myosin cross bridge is released to cause the cross bridge bending responsible for the power stroke that pulls the thin actin filament inward.
- The contractile process is turned off when the calcium is returned to the lateral sacs upon cessation of local electrical activity.
- When acetylcholinesterase removes ACh from the neuromuscular junction, the muscle fiber action potential ceases.
- Removal of cytosolic calcium allows the troponin-tropomyosin complex to slip back into its blocking position.
- Relaxation has occurred.

Contractile activity far outlasts the electrical activity that initiated it.

- The action potential that induced contractions ends before contraction begins.
- The time from the onset of contraction until peak tension is developed—the "contraction time."

- The time from peak tension until relaxation is complete is called the "relaxation time."

SKELETAL MUSCLE MECHANICS
Whole muscles are groups of muscle fibers bundled together and attached to bones.
- Groups of muscle fibers are organized into whole muscles.
- Each muscle is covered by a sheath of connective tissue that penetrates from the surface into the muscle to envelop each individual fiber and divide the muscle into columns or bundles.
- The connective tissue further extends beyond the ends of the muscle to form tough, collagenous tendons that attach the muscle to bones.

Muscle tension is transmitted to bone as the contractile component tightens the series-elastic component.
- Tension is produced internally within the sarcomeres, considered the contractile component of the muscle.
- The noncontractile tissues are referred to as the "series-elastic component."
- Typically, a muscle is attached to at least two different bones across a joint by means of tendons that extend from each end of the muscle.

The two primary types of contraction are isotonic and isometric.
- Tension is produced internally within the sarcomeres, the contractile component of the muscle.
- Tension generated by the contractile components must be transmitted to the bone via connective tissue and tendons.
- The end of the muscle attached to the more stationary part of the skeletal is called the "origin," and the end attached to the skeletal part that moves is referred to as the "insertion."
- The load is the weight of an object being lifted.
- In an isotonic contraction, muscle tension remains constant as the muscle changes length.

- In an isometric contraction, the muscle is prevented from shortening, so tension develops at constant muscle length.
- With concentric contractions the muscle shortens, whereas with eccentric contractions the muscle lengthens because it is being stretched by an external force while contracting.

The velocity of shortening is related to the load.
- The greater the load, the lower the velocity at which the muscle fiber shortens during an isotonic tetanic contraction.

Contractions of whole muscles can be of varying strength.
- A single action potential in a muscle fiber produces a brief, weak contraction known as a "twitch."
- Muscle fibers are arranged into whole muscles where they can function cooperatively to produce contractions of variable grades of strength stronger than a twitch.
- Two primary factors can be adjusted to accomplish gradation of whole-muscle tension: (1) the number of muscle fibers contracting within a muscle and (2) the tension developed by each contracting fiber.

The number of fibers contracting within a muscle depends on the extent of motor unit recruitment.
- Larger muscles consisting of more muscle fibers are obviously capable of generating more tension than are smaller muscles with fewer fibers.
- One motor neuron innervates a number of muscle fibers, but each muscle fiber is supplied by only one motor neuron.
- This functional unit—one motor neuron plus all of the muscle fibers it innervates—is called a "motor unit."
- For a weak contraction of the whole muscle, only one or a few of its motor units is activated.

- For stronger and stronger contractions, more and more motor units are recruited, or stimulated to contract, a phenomenon known as "motor unit recruitment."
- To delay or prevent fatigue during a sustained contraction involving only a portion of a muscle's motor units, as is necessary in muscles supporting the weight of the body against the force of gravity, asynchronous recruitment of motor units takes place.
- The body alternates motor unit activity to give motor units that have been active an opportunity to rest while others take over.

The frequency of stimulation can influence the tension developed by each muscle fiber.
- Various factors influence the extent to which tension can be developed including: (1) the frequency of stimulation, (2) the length of the fiber at the onset of contraction, (3) the extent of fatigue, and (4) the thickness of the fiber.
- Even though a single action potential in a muscle fiber produces only a twitch, greater tension can be achieved by repetitive stimulation of the fiber.
- If the muscle fiber is stimulated a second time before it has completely relaxed from the first twitch, a second action potential occurs that causes a second contractile response.
- The two twitches resulting from the two action potentials add together, or sum, to produce greater tension than produced by a single action potential.
- This is known as "twitch summation."
- If the muscle fiber is stimulated so rapidly that it does not have a chance to relax between stimuli, a smooth, sustained contraction of maximal strength known as "tetanus" occurs.

Twitch summation results from a sustained elevation of cytosolic calcium.
- Twitch summation depends on how long sufficient calcium is available.

- When stimulations occur rapidly, cytosolic calcium levels rise and stay high for extended times.
- This allows maximal cross-bridge cycling for maximal force.
- Additional factors not directly under nervous control also influence the tension developed during contraction.

There is an optimal muscle length at which maximal tension can be developed.
- For every muscle there is an optimal length at which maximal force can be achieved upon a subsequent tetanic contraction.
- In the body, the muscles are so positioned that their relaxed length is approximately their optimal length.
- At lengths greater or less than the optimal length, less tension can be developed.

SKELETAL MUSCLE METABOLISM AND FIBER TYPES
Muscle fibers have alternate pathways for forming ATP.
- Three different steps in the contraction-relaxation process require ATP: (1) the splitting of ATP by myosin ATPase; (2) the binding of a fresh molecule of ATP to myosin; and (3) the active transport of calcium back into the sarcoplasmic reticulum during relaxation depends on energy derived from the breakdown of ATP.
- Only limited stores of ATP are immediately available in muscle, but three pathways supply additional ATP as needed during muscle contraction: (1) transfer of a high-energy phosphate from creatine phosphate to ADP, (2) oxidative phosphorylation, and (3) glycolysis.
- Creatine phosphate is the first energy storehouse tapped at the onset of contractile activity.
- The energy released from the hydrolysis of creatine phosphate, along with the phosphate, can be donated directly to ADP to form ATP.

- Oxidation phosphorylation takes place within the muscle mitochondria if sufficient oxygen is present.
- During light exercise to moderate exercise, muscle cells are able to form sufficient amounts of ATP through oxidative phosphorylation to keep pace with the modest energy demands of the contractile machinery for prolonged periods of time.
- Activity that can be supported in this way is known as "endurance type exercise" or "aerobic exercise."
- The oxygen required for oxidative phosphorylation is primarily delivered by the blood.
- Some types of muscle fibers have an abundance of myoglobin, which increases the rate of oxygen transfer from the blood into muscle fibers.
- When oxygen delivery or oxidative phosphorylation cannot keep up with the demand for ATP formation as the intensity of exercise increases, the muscle fibers rely increasingly on glycolysis to generate ATP.
- Glycolysis alone has two advantages over the oxidative-phosphorylation pathway: (1) glycolysis can form ATP in the absence of oxygen, and (2) it can proceed more rapidly than oxidative phosphorylation because it requires fewer steps.
- Even though anaerobic glycolysis provides a means of performing intense exercise, using this pathway has two consequences: (1) large amounts of nutrient fuel must be processed, and (2) the end product of anaerobic glycolysis, pyruvic acid, is converted to lactic acid when it cannot be further processed by the oxidative-phosphorylation pathway.
- High-intensity or anaerobic exercise can be sustained for only a short duration in contrast to the body's prolonged ability to sustain aerobic activities.

Increased oxygen consumption is necessary to recover from exercise.
- The best known reason of elevated oxygen uptake during recovery is repayment of an oxygen debt that was incurred during exercise.
- During exercise, the creatine phosphate stores of active muscles are reduced, lactic acid may accumulate, and glycogen stores may be tapped.
- During the recovery period, fresh supplies of ATP are formed by oxidative phosphorylation using the newly acquired oxygen.
- This ATP is used to resynthesize creatine phosphate to restore its reserves.
- Any accumulated lactic acid is converted back into pyruvic acid, part of which is used in oxidative phosphorylation for ATP production.
- The remainder of the pyruvic acid is converted back into glucose by the liver.
- The glucose is used to replenish glycogen stores drained from the muscles and liver during exercise.
- These biochemical transformations involving pyruvic acid require oxygen.
- Part of the extra oxygen uptake during recovery is not directly related to the repayment of energy stores but instead is the result of a general metabolic disturbance following exercise.

There are three types of skeletal muscle fibers based on differences in ATP hydrolysis and synthesis.
- Based on their biochemical capacities, there are three major types of muscle fibers: (1) slow-oxidative (type I) fibers, (2) fast-oxidative (type IIa) fibers, and (3) fast-glycolytic (type IIb) fibers.
- Fast fibers have higher myosin ATPase activity than slow fibers.
- Fibers also differ in their ATP-synthesizing ability.
- Oxidative types of muscle fibers are more resistant to fatigue than are glycolytic fibers.

- Oxidative fibers also have a high myoglobin content and are referred to as "red fibers."
- The glycolytic fibers contain very little myoglobin and therefore are pale in color, so they are sometimes called "white fibers."
- In humans, most of the muscles contain a mixture of all three fiber types.

Muscle fibers adapt considerably in response to the demands placed on them.

- Two types of changes can be induced in muscle fibers: changes in their ATP-synthesizing capacity and changes in their diameter.
- Regular endurance exercise induces metabolic changes within the oxidative fibers, which are the ones primarily recruited during aerobic exercises.
- Muscles so adapted are better able to endure prolonged activity without fatiguing, but they do not change in size.
- The actual size of muscles can be increased by regular bouts of anaerobic, short duration, high-intensity resistance training, such as weight lifting.
- The resulting muscle enlargement comes primarily from an increase in diameter of the fast-glycolytic fibers that are called into play during such powerful contractions.
- The resultant bulging muscles are better adapted to activities that require intense strength for brief periods, but endurance has not been improved.
- Adaptive changes that take place in skeletal muscle gradually reverse to their original state over a period of months if the regular exercise program that induced these changes is discontinued.
- With age a person experiences a slow, continuous reduction in skeletal muscle mass and an accompanying loss of strength. Muscle loss is called atrophy.

CONTROL OF MOTOR MOVEMENT
Multiple neural inputs influence motor unit output.

- Three levels of input control motor neuron output: (1) input from afferent neurons, usually through intervening interneurons, at the level of the spinal cord—that is, spinal reflexes; (2) input from the primary motor cortex, the corticospinal (or pyramidal) motor system; and (3) input from the multineuronal (or extrapyramidal) motor system.
- The corticospinal system primarily mediates performances of fine, discrete, voluntary movements of the hands and fingers.
- The multineuronal system is primarily concerned with regulation of overall body posture involving involuntary movements of large muscle groups of the trunks and limbs.
- Premotor and supplementary motor areas, with input from the cerebrocerebellum, plan the voluntary motor command that is issued to the appropriate motor neurons by the primary motor cortex through this descending system. Defective motor control can result in spastic paralysis, flaccid paralysis, and loss of ability to generate smooth, coordinated movements.

Muscle receptors provide afferent information needed to control skeletal muscle activity.

- The CNS must know the starting position of your body to appropriately program muscle activity.
- The CNS must be constantly apprised of the progression of movement it has initiated so that it can make adjustments as needed.
- Your brain receives this information, which is known as "proprioceptive input," from receptors in your eyes, joints, vestibular apparatus, and skin, as well as from the muscles themselves.
- Muscle length is monitored by muscle spindles, whereas changes in muscle tension are detected by Golgi tendon organs.

- Muscle spindles consist of collections of specialized muscle fibers known as "intrafusal fibers," which lie within spindle-shaped connective tissue capsules parallel to the "ordinary" extrafusal fibers.
- The efferent neuron that innervates a muscle spindle's intrafusal fibers is known as a "gamma motor neuron," whereas the motor neurons that supply the ordinary extrafusal fibers are designated as "alpha motor neurons."
- Whenever the whole muscle is passively stretched, the intrafusal fibers are likewise stretched, increasing the rate of firing in the afferent nerve fibers.
- The afferent neuron directly synapses on the alpha motor neuron that innervates the extrafusal fibers of the same muscle, resulting in contraction of that muscle.
- This stretch reflex serves as a local negative-feedback mechanism to resist any passive changes in muscle length so that optimal resting length can be maintained.
- The primary purpose of the stretch reflex is to resist the tendency for the passive stretch of extensor muscles caused by gravitational forces when a person's standing upright.
- Golgi tendon organs are located in the tendons of the muscle, where they are able to respond to changes in the muscle's externally applied tension rather than to changes in its length.
- It is essential that motor control systems be apprised of the tension actually achieved so that adjustments can be made if necessary.
- The Golgi tendon organs consist of endings of afferent fibers entwined within bundles of connective tissue fibers that make up the tendon.
- When the extrafusal muscle fibers contract, the entwined Golgi organ afferent receptor endings are stretched, causing the afferent fibers to fire; the frequency of firing is directly related to the tension developed.

- Input from the activated Golgi tendon organs counterbalances excitatory inputs to the alpha motor neurons.
- This inhibitory response halts further contraction and brings about sudden reflex relaxation, thus helping prevent damage to muscle or tendons from excessive, tension-developing muscle contractions.

SMOOTH AND CARDIAC MUSCLE
- Smooth muscle and cardiac muscle share some basic properties with skeletal muscle, but each also displays unique characteristics.

Smooth muscle cells are small and unstriated.
- The majority of smooth muscle cells are found in the walls of hollow organs and tubes.
- Smooth muscle cells are spindle-shaped, have a single nucleus, and are considerably smaller than skeletal muscle cells.
- Also, a single smooth muscle cell does not extend the full length of a muscle.
- Three types of filaments are found in a smooth muscle cell: (1) thick myosin filaments, (2) thin actin filaments, and (3) unique filaments of intermediate size.
- Lacking sarcomeres, smooth muscle does not have Z lines, but dense bodies containing the same protein constituent found in Z lines are present.
- Dense bodies are positioned throughout the smooth muscle cell as well as being attached to the internal surface of the plasma membrane.
- Considerably more actin is present in smooth muscle cells than in skeletal muscle cells, with 10 to 15 thin filaments for each thick myosin filament in smooth muscle compared to 2 thin filaments for each thick filament in skeletal muscle.
- The actin filaments are anchored to the dense bodies.

- Because of the arrangement of myosin and actin the surrounding thin filaments can be pulled along the thick filaments for longer distances than in skeletal muscle

Smooth muscle cells are turned on by calcium-dependent phosphorylation of myosin.
- Smooth muscle myosin is able to interact with actin only when the myosin is phosphorylated.
- Smooth muscle calcium binds with calmodulin.
- This calcium-calmodulin complex binds to and activates myosin kinase, which in turn phosphorylates myosin.
- Phosphorylated myosin then binds with actin so that cross-bridge cycling can begin.
- When calcium is removed the muscle relaxes.

Multiunit smooth muscle is neurogenic.
- Smooth muscle is grouped into two categories—multiunit and single-unit smooth muscle—based on differences in how the muscle fibers become excited.
- Contractile activity in both skeletal muscle and multiunit smooth muscle is neurogenic.
- Multiunit smooth muscle is supplied by the involuntary autonomic nervous system.
- Multiunit smooth muscle is found: (1) in the walls of large blood vessels, (2) in large airways to the lungs, (3) in the muscle of the eye that adjusts the lens, (4) in the iris of the eye, and (5) at the base of hair follicles.

Single-unit smooth muscle cells form functional synctia.
- Most smooth muscle is of the single-unit variety.
- Single-unit is alternately called "visceral smooth muscle" because it is found in the walls of the hollow organs or viscera (for example, the digestive,

reproductive, and urinary tracts and small blood vessels).
- The muscle fibers in single-unit smooth muscle are electrically linked by gap junctions.
- When an action potential occurs anywhere within a sheet of single-unit smooth muscle, it is quickly propagated throughout the entire group of interconnected cells, which then contract as a single coordinated unit, known as a "functional syncytium."

Single-unit smooth muscle is myogenic.
- Single-unit smooth muscle is self-excitable rather than requiring nervous stimulation for contraction.
- There are two major types of spontaneous depolarizations displayed by self-excitable cells: pacemaker activity and slow-wave potentials.
- In pacemaker activity the membrane potential gradually depolarizes on its own.
- When the membrane has depolarized to threshold, an action potential is initiated.
- Self-generated action potentials are cyclically produced.
- Slow-wave potentials are gradually alternating hyperpolarizing and depolarizing swings in the potential caused by automatic cyclical changes in the rate at which sodium ions are actively transported across the membrane.
- If threshold is reached, a burst of action potentials occurs at the peak of a depolarizing swing.
- Such nerve-independent contractile activity initiated by the muscle itself is called "myogenic activity."

Gradation of single-unit smooth muscle contraction differs from that of skeletal muscle.
- In smooth muscle, the gap junctions assure that an entire smooth muscle mass contracts as a single unit, making it impossible to vary the number of muscle fibers contracting.

- The portion of cross bridges activated and the tension subsequently developed in single-unit smooth muscle can be graded by varying the cytosolic calcium concentration.
- Many single-unit smooth muscle cells have sufficient levels of cytosolic calcium to maintain a low level of tension, or tone.
- Smooth muscle is typically innervated by both branches of the autonomic nervous system.
- In single-unit smooth muscle, this nerve supply does not initiate contraction, but it can modify the rate and strength of contraction, either enhancing or retarding the inherent contractile activity of a given organ.
- The transmitter diffuses to the many receptor sites specific for it on the cells underlying the terminal.
- A given smooth muscle cell can be influenced by more than one type of neurotransmitter, and each autonomic terminal can influence more than one smooth muscle cell.
- In addition to autonomic neurotransmitters, smooth muscle activity can be influenced by mechanical stretch, certain hormones and local metabolites.

Smooth muscle can still develop tension when stretched.

- The ability of a considerably stretched smooth muscle fiber to still develop tension is important, because the smooth muscle fibers within the wall of a hollow organ are progressively stretched as the volume of the organ's contents increases.
- When smooth muscle is suddenly stretched, the muscle quickly adjusts to this new length and inherently relaxes to the tension level prior to the stretch.

Smooth muscle is slow and economical.

- A smooth muscle contractile response proceeds at a more leisurely pace than does a skeletal muscle twitch.

- Because of the low rate of cross bridge cycling, cross bridges are maintained in the attached state for a longer period of time during each cycle; that is, the cross bridges "latch onto" the thin filaments for a longer time each cycle.
- This so-called latch phenomenon enables smooth muscle to maintain tension with comparatively less ATP consumption.
- A single smooth muscle contraction may last as long as 3 seconds (3000 msec), compared to the maximum of 100 msec for a single contractile response in skeletal muscle

Cardiac muscle blends features of both skeletal and smooth muscle.

- Cardiac muscle, found only in the heart, shares structural and functional characteristics with both skeletal and single-unit smooth muscle.
- Cardiac muscle is striated.
- Cardiac thin filaments contain troponin and tropomyosin.
- Cardiac muscle has a clear-cut length-tension relationship.
- Cardiac muscle cells have an abundance of mitochondria and myoglobin and possess T tubules.
- As in smooth muscle, calcium enters the cytosol from both the ECF and the sarcoplasmic reticulum during cardiac excitation.
- Like single-unit smooth muscle, the heart displays pacemaker activity.
- Cardiac cells are interconnected by gap junctions.
- The heart is innervated by the autonomic nervous system, which along with certain hormones and local factors, can modify the rate and strength of contraction.
- Unique to cardiac muscle, the cardiac fibers are joined together in a branching network.

Key Terms

A-band
Actin
Aerobic exercise
Alpha motor neurons
Atrophies
Calmodulin
Cardiac muscle
Central fatigue
Contraction time
Creatine phosphate
Cross bridges
Dense bodies
Endurance exercise
Excitation-contraction coupling
Extrafusal fibers
Fatigue
Flaccid paralysis
Foot proteins
Force
Glycolysis
Golgi tendon organs
H zone
Hyperplasia
Hypertrophy
I band
Isometric contraction
Isotonic contraction
Intrafusal fibers
Latch phenomenon
Latent period
Lateral sacs
Lever
Load arm
M line
Motor unit
Motor unit recruitment
Multiunit smooth muscle
Muscle fatigue
Muscle fiber
Muscle spindle
Muscular dystrophy
Myofibrils
Myogenic activity
Myoglobin
Myosin
Myosin kinase
Neuromuscular fatigue

Optimal length
Oxidative phosphorylation
Oxygen debt
Pacemaker potential
Paraplegia
Patella tendon reflex
Power arm
Power stroke
Pyruvic acid
Quadriplegia
Regulatory proteins
Relaxation time
Rigor mortis
Sarcomere
Sarcoplasmic reticulum
Series elastic component
Skeletal muscle
Slow wave potential
Smooth muscle
Stretch reflex
Summation
Tendons
Tension
Tetanus
Thick filaments
Thin filaments
Transverse tubule
Tropomyosin
Troponin
Twitch
Twitch summation
White fibers
Work
Z line

Review Exercises

Answers are in the appendix.

True/False-if an answer is false, underline the incorrect portion of the statement and rewrite it to make the statement true.

_____ 1. The corticospinal system primarily mediates performance of fine, discrete, voluntary movements of the hands and fingers.

_____ 2. Muscle spindles consist of specialized muscle fibers known as "extrafusal fibers."

_____ 3. Gamma motor neurons are function to stimulate skeletal muscle contractions.

_____ 4. The stretch reflex serves as a local positive-feedback mechanism to resist any passive changes in muscle length.

_____ 5. Golgi tendon organs respond to changes in muscle length.

_____ 6. After exercise, oxygen is needed for skeletal muscle recovery.

_____ 7. Ordinary muscle fibers are known as "extrafusal fibers."

_____ 8. Smooth muscle cells are spindle-shaped with multiple nuclei.

_____ 9. In smooth muscle, calcium binds to troponin to promote contraction.

_____ 10. Smooth muscle calcium binds with calmodulin, an intracellular protein similar to troponin.

_____ 11. Contractile activity initiated by the muscle, itself, is called "myogenic activity."

_____ 12. A single smooth muscle contraction may last as long as 3 seconds

_____ 13. The portion of cross bridges activated and the tension subsequently developed in a single-unit smooth muscle can be graded by varying the cytosolic magnesium concentration.

_____ 14. The splitting of ATP by actin ATPase provides the energy for the power stroke of the cross bridge.

_____ 15. Creatine phosphate is the first energy storehouse tapped at the onset of contractile activity.

_____ 16. Hemoglobin increases the rate of oxygen transfer from the blood into the muscle fiber.

_____ 17. The end product of anaerobic glycolysis is pyruvic acid.

_____ 18. Smooth muscle contains more actin than skeletal muscle.

_____ 19. Glycolytic fibers are more resistant to fatigue.

_____ 20. A single skeletal muscle cell is known as a "myofibril."

_____ 21. Thick filaments are special assemblies of the protein myosin.

_____ 22. Each myofibril consists of a regular arrangement of the proteins myosin and actin.

_____ 23. In addition to autonomic neurotransmitters, smooth muscle activity can be influenced by mechanical stretch, hormones and local metabolites

_____ 24. The globular heads of the myosin molecule form the cross bridges between the thick and thin filaments.

_____ 25. The area between two Z lines is called a "sarcomere."

_____ 26. When magnesium binds to troponin, the shape of this protein is changed in such a way that tropomyosin is allowed to slide away from its blocking position.

_____ 27. One motor neuron plus all of the muscle fibers it innervates is called a "motor unit."

_____ 28. A single action potential in a muscle fiber produces a brief, weak contraction known as a "twitch."

_____ 29. Smooth muscle contracts faster than skeletal muscle.

_____ 30. The asynchronous recruitment of motor units produces muscle spasms.

_____ 31. The thickness of the fiber influences the extent to which tension can be developed.

_____ 32. In skeletal muscle, at lengths greater or less than the optimal length, less tension is developed

_____ 33. A sustained contraction of maximal strength is known as "tetanus."

_____ 34. The end of the muscle attached to the more stationary part of the skeleton is known as the "insertion."

_____ 35. In eccentric contractions the muscle shortens.

_____ 36. At each junction of an A band and an I band, the surface membrane dips into the muscle fiber to form a transverse tubule.

_____ 37. The sarcoplasmic tubule is a modified endoplasmic reticulum.

_____ 38. T tubules store calcium.

_____ 39. In skeletal muscle, magnesium must be attached to ATP before myosin ATPase can split the ATP.

_____ 40. The time delay between stimulation and the onset of contraction is known as the "contraction time."

_____ 41. The contractile process is turned off when calcium is returned to the latent sacs upon cessation of local electrical activity.

_____ 42. Removal of cytosolic potassium allows the troponin-tropomyosin complex to slip back into its blocking position.

Fill in the Blank

43. _____ refers to the series of events linking muscle excitation to muscle contraction.

44. At each junction of an A band and an I band, the surface membrane dips into the muscle fiber to form a(n) _____, which runs perpendicularly from the surface of the muscle cell membrane into the central portions of the muscle fiber.

45. The _____ is a modified endoplasmic reticulum that consist of a fine network of interconnected tubules surrounding each myofibril.

46. The ends of each sarcoplasmic reticulum segment expand to form saclike regions, the _____, which lie in close proximity to the adjacent T tubules.

47. A myosin cross bridge has two special sites, a(n) _____ site and a(n) _____ site.

48. The necessity for ATP in the separation of myosin and actin is amply demonstrated by the phenomenon of _____.

49. Loss of function of alpha motor neurons produces _____ paralysis.

50. _____ is an intracellular protein found in most cells that is structurally similar to troponin.

51. _____ exhibits properties pathway between skeletal muscle and single unit smooth muscle.

52. _____ means they depend on their nerve supply to initiate contraction. Examples are skeletal muscle and multiunit smooth muscle.

53. A group of interconnected muscle cells that function electrically and mechanically as a unit is known as a(n)_____.

54. _____ are gradually alternating hyperpolarizing and depolarizing swings in potential caused by automatic cyclical changes in the rate at which sodium ions are actively transported across the membrane.

55. _____ enables smooth muscle to maintain tension with comparatively less ATP consumption, because each cross-bridge cycle uses up one molecule of ATP.

56. A single skeletal muscle cell is known as _____.

57. The most predominant structural feature of a skeletal muscle fiber is the presence of numerous _____.

58. A(n) _____ consists of a stacked set of thick filaments along with the portions of the thin filaments that overlap on both ends of the thick filaments.

59. The area between two Z lines is called a(n) _____.

60. A(n) _____ molecule is a protein consisting of two identical subunits, each shaped somewhat like a golf club. _____ molecules are spherical in shape.

61. _____ molecules are threadlike proteins that lie end to end alongside the groove of the actin spiral.

62. _____ and _____ are often referred to as "regulatory proteins."

63. _____ is paralysis of one side of the body.

64. _____ is paralysis of all four limbs.

65. _____ is paralysis of the legs.

66. The efferent neuron that innervates a muscle spindle's intrafusal fibers is known as a(n) _____, whereas the motor neurons that supply the ordinary extrafusal fibers are designated as _____.

67. _____ are located in the tendons of the muscle, where they are able to respond to changes in the muscle's externally applied tension rather than to changes in its length.

68. One motor neuron plus all of the muscle fibers it innervates is called a(n) _____.

69. Cross-bridge activity produces _____ within the sarcomeres.

70. The noncontractile tissues are referred to as the _____ component of the muscle.

71. The end of the muscle attached to the more stationary part of the skeleton is called the _____, whereas the end attached to the skeletal part that moves is referred to as the _____.

72. In a(n) _____ contraction, muscle tension remains constant as the muscle changes length.

73. There are actually two types of isotonic contractions, _____ and _____.

74. _____ muscles guard the exit of urine and feces from the body by isotonically contracting.

75. _____ paralysis occurs when neurons within the brain are damaged.

76. A(n) _____ is a rigid structure capable of moving around a pivot point known as a(n) _____.

77. _____ is the first energy storehouse tapped at the onset of contractile activity.

78. _____ takes place within the muscle mitochondria if sufficient oxygen is present.

79. _____ increases the rate of oxygen transfer from the blood into muscle fibers.

80. _____ occurs when a muscle is not used for a long period of time.

81. _____ is a hereditary pathological condition characterized by progressive degeneration of contractile elements, which are ultimately replaced by fibrous tissue.

Matching
Match the characteristic to the type of fiber.

a. fast-glycolytic type IIb
b. fast-oxidative type IIa
c. slow-oxidative type I

_____ 82. Intermediate intensity of contraction
_____ 83. Few capillaries
_____ 84. High resistance to fatigue
_____ 85. Mostly anaerobic production of ATP
_____ 86. Small fiber diameter
_____ 87. Low oxidative phosphorylation capacity
_____ 88. Slow speed of contraction
_____ 89. White fibers
_____ 90. Intermediate amount of enzymes for anaerobic glycolysis
_____ 91. Intermediate glycogen content
_____ 92. Low ability to store oxygen
_____ 93. Low myosin-ATPase activity

Multiple Choice

94. Which of the following consists of a stacked set of thick filaments along with the portions of the thin filaments that overlap on both ends of the thick filaments?
    a. Z line
    b. M line
    c. A band
    d. I band
    e. H zone

95. Which of the following are threadlike proteins that lie end to end alongside the groove of the actin spiral?
    a. tropomyosin
    b. troponin
    c. myosin
    d. tubulin
    e. myoglobin

96. Which of the following ions must be attached to ATP before myosin ATPase can split the ATP in skeletal muscle?
    a. magnesium
    b. calcium
    c. potassium
    d. sodium
    e. calmodulin

97. Which of the following represents the region where only the central portions of the thick filaments are found?
    a. Z line
    b. M line
    c. A band
    d. I band
    e. H zone

98. Which of the following pairs of cross-bridge sites is crucial to the contractile process?
    a. troponin binding site and myosin ATPase site
    b. tropomyosin binding site and actin ATPase site
    c. myosin binding site and myosin ATPase site
    d. actin binding site and myosin ATPase site
    e. myoglobin binding site and troponin ATPase site

99. Which of the following ions binds to calcium in smooth muscles?
    a. magnesium
    b. calmodulin
    c. ATPase
    d. sodium
    e. troponin

100. Which of the following types of muscle has no gap junctions?
    a. multiunit smooth
    b. skeletal
    c. single-unit smooth
    d. cardiac
    e. all of the above

101. Which of the following types of muscle has cross bridges turned on by calcium?
    a. multiunit smooth
    b. skeletal
    c. single-unit smooth
    d. cardiac
    e. all of the above

102. Which of the following is innervated by the somatic nervous system?
    a. multiunit smooth
    b. skeletal
    c. single-unit smooth
    d. cardiac
    e. all of the above

103. Which of the following is the functional unit of a skeletal muscle?
    a. Golgi organ
    b. thick filaments
    c. sarcomere
    d. sliding filaments
    e. transverse-contractile coupler

104. Which of the following is not characteristic of smooth muscle?
    a. contracts slowly   b. is self-excitable   c. has a striated appearance
    d. contains large amounts of actin
    e. can contain gap junctions

105. Which type of fatigue occurs due to an accumulation of lactic acid?
    a. muscle fatigue
    b. neuromuscular fatigue
    c. central fatigue
    d. all of the above
    e. none of the above

106. Thin filaments in skeletal muscle are composed of:
    a. actin only
    b. tropomyosin only
    c. troponin only
    d. b and c only
    e. a, b and c

Modified Multiple Choice
Indicate which type(s) of muscle are associated with the property in question by writing the appropriate letter in the blank using the code below.

> A = skeletal muscle only
> B = single-unit smooth muscle only
> C = cardiac muscle only
> D = skeletal muscle and cardiac muscle
> E = skeletal muscle and single-unit smooth muscle
> F = single-unit smooth muscle and cardiac muscle
> G = skeletal, single-unit smooth, and cardiac muscle

107. _____ contains actin, myosin, troponin, and tropomyosin
108. _____ contains gap junctions
109. _____ activity influenced by the supplemental motor area
110. _____ is self excitable
111. _____ maintains a constant membrane potential unless stimulated
112. _____ innervated by the autonomic nervous system
113. _____ relaxation caused by acetylcholinesterase
114. _____ considered to be involuntary
115. _____ thick and thin filaments are highly organized into a banding pattern
116. _____ found in the heart
117. _____ can exist over a variety of lengths with little change in tension
118. _____ is striated
119. _____ is found in the walls of hollow organs such as the digestive tract, bladder, uterus, and blood vessels
120. _____ cells connected together by gap junctions
121. _____ is under voluntary control
122. _____ has a clear-cut length tension relationship
123. _____ basis of contraction is cross bridge interaction between actin and myosin
124. _____ contraction is triggered as calcium physically pulls troponin and tropomyosin from its blocking position over actin's binding sites for cross bridges
125. _____ myosin must be phosphorylated before it can bind with actin
126. _____ contains T tubules
127. _____ displays pacemaker potentials and slow wave potentials
128. _____ is neurogenic

Indicate which bands are being described in each statement by writing the appropriate letter(s) in the blank using the answer code below. Note that more than one answer may apply.

A = A band
B = I band
C = H zone

129. _____ composed of thin filaments only
130. _____ composed of thick filaments only
131. _____ composed of both thick and thin filaments
132. _____ shortens during muscular contraction
133. _____ remains the same size during muscular contraction

Points to Ponder

1. Can you think of an example where involuntary muscles are stimulated to contract voluntarily?

2. How can an average-sized woman beat a contestant for Mr. Universe in arm wrestling?

3. How does weight-lifting shape up as a healthy form of exercise?

4. Why do you think the average life expectancy for NFL players is only 57 years?

5. Why does skeletal muscle constitute about 40-50% of an adult male's body weight and 30-40% of an adult female's body weight?

6. What makes one muscle larger than another muscle?

7. What causes your foot to "fall asleep"?

8. What causes rigor mortis to occur after death?

9. While running on a hot day, you develop muscle cramps. Which of the following will help the muscle cramps: drinking a lot of water, massaging the muscles, stretching the cramped muscles, or eating a banana as a source of potassium?

10. 35% of each cardiac muscle cell is made up of mitochondria, compared with 2% in skeletal muscle cells. What does this reflect physiologically?

Clinical Perspectives

1. Explain why heat and/or cold packs are used to treat the so-called "pulled muscle."

2. How does rigor mortis help the medical examiner determine the time of death?

3. Why does Duchenne muscular dystrophy only occur in males?

4. Rapid stretching of skeletal muscles produces very forceful muscle contractions. This can result in painful muscle spasms. How can these muscle spasms be prevented?

5. How does the disease polio cause flaccid paralysis, reduced muscle tone, depressed stretch reflexes, and atrophy?

6. Parkinson's disease causes a resting tremor to occur. This "shaking of the limbs" tends to disappear during voluntary movements and then reappears when the limbs are at rest again. Why does this occur?

7. The enzyme that transfers phosphate between creatine and ATP is called "creatine kinase." Skeletal muscle and heart muscle have two different forms of this enzyme (isozymes). What diseases can these isozymes detect in muscle tissue?

8. How do calcium channel blocking drugs work in the walls of blood vessels to help reduce blood pressure in those patients who have hypertension?

9. In some patients, calcium channel blocking drugs can cause fatigue and muscle pain. Explain why this occurs physiologically.

10. From 12-48 hours after strenuous exercise, skeletal muscles often become sore. Such delayed onset of muscle soreness is accompanied by stiffness, tenderness, and swelling. Why do these symptoms occur?

Experiments of the Day

1. Go to the meat department of your favorite grocery store and explain the appearance of the meats in terms of muscle structure.

2. Go to the vitamin supplement department of your favorite pharmacy and find supplements that claim to increase muscle mass. Are these the same as metabolic steroids? Are they good for you? Explain.

3. The next time you eat a steak, try to identify the anatomical parts of the steak.

9

Cardiac Physiology

Chapter Overview

The importance of the circulatory system, particularly the heart, in maintaining homeostasis has been known to some degree by most of us since our childhood days. This heart-homeostasis relationship is probably the best known aspect of homeostasis. From the bending and twisting of an embryonic blood vessel an organ develops that will pump the blood. Complete with muscular walls and valves, the heart responds to the autorhythmicity of the sinoatrial node. Using specialized conductive cells, action potentials are spread throughout the organ and coordinated contractions occur. After a brief period of relaxation, the cycle begins again.

The electrical and mechanical activities of the heart produce sounds, pressures, and electrical changes that can be detected on the surface of the body. Electrocardiography is an important tool in monitoring cardiac activities.

To meet the changing homeostatic needs of all the cells, the heart must be able to change its output. Cardiac output is controlled through changes in heart rate and stroke volume. While the heart is working for all the cells it must also nourish itself. This is one of the most frequent areas of complications in the circulatory system.

Even though the heart is emphasized in science and culture, in reality it is but a part of an organ system controlled by the nervous and endocrine systems. The remarkable properties of the heart are specializations of the general properties and functions of all cells. The circulatory system aids in maintaining homeostasis, which is essential for the survival of cells.

Chapter Outline

INTRODUCTION
- The circulatory system consists of the heart, the blood vessels, and the blood. The heart serves as the pump; the blood vessels serve as pathways for the blood and the blood serves as the transport medium.
- The pulmonary circulation consists of a closed loop of vessels carrying blood between the heart and lungs, whereas the systemic circulation consists of a circuit of vessels carrying blood between the heart and organ systems.

ANATOMY OF THE HEART
- The heart is situated at an angle under the sternum so that its base lies predominately to the right and the apex to the left of the sternum.
- The fact that the heart is positioned between two bony structures, the sternum and vertebrae, make it possible to manually drive blood out of the heart which is part of CPR.

The heart is a dual pump.
- Even though anatomically the heart is a single organ, the right and left sides of the heart function as two separate pumps.

- The upper chambers, the atria, receive blood returning to the heart and transfer it to the lower chambers, the ventricles, which pump the blood from the heart.
- The vessels that return blood from the tissues to the atria are veins, and those that carry blood away from the ventricles to the tissues are arteries.
- The right side of the heart pumps blood into the pulmonary circulation.
- The right ventricle pumps blood out through the pulmonary artery to the lungs.
- The left side of the heart pumps blood into the systemic circulation.
- The large artery carrying blood away from the left ventricle is the aorta.
- Both sides of the heart simultaneously pump equal amounts of blood.
- The volume of oxygen-poor blood being pumped to the lungs by the right side of the heart soon becomes the same volume of oxygen-rich blood being delivered to the tissues by the left side of the heart.
- The pulmonary circulation is a low-pressure, low-resistance system, whereas the systemic circulation is a high-pressure, high-resistance system.

Pressure-operated heart valves ensure that the blood flows in the right direction through the heart.
- Blood flows through the heart in one fixed direction from veins to atria to ventricles to arteries.
- Two of the heart valves, the right and left atrioventricular (AV) valves, are positioned between the atrium and the ventricle on the right and left sides, respectively. Chordae

tendineae and papillary muscles are associated with AV valves and prevent the valves from everting.
- The two remaining heart valves, the aortic and pulmonary valves, are located at the junction where the major arteries leave the ventricles.
- They are known as "semilunar valves" because they are composed of three cusps, each resembling a shallow half-moon shaped pocket.

The heart walls are composed primarily of spirally arranged cardiac muscle fibers.
- The heart wall consists of three distinct layers: (1) The endocardium is a thin inner layer of endothelium that lines the entire circulatory system. (2) The myocardium, the middle layer composed of cardiac muscle, constitutes the bulk of the heart wall. (3) The epicardium is a thin external membrane covering the heart.

Cardiac muscle fibers are interconnected by intercalated discs and form functional synctia.
- The individual cardiac muscle cells are interconnected to form branching fibers, with adjacent cells joined end to end at specialized structures known as "intercalated discs."
- Within an intercalated disc, there are two types of membrane junctions: desmosomes and gap junctions.
- Cardiac muscle is capable of generating action potentials without nervous stimulation.
- There are no gap junctions between atrial and ventricular contractile cells, and furthermore, these muscle masses are separated by the electrically nonconductive fibrous skeleton that surrounds the valves.
- An important specialized conducting system is present to facilitate and coordinate the transmission of electrical excitation from the atria to the ventricles.
- Either all of the cardiac muscle fibers contract or none of them do.
- Gradation of cardiac contraction is accomplished by varying the strength of contraction of all the cardiac muscle cells.

ELECTRICAL ACTIVITY OF THE HEART

Cardiac autorhythmic cells display pacemaker activity.

- The heart contracts, or beats, rhythmically as a result of action potentials that it generates by itself, a property known as "autorhythmicity."
- There are two types of cardiac muscle cells: (2) 99% of the cardiac muscle cells are contractile cells. (2) The autorhythmic cells do not contract, but instead are specialized for initiating and conducting the action potentials responsible for contraction.
- The cardiac autorhythmic cells display pacemaker activity.
- A unique voltage-gated channel found only in cardiac autorhythmic cells plays a key role in producing the pacemaker potential.
- These unusual channels open when the potential becomes more negative (hyperpolarizes) at the end of repolarization from the previous action potential. Because of this funny behavior, they are sometimes called funny channels. When one action potential ends and the funny channels open, the resultant depolarizing net inward Na$^+$ movement starts immediately moving the pacemaker cell's potential toward threshold once again.
- Once threshold is reached, the rising phase of the action potential occurs in response to opening of voltage-gated Ca^{2+} channels and the resulting entry of Ca^{2+}. The falling phase occurs as usual by K$^+$ leaving the pacemaker cell as a result of opening of voltage-gated K$^+$ channels

- The cardiac cells capable of autorhythmicity arc found in the following specific locations: (1) the sinoatrial node, (2) the atrioventricular node, (3) the bundle of His, and (4) the Purkinje fibers.
- Various autorhythmic cells differ in the rates at which they are normally capable of generating action potentials.

Normal pacemaker activity.

- The SA node, which normally exhibits the fastest rate of autorhythmicity, is known as the pacemaker of the heart.
- The non-SA nodal autorhythmic tissues are latent pacemakers.

The spread of cardiac excitation is coordinated to ensure efficient pumping.

- For efficient cardiac function, the spread of excitation should satisfy three criteria: (1) Atrial excitation and contraction should be complete before the onset of ventricular contraction. (2) Excitation of cardiac muscle fibers should be coordinated to ensure that each heart chamber contracts as a unit to accomplish efficient pumping. (3) The pair of atria and pair of ventricles should be functionally coordinated so that both members of the pair contract simultaneously.
- Simultaneous contraction permits synchronized pumping of blood into the pulmonary and systemic circulation.
- The normal spread of cardiac excitation is carefully orchestrated to ensure that these criteria are met and the heart functions efficiently.
- An action potential originating in the SA node first spreads throughout both atria, primarily from cell to cell via gap junctions.
- The interatrial pathway extends from the SA node within the right atrium to the left atrium.
- This pathway ensures that both atria become depolarized to contract more or less simultaneously.
- The internodal pathway extends from the SA node to the AV node.

- The internodal conduction pathway directs the spread of an action potential originating at the SA node to the AV node to ensure sequential contraction of the ventricles following atrial contraction.
- The action potential is conducted relatively slowly through the AV node, called the AV nodal delay.
- The slowness is advantageous because it allows time for complete ventricular filling to occur.
- The impulse rapidly travels down the bundle of His and throughout the ventricular myocardium via the Purkinje fibers.
- The ventricular conduction system is more highly organized and more important than the interatrial and internodal conduction pathways.
- The rapid conduction of the action potential down the bundle of His and its swift, diffuse distribution throughout the Purkinje network lead to almost simultaneous activation of the ventricular myocardial cells in both ventricular chambers, which ensures a single, smooth, coordinated contraction that can efficiently eject blood into both the systemic and pulmonary circulations at the same time.

The action potential of contractile cardiac muscle cells shows a characteristic plateau.

- Unlike autorhythmic cells, the membrane of contractile cells remains essentially at rest at about -90mV until excited by electrical activity propagated from the pacemaker.
- Once the membrane is excited, an action potential is generated by a complicated interplay of permeability changes and membrane potential changes.
- During the rising phase of the action potential, the membrane potential rapidly becomes reversed to a positive value of +30mV as a result of an explosive sodium influx.
- The membrane potential is maintained at this positive level for several hundred milliseconds, producing a plateau phase of the action potential.
- Two voltage-dependent permeability changes are responsible for maintaining this plateau: activation of "slow" calcium channels and a marked decrease in potassium permeability.
- The rapid falling phase of the action potential results from inactivation of the calcium channels and activation of potassium channels.
- The mechanism by which an action potential in a cardiac muscle fiber brings about contraction of that fiber is quite similar to the excitation—contraction coupling process of skeletal muscle and smooth muscle.

Calcium entry from the ECF induces a much larger calcium release from the sarcoplasmic reticulum.

- Calcium diffuses into the cytosol across the T tubule membrane during cardiac action potential.
- Entering calcium triggers release of calcium from the sarcoplasmic reticulum.
- Calcium also diffuses into the cytosol across the plasma membrane from the ECF during a cardiac action potential.
- This extra supply of calcium is not only the major factor responsible for the prolongation of the cardiac action potential but is also responsible for the subsequent lengthening of the period of cardiac contraction.
- In cardiac muscle the extent of cross-bridge activity varies with the amount of cytosolic calcium.
- A long refractory period prevents tetanus of cardiac muscle .

- During the refractory period, which occurs immediately after the initiation of an action potential, an excitable membrane's responsiveness is totally abolished.
- Cardiac muscle has a long refractory period.
- Consequently, cardiac muscle cannot be restimulated until contraction is almost over, making summation of contractions and tetanus of cardiac muscle impossible.

The ECG is a record of the overall spread of electrical activity through the heart.

- An ECG is a recording of that portion of the electrical activity induced in the body fluids by the cardiac impulse that reaches the surface of the body, not a direct recording of the actual electrical activity of the heart.
- The ECG is a complex recording representing the overall spread of activity throughout the heart during depolarization and repolarization.
- The ECG is not a recording of a single action potential in a single cell.
- The recording represents comparisons in voltage detected by electrodes at two different points on the body surface, not the actual potential.
- To provide standard comparisons, ECG records routinely consist of twelve conventional electrode systems or leads.

Different components of the ECG record can be correlated to specific cardiac events.

- A normal ECG exhibits three distinct waveforms: the P wave, the QRS complex, and the T wave.
- The P wave represents atrial depolarization.
- The QRS complex represents ventricular depolarization.
- The T wave represents ventricular repolarization.
- The electrical activity associated with atrial repolarization normally occurs simultaneously with ventricular depolarization and is marked by the QRS complex.
- The P wave is much smaller than the QRS complex because the atria have a much smaller muscle mass than the ventricles.
- There are three times when no current is flowing in the heart musculature and the ECG remains at baseline, during: (1) the AV nodal delay, (2) the ST segment, and (3) the TP interval.

The ECG can be useful in diagnosing abnormal heart rates, arrhythmias, and damage of heart muscle.

- Because electrical activity triggers mechanical activity, abnormal electrical patterns are usually accompanied by abnormal contractile activity of the heart.
- The principle deviations from normal that can be ascertained through electrocardiography are as follows: tachycardia, bradycardia, arrhythmia, atrial flutter, atrial fibrillation, ventricular fibrillation, and heart block.
- Abnormal ECG waves are also important in the recognition and assessment of cardiac myopathies, such as myocardial ischemia, necrosis, coma, and acute myocardial infarction.

MECHANICAL EVENTS OF THE CARDIAC CYCLE
The heart alternately contracts to empty and relaxes to fill.

- The cardiac cycle consists of alternate periods of systole and diastole.
- The atria and ventricles go through separate cycles of systole and diastole.

- Because of the continuous inflow of blood from the venous system into the atrium, atrial pressure slightly exceeds ventricular pressure even though both chambers are relaxed.
- The AV valve opens and blood flows directly from the atrium into the ventricle throughout ventricular diastole.
- Ventricular volume slowly continues to rise even before atrial contraction takes place.
- Late in ventricular diastole, the SA node fires.
- The impulse spreads throughout the atria (P wave).
- Atrial depolarization brings about atrial contraction.
- Throughout atrial contraction, atrial pressure still slightly exceeds ventricular pressure, so the AV valve remains open.
- Ventricular diastole ends at the onset of ventricular contraction.
- The volume of blood in the ventricle at the end of diastole is known as the "end-diastolic volume" (EDV).
- Following atrial excitation, the impulse passes through the AV node and specialized conducting system to excite the ventricle.
- Simultaneously, atrial contraction is occurring.
- By the time ventricular activation is complete, atrial contraction is already accomplished.
- The QRS complex represents this ventricular excitation, which induces ventricular contraction.
- As ventricular contraction begins, ventricular pressure immediately exceeds atrial pressure.
- The backward pressure differential forces the AV valve closed.

- There is a brief period of time between closure of the AV valve and opening of the aortic valve when the ventricle remains a closed chamber.
- This interval is termed the "period of isovolumetric ventricular contraction."
- When ventricular pressure exceeds aortic pressure, the aortic valve is forced open and ejection of blood begins.
- Ventricular systole includes both the period of isovolumetric contraction and the ventricular ejection phase.
- About half of the blood contained within the ventricle at the end of diastole is pumped out during the subsequent systole.
- The amount of blood remaining in the ventricle at the end of ventricular systole is known as "end-systolic volume" (ESV).
- The amount of blood pumped out of each ventricle with each contraction is known as the "stroke volume," the difference between the volume of blood in the ventricle before contraction and the volume after contraction (EDV-ESV).
- The T wave signifies ventricular repolarization occurring at the end of ventricular systole.
- As the ventricle starts to relax, the aortic valve closes.
- Closure of the aortic valve produces a disturbance or notch on the aortic pressure curve.
- All valves are once again closed for a brief period of time known as "isovolumetric ventricular relaxation."
- When the ventricular pressure falls below the atrial pressure, the AV valve opens and ventricular filling occurs once again.

Two normal heart sounds are associated with valve closures.
- Two major heart sounds normally can be heard during the cardiac cycle when listening with a stethoscope.
- The first sound is associated with the closure of the AV valves.
- Closure of the AV valves occurs at the onset of ventricular contraction.

- Closure of the semilunar valves occurs at the onset of ventricular relaxation and produces the second heart sound.

Turbulent blood flow produces heart murmurs.
- Abnormal heart sounds, or murmurs, are usually associated with cardiac disease.
- Blood normally flows in a laminar fashion and does not produce any sound.
- When blood flow becomes turbulent, however, a sound can be heard. The most common cause of turbulence is valve malfunction, either a stenotic (a stiff, narrowed valve) or an insufficient valve (one that cannot close completely).
- Most often, both valvular stenosis and insufficiency are caused by rheumatic fever.
- The valve involved and the type of defect can usually by detected by the location and timing of the murmur.
- A murmur occurring between the first and second heart sounds signifies a systolic murmur.
- A diastolic murmur occurs between the second and first heart sounds.
- The sound of the murmur characterizes it as either a stenotic (whistling) murmur or an insufficient (swishy) murmur.

CARDIAC OUTPUT AND ITS CONTROL
Cardiac output depends on the heart rate and the stroke volume.
- Cardiac output is the volume of blood pumped by each ventricle per minute.
- The two determinants of cardiac output are heart rate and stroke volume.

Heart rate is determined primarily by autonomic influences on the SA node.
- The heart is innervated by both divisions of the autonomic nervous system, which can modify the rate of contraction, even though nervous stimulation is not required to initiate contraction.
- The parasympathetic nerve to the heart, the vagus nerve, primarily supplies the atrium, especially the SA and AV nodes.
- There is no significant parasympathetic innervation to the ventricles.
- The cardiac sympathetic nerves also supply the atria, including the SA and AV nodes, and richly innervate the ventricles as well.
- The parasympathetic nervous system's influence on the SA node is to decrease the heart rate.
- Parasympathetic influence on the AV node decreases the node's excitability, prolonging the AV nodal delay.
- Parasympathetic stimulation of the atrial contractile cells shortens the action potential, that is, the plateau phase is reduced and atrial contraction is weakened.
- The sympathetic nervous system speeds up the heart rate through its effect on the pacemaker tissue.
- Sympathetic stimulation of the AV node reduces the AV nodal delay.
- Similarly, sympathetic stimulation speeds up the spread of the action potential throughout the specialized conducting pathway.
- Sympathetic stimulation increases contractile strength so that the heart beats more forcefully and squeezes out more blood. It does so by increase calcium permeability allowing more calcium to act in excitation-contraction coupling.

Control of Heart Rate
- Under resting conditions, parasympathetic discharge dominates because acetylcholine suppresses sympathetic activity by inhibiting the release of norepinephrine from neighboring sympathetic nerve endings

Stroke volume is determined by the extent of venous return and by sympathetic activity.
- Two types of controls influence stroke volume: (1) intrinsic control related to the extent of venous return and (2) extrinsic control related to the extent of sympathetic stimulation of the heart.

Increased end-diastolic volume results in increased stroke volume.
- The direct correlation between end-diastolic volume and stroke volume constitutes the intrinsic control of stroke volume, which refers to the heart's inherent ability to vary the stroke volume.
- For cardiac muscle, the resting cardiac muscle fiber length is <u>less</u> than optimal length.
- An increase in cardiac muscle fiber length, by moving closer to the optimal length, increases the contractile tension of the heart on the following systole.
- The main determinant of cardiac muscle fiber length is the degree of diastolic filling.
- This intrinsic relationship between end-diastolic volume and stroke volume is known as the Frank-Starling law of the heart.
- Stated simply, the law says that the heart normally pumps all the blood returned to it.

Sympathetic stimulation increases the contractility of the heart. .
- Stroke volume is also subject to extrinsic control.
- Sympathetic stimulation and epinephrine enhance the heart's contractility.
- Sympathetic stimulation increases stroke volume by enhancing venous return.

- Sympathetic stimulation constricts the veins, which squeezes more blood forward from the veins to the heart, increasing the end-diastolic volume.
- Sympathetic stimulation also increase calcium influx resulting in a stronger contraction.
- The strength of cardiac contraction and the stroke volume can thus be graded by: (1) varying the initial length of the muscle fibers and (2) varying the extent of sympathetic stimulation.

High blood pressure increases the workload of the heart.
- When the ventricles contract, they must generate sufficient pressure to exceed the blood pressure in the major arteries in order to force open the semilunar valves.
- The arterial blood pressure is referred to as the "afterload" because it is the workload imposed on the heart.
- If the arterial blood pressure is chronically elevated, the ventricle has to generate more pressure to eject blood.
- A chronically elevated afterload is one of the two major factors that cause heart failure.

In heart failure the contractility of the heart decreases.
- Heart failure refers to the inability of the cardiac output to keep pace with the body's demands for supplies and removal of wastes.
- Heart failure may occur for a variety of reasons, but the two most common are: (1) damage to the heart muscle and (2) prolonged pumping against a chronically increased afterload.
- The prime defect in heart failure is a decrease in cardiac contractility.
- A failing heart will pump out a smaller stroke volume than a normal, healthy heart.
- Two major compensatory measures help restore the stroke volume to normal; sympathetic activity to the heart is reflexively increased, and when cardiac output is reduced. the kidneys, in a compensatory attempt to improve their reduced blood flow, retain extra salt and

water in the body during urine formation to expand the blood volume.

- Forward failure is when the heart fails to pump an adequate amount of blood forward to the tissues because the stroke volume becomes progressively smaller. Backward failure occurs simultaneously as blood that cannot enter and be pumped out by the heart continues to dam up in the venous system. The congestion in the venous system is the reason why this condition is sometimes termed congestive heart failure.

NOURISHING THE HEART MUSCLE
The heart receives most of its own blood supply through the coronary circulation during diastole.

- Cardiac muscle cells contain an abundance of mitochondria, the O_2-dependent energy organelles. In fact, up to 40% of the cell volume of cardiac muscle cells is occupied by mitochondria, indicative of how much work the heart muscle does.
- The heart muscle must receive blood through blood vessels, specifically by means of the coronary circulation.
- The coronary arteries branch from the aorta just beyond the aortic valve, and the coronary veins empty into the right atrium.
- Most coronary arterial flow occurs during diastole.
- Coronary blood flow is adjusted primarily in response to changes in the heart's oxygen requirements.
- The link that coordinates coronary blood flow with

myocardial oxygen needs is adenosine.

Atherosclerotic coronary artery disease can deprive the heart of essential oxygen.

- Coronary artery disease (CAD) can cause myocardial ischemia by the three following mechanisms: (1) profound vascular spasm of the coronary arteries, (2) the formation of atherosclerotic plaques, and (3) thromboembolism.
- Vascular spasm is an abnormal spastic constriction that transiently narrows the coronary vessels.
- Atherosclerosis is a progressive, degenerative arterial disease that leads to occlusion of affected vessels.
- Potential complications of coronary atherosclerosis are angina pectoris and thromboembolism.
- The amount of "good" cholesterol versus "bad" cholesterol in the blood is linked to atherosclerosis.
- In response to the presence of LDL and/or other irritants, the endothelial cells produce chemicals that attract monocytes, a type of white blood cell, to the site. These immune cells trigger a local inflammatory response. This inflammatory response accelerates the progression of the disease.

Key Terms

Acute myocardial infarction
After load
Angina pectoris
Arrhythmia
Arteries
Atherosclerosis
Atrial fibrillation
Atrial flutter
Atrioventricular (AV) node
Atrioventricular (AV) valves
Atrium
Autorhythmicity
AV nodal delay
Bicuspid valve
Blood
Blood vessels
Bradycardia
Bundle of His (atrioventricular bundle)
Cardiac myopathies
Cardiac output (CO)
Cardiopulmonary resuscitation (CPR)
Chordae tendineae
Complete heart block
Congestive heart failure
Contractility
Coronary artery disease (CAD)
Diastole
Diastolic murmur
Electrocardiogram
Electrical defibrillation
End-diastolic volume
End-systolic volume
Endocardium
Endothelium
Epicardium
Extrasystole (premature beat)
Fibrillation
Fibrous skeleton
First heart sound
Foam cells
Frank-Starling Law of the heart
Functional murmurs
Heart
Heart attack
Heart block

High density lipoproteins
Homocysteine
Intercalated discs
Internodal pathway
Isovolumetric ventricular contraction
Lead
Low density lipoproteins
Mitral valve
Myocardial infarction
Myocardial ischemia
Myocardium
Necrosis
Nitric oxide
Pacemaker
Pacemaker activity
Pacemaker potential
Papillary muscle
Pericardial fluid
Pericardial sac
Pericarditis
Plaque
Preload
Pulmonary circulation
Purkinje fibers
P wave
QRS complex
Regurgitation
Second heart sound
Semilunar valves
Septum
Sinoatrial (SA) node
Stenotic valve
Sternum
Stroke volume
Systemic circulation
Systole
Systolic murmur
Tachycardia
Thoracic cavity
Thromboembolism
Thrombus
Tricuspid valve
T wave
Ventricular fibrillation

Review Exercises

Answers are in the appendix.

True/False-if an answer is false, underline the incorrect portion of the statement and rewrite it to make the statement true.

_____ 1. 99% of the cardiac muscle cells are autorhythmic cells.

_____ 2. The SA node, which normally exhibits the fastest rate of autorhythmicity, is known as the "pacemaker of the heart."

_____ 3. An action potential originating in the AV node first spreads throughout both atria, primarily from cell to cell via gap junctions.

_____ 4. The internodal conduction pathway directs the spread of an action potential originating at the SA node to the AV node to assure sequential contraction of the ventricles following atrial contraction.

_____ 5. In cardiac pacemaker cells, the rising phase of the action potential is due to an inward sodium current.

_____ 6. Cardiac muscle has a short refractory period.

_____ 7. A normal ECG exhibits three distinct wave forms: the P wave, the QRS complex, and the T wave.

_____ 8. The QRS complex represents atrial depolarization.

_____ 9. The two determinants of cardiac output are heart rate and stroke volume.

_____ 10. The parasympathetic nervous system's influence on the SA node is to increase the heart rate.

_____ 11. The parasympathetic nervous system stimulation of the AV node reduces the AV nodal delay.

_____ 12. Sympathetic stimulation increases stroke volume by enhancing venous return.

_____ 13. The prime defect in heart failure is a decrease in cardiac contractility.

_____ 14. Cardiac muscle cell have less mitochondria compared to skeletal muscle cells.

_____ 15. Forward failure heart failure is when the heart fails to pump an adequate amount of blood forward to the tissues because the stroke volume becomes progressively smaller.

_____ 16. The liver has a primary role in determining total blood cholesterol levels.

_____ 17. Under normal conditions, sympathetic nervous system controls resting heart rate.

_____ 18. The cardiac cycle consists of alternate periods of systole and diastole.

_____ 19. Early in ventricular diastole, the SA node reaches threshold and fires.

_____ 20. Atrial depolarization brings about atrial contraction.

_____ 21. Ventricular systole includes both the period of isovolumetric contraction and the ventricular ejection phase.

_____ 22. The first heart sound is associated with closure of the AV valve.

_____ 23. Laminar flow produces a faint sound.

_____ 24. The sound of the murmur characterizes it as either a stenotic murmur or an insufficient murmur.

_____ 25. There are no gap junctions between the atrial and ventricular contractile cells.

_____ 26. Either all the cardiac muscle fibers contract or none of them do.

_____ 27. Contract of cardiac muscle cells depends on calcium from both the sarcoplasmic reticulum and ECF.

_____ 28. Sympathetic regulation of heart activity results in stronger contractions.

_____ 29. Blood serves as the transport medium.

_____ 30. Pulmonary circulation consists of a circuit of vessels carrying blood between the heart and organ systems.

_____ 31. Anatomically the heart is a single organ; though it has a right and left side, they still function as a single pump.

_____ 32. The left side of the heart pumps blood into the pulmonary circulation.

_____ 33. Both sides of the heart simultaneously pump equal amounts of blood under equal pressure.

_____ 34. The systemic circulation is a low-pressure, low-resistance system.

_____ 35. Blood flows through the heart in one fixed direction from veins to atria to ventricles to arteries.

_____ 36. Cardiac muscle is capable of generating action potentials without any nervous stimulation.

Fill in the Blank

37. The _____ allows time for the ventricles to fill with blood before they contract.

38. _____ are connective tissue structures that prevent the AV valves from averting.

39. _____ channels found within the cardiac pacemaker cells open when the potential becomes more negative (hyperpolarizes).

40. _____ exists when small terminal branches from adjacent blood vessels nourish the same area.

41. A long _____ prevents tetanus of cardiac muscle

42. Random, uncoordinated excitation and contraction of the cardiac cells is known as _____.

43. The _____ pathway extends from the SA node within the right atrium to the left atrium.

44. The _____ represents ventricular repolarization.

45. A rapid heart rate of more than 100 beats per minute is known as _____.

46. A slow heart rate of fewer than 60 beats per minute is known as _____.

47. _____ fibrillation is characterized by rapid, irregular, uncoordinated depolarization of the atria with no definite P waves.

48. The heart has a broad base at the top and tapers to a pointed tip known as the _____ at the bottom.

49. CPR is short for _____.

50. _____ pump blood from the heart.

51. The vessels that return blood from the tissues to the atria are _____.

52. The large artery carrying blood away from the left ventricle is the _____.

53. The right AV valve is also called the _____.

54. The _____ is a thin external membrane covering the heart.

55. Up to _____ percent of the cell volume of cardiac muscle cells is occupied by mitochondria

56. _____ is the volume of blood pumped by each ventricle per minute.

57. The difference between the cardiac output at rest and the maximum volume of blood the heart is capable of pumping per minute is known as the _____.

58. The direct correlation between end-diastolic volume and stroke volume constitutes the _____ of stroke volume, which refers to the heart's inherent ability to vary the stroke volume.

59. The intrinsic relationship between end-diastolic volume and stroke volume is known as the _____.

60. _____ refers to the inability of the cardiac output to keep pace with the body's demands for supplies and removal of wastes.

61. The volume of blood in the ventricle at the end of diastole is known as the _____.

62. When all the valves are closed and no blood can enter or leave the ventricle during this time, this interval is termed the period of _____.

63. The amount of blood remaining in the ventricle at the end of systole when ejection is complete is known as the _____.

64. Abnormal heart sounds are called _____.

65. An abnormal heart sound occurring between the first and second heart sounds signifies a(n) _____.

Matching
Match the division of the autonomic nervous system to the effect it exerts on the heart or structures that influence the heart.

a. sympathetic
b. parasympathetic
c. neither sympathetic or parasympathetic

_____ 66. Influences skeletal muscle activity
_____ 67. Increases contractility of the atrial muscle and strengths contraction
_____ 68. Decreases excitability; increases AV nodal delay
_____ 69. Has no effect on the ventricular conduction pathway
_____ 70. Increases rate of depolarization to threshold of SA node; increases heart rate
_____ 71. Promotes adrenal medulla secretion of epinephrine
_____ 72. Increases contractility of the ventricular muscle; strengthens contraction
_____ 73. Increases venous return
_____ 74. Decreases contractility of atrial muscle
_____ 75. Increases calcium influx

Multiple Choice

76. This cardiac disorder is most often caused by a streptococcus bacterial infection.
    a. myocardial infarction
    b. myocardial ischemia
    c. atherosclerosis
    d. congestive heart failure
    e. valvular stenosis

77. This cardiac complication can be caused by prolonged pumping against a chronically increased afterload.
    a. myocardial infarction
    b. myocardial ischemia
    c. atherosclerosis
    d. congestive heart failure
    e. valvular stenosis

78. This chest pain is associated with myocardial ischemia.
    a. myocardial nociceptosis
    b. angina pectoris
    c. atheroma
    d. valvular stenosis
    e. heartburn

79. Influx of this ion, associated with the plateau phase of a cardiac muscle cell's action potential prevents tetanus
    a. sodium
    b. potassium
    c. calcium
    d. chloride
    e. both a and b

80. This cardiac disorder can be caused by profound vascular spasm of the coronary arteries.
    a. myocardial infarction
    b. myocardial ischemia
    c. atherosclerosis
    d. congestive heart failure
    e. valvular stenosis

81. This is a progressive, degenerative arterial disease that leads to occlusion of affected vessels.
    a. myocardial infarction
    b. myocardial ischemia
    c. atherosclerosis
    d. congestive heart failure
    e. valvular stenosis

82. This heart disorder can be caused by thromboembolism.
    a. myocardial infarction
    b. myocardial ischemia
    c. atherosclerosis
    d. congestive heart failure
    e. valvular stenosis

83. This disease has a close relationship with cholesterol.
    a. myocardial infarction
    b. myocardial ischemia
    c. atherosclerosis
    d. congestive heart failure
    e. valvular stenosis

84. This disease is characterized by abnormally large amounts of blood being dammed up in the venous system.
    a. myocardial infarction
    b. myocardial ischemia
    c. atherosclerosis
    d. congestive heart failure
    e. valvular stenosis

85. This condition is commonly known as a heart attack.
    a. myocardial infarction
    b. myocardial ischemia
    c. atherosclerosis
    d. congestive heart failure
    e. valvular stenosis

86. Which of the following compounds reduces the force of cardiac contraction by blocking calcium influx during an action potential?
    a. digitalis
    b. nicotine
    c. verapamil
    d. caffeine
    e. all of the above

Modified Multiple Choice

Indicate the proper order of the events during the cardiac cycle by placing the numbers in the blank preceding the events in sequence. The first and last events are already so indicated as a guide.

87. __1__ AV vale open; aortic valve closed; ventricular filling occurring
88. _____ blood ejected from the ventricle
89. _____ isovolumetric ventricular relaxation
90. _____ atrial contraction
91. __12_ AV valves open; ventricular filling occurs again; one cardiac cycle is complete
92. _____ aortic valve opens
93. _____ SA node discharges
94. _____ ventricular filling complete
95. _____ ventricular relaxation begins

96. _____ aortic valve closes
97. _____ isovolumetric ventricular contraction
98. _____ ventricular contraction begins; AV valves close

Indicate which valve abnormality is being described using the answer code below.

        A = valvular stenosis
        B = valvular insufficiency

99. _____ produces a "gurgling" murmur
100. _____ produces a "whistling" murmur
101. _____ valve does not close completely
102. _____ valve does not open completely

Indicate which ion is involved in each event being described by using the following answer code:

$$A = K^+$$
$$B = Na^+$$
$$C = Ca^{2+}$$

103. Inactivity of _____ channels brings about the slow drift of membrane potential to threshold in the cardiac autorhythmic cells.
104. Explosive increase in membrane permeability to _____ brings about the rapidly rising phase of the action potential in contractile cardiac cells.
105. Slow inward diffusion of _____ is largely responsible for the plateau portion of the cardiac action potential.
106. The rapid falling phase of the cardiac action potential is brought about primarily by the outward diffusion of _____.
107. Changes in cytosolic _____ concentration bring about changes in the strength of cardiac contraction.
108. Parasympathetic stimulation increases the permeability of the SA node to _____, whereas sympathetic stimulation decreases the permeability to this same ion.

Complete the following discussion by circling the correct phrase within the parentheses.

If the venous return increases, at the end of diastole the ventricular volume will be (109. increased, decreased, the same as before). Therefore, the length of the cardiac muscle cells will be (110. increased, decreased, the same as before). Consequently, during the next contraction the tension developed by the heart will be (111. greater than before, less than before, the same as before). The amount of blood pumped out as a result of this contraction will be (112. more than, less than, the same as) the amount pumped out by the contraction prior to the increase in venous return. Therefore, as venous return to the heart increases, the stroke volume ejected by the heart (113. increases, decreases, remains unchanged).

Use the following answer code to indicate which factor involved in the initiation and spread of cardiac excitation is being identified.

A = SA node
B = AV node
C = bundle of His and Purkinje system
D = gap junctions

114. _____ has the fastest rate of pacemaker activity
115. _____ allows the impulse to spread from cell to cell
116. _____ delays conduction of the impulse
117. _____ only point of electrical contact between the atria and ventricles
118. _____ normal pacemaker of the heart
119. _____ rapidly conducts the impulse down the ventricular septum and throughout much of the ventricular musculature
120. _____ firing of this structure produces the P wave seen on an ECG
121. _____ firing of this structure produces the QRS complex seen on an ECG

Points to Ponder

1. Congenital heart defects are routinely corrected through the miracles of modern surgery. What happens to the number of people who have such disorders as a result of these surgical techniques? What is the solution to this problem?

2. After carefully studying the sections of this chapter involving cholesterol and lipids, why do you suppose a good physician checks the triglycerides in the blood?

3. Speculate as to how much more work the heart of a physically inactive person has to perform than that of a physically active person.

4. Do adults and babies have the same heart rate? Blood pressure? Explain your answers.

5. How do the AV valves ensure a one-way flow of blood?

6. How does aerobic exercise help the heart?

7. How would you differentiate between heart flutter and fibrillation?

8. How does a defibrillator work physiologically?

9. What risk factors for heart disease can be modified? Which can't be modified?

Clinical Perspectives

1. With respect to the heart, why do people generally sleep in a prone position?

2. What happens to the heart in these two situations?
   A. You sense you are about to sneeze, so you close your mouth and hold your nose to muffle the sneeze. Then you sneeze.
   B. You are extremely constipated. In an effort to have a bowel movement you strain very hard.

3. Why do you suppose the sympathetic nervous system has an effect on the adrenal medulla while the parasympathetic has no effect at all?

4. Why is the nitroglycerin tablet placed under the tongue in cardiac patients?

5. Why does a physician or nurse practitioner tap a patient's chest wall during a physical examination?

6. Why is it that individuals who have atrial fibrillation do not appear to have a higher mortality rate than those who have normal functioning atria?

7. What is meant by the phrase "splitting the heart sounds"?

8. How would you explain to a patient the relationship between plasma lipids and heart disease?

9. If a patient has premature ventricular contractions, how would this affect stroke volume?

The Blood Vessels and Blood Pressure

Chapter Overview

The circulatory system contributes to homeostasis by serving as the body's transport system. This aids in the performance of many vital functions. The circulatory contribution to all these functions is flow. Flow depends on pressure and resistance. The circulatory system includes arteries, arterioles, capillaries, venules, and veins. It is the flow of blood through these vessels that is so important to homeostasis. For this system to function, pressure and resistance are essential. The heart and the blood vessels provide the pressure. The pressure is monitored by baroreceptors, which provide the CNS with the information necessary to control the blood flow. Arteries and veins play a major role in the gross transport of blood. The cardiac output is distributed throughout the body in a highly variable pattern. In this capacity it is the arterioles and capillaries that play the major role. Through changes in resistance the arterioles provide the mechanism for the distributional changes. The distribution of cardiac output is adjusted to maintain homeostasis. Consider all the substances released into or removed from the blood. Many of these molecules have typical cells as their destinations, but most are being transported to specific tissues. The capillaries are the sites of exchange. Arterioles, which respond to intrinsic and extrinsic controls, provide the proper pressure for the exchanges to occur. The circulatory system, like the other body systems, functions to maintain homeostasis. Again, all levels of organization are performing to maintain homeostasis for the survival of cells.

Chapter Outline

PATTERNS AND PHYSICS OF BLOOD FLOW

To maintain homeostasis, reconditioning organs receive blood flow in excess of their own needs.

- The majority of the body's cells are not in direct contact with the external environment, yet these cells must make exchanges with the environment, such as picking up oxygen and nutrients and eliminating wastes.
- All blood pumped by the right side of the heart passes through the lungs for oxygen pickup and carbon dioxide removal.
- The blood pumped by the left side of the heart is parceled out in various proportions to the systemic organs through a parallel arrangement of vessels that branch from the aorta.
- Blood is constantly "reconditioned" so that its composition remains relatively constant despite an ongoing drain of supplies to support metabolic activities and the continual addition of wastes from the tissues.
- The organs that recondition the blood (digestive tract, kidneys, and skin) normally receive substantially more blood than is necessary to meet their basic metabolic needs so that they can perform homeostatic adjustments on the blood.

- Reconditioning organs can withstand temporary reductions in blood flow.
- In contrast, the brain can least tolerate a disruption in its blood supply and a constant supply of blood to the brain is a high priority for the circulatory system

Blood flow through vessels depends on the pressure gradient and vascular resistance.

- Arteries carry blood from the heart.
- The flow rate of blood through a vessel is directly proportional to the pressure gradient and inversely proportional to vascular resistance.
- The pressure gradient—the difference in pressure between the beginning and end of a vessel—is the main driving force for flow through the vessel.
- The greater the pressure gradient forcing blood through a vessel, the greater the rate of flow through that vessel.
- Resistance is a measure of the hindrance to blood flow through a vessel caused by friction between the moving fluid and the stationary vascular walls.
- Resistance to blood flow depends on three factors: (1) viscosity of the blood; (2) vessel length, and (3) vessel radius.
- Resistance to blood flow is directly proportional to viscosity, directly proportional to vessel length and inversely proportion to vessel radius.
- Of these three factors, vessel radius is by far the most important.
- Viscosity refers to the friction developed between the molecules of a fluid as they slide over each other as the fluid flows. A greater viscosity the greater the resistance to flow.
- The greater the surface area of a blood vessel, the more contact blood has with the surface so the greater resistance to flow there is.
- Both length and radius of a vessel affect the surface area of the vessel, but of the two, length of vessels does not change in the body so radius is the most important variable.
- Vessel radius can be controlled in certain vessels to influence resistance.

The vascular tree consists of arteries, arterioles, capillaries, venules, and veins.

- The systemic and pulmonary circulation each consist of a closed system of vessels called "arteries," "arterioles," "capillaries," "venules," and "veins."
- When a small artery reaches the organ it is supplying, it branches into numerous arterioles.
- Arterioles branch further within the organs into capillaries.
- Capillaries rejoin to form small venules, which further merge to form small veins.
- The small veins unite to form larger veins.
- The arterioles, capillaries, and venules are collectively referred to as the "microcirculation."

ARTERIES

Arteries serve as rapid-transit passageways to the organs and as a pressure reservoir.

- Arteries are specialized to serve as rapid-transit passageways for blood from the heart to the tissues and to act as a pressure reservoir to provide the driving force for blood when the heart is relaxing.
- Capillary flow does not fluctuate between cardiac systole and diastole; blood flow is continuous through the capillaries supplying the tissues.
- The driving force for the continued flow of blood to the tissues during cardiac relaxation is provided by the elastic properties of the arterial wall.
- The arteries' elasticity enables them to expand to temporarily hold an excess volume of ejected blood, storing some of the pressure energy imparted by the cardiac contraction.
- When the heart relaxes and temporarily stops pumping blood into the arteries, the stretched arterial walls passively recoil and the recoil pushes the excess blood contained in the arteries into the vessels downstream.

Arterial pressure fluctuates in relation to ventricular systole and diastole.

- The maximum pressure exerted in the arteries when the blood is ejected into them during systole, the systolic pressure, averages 120 mm Hg.
- The minimum pressure within the arteries when blood is draining off into the remainder of the vessels during diastole, the diastolic pressure, averages 80 mm Hg.
- The pressure difference between systolic and diastolic pressures is known as the "pulse pressure."

Blood pressure can be measured indirectly by using a sphygmomanometer.

- It is convenient and reasonably accurate to measure the pressure indirectly through the use of a sphygmomanometer, an externally applied inflatable cuff attached to a pressure gauge.
- During the determination of blood pressure, a stethoscope is placed over the brachial artery.
- At the onset of a blood pressure determination, the cuff is inflated to a pressure greater than systolic blood pressure so that the brachial artery collapses.
- The pressure in the cuff is gradually reduced.
- The highest cuff pressure at which the first sound can be heard is indicative of the systolic blood pressure.
- The highest cuff pressure at which the last sound can be detected is indicative of the diastolic pressure.
- The sounds heard when determining blood pressure are distinct from the heart sounds associated with valve closure heard when listening to the heart with a stethoscope.

Mean arterial pressure is the main driving force for blood flow.

- The mean arterial pressure is the average pressure responsible for driving blood forward into the tissues throughout the cardiac cycle.
- The mean arterial pressure equals the diastolic pressure plus one-third the pulse pressure.
- The mean arterial pressure averages 93 mm Hg.
- Arterial pressure is essentially the same throughout the arterial tree.

ARTERIOLES

Arterioles are the major resistance vessels.

- The arterioles are the major resistance vessels in the vascular tree, even though the capillaries have smaller radii than the arterioles.
- The radii of arterioles supplying individual organs can be adjusted independently to determine the distribution of cardiac output and to regulate arterial blood pressure.
- Arteriolar walls have a thick layer of smooth muscle that is richly innervated by sympathetic nerve fibers.
- The smooth muscle layer runs circularly around the arteriole, so when it contracts, the vessel's circumference becomes smaller, (vasoconstriction) thus increasing resistance and decreasing the flow through that vessel. When the muscle relaxes, the circumference enlarges thus decreasing resistance and increasing blood flow through that vessel.
- Arteriolar smooth muscle normally displays a state of partial constriction known as **vascular tone,** which establishes a baseline of arteriolar resistance
- Local (intrinsic) controls and extrinsic controls influence the level of contractile activity in arteriolar smooth muscle.

Local control of arteriolar radius is important in determining distribution of cardiac output.

- Blood is delivered to all tissues at the same mean arterial pressure.

- The distribution of cardiac output can be varied by differentially adjusting arteriolar resistance in the various vascular beds.
- Local controls are changes within a tissue that alter the radii of the vessels and hence adjust blood flow through the tissue by directly affecting the smooth muscle of the tissue's arterioles.

Local metabolic influences on arteriolar radius help match blood flow with the organ's needs.
- Local influences may be either chemical or physical in nature.
- Local chemical influences on arteriolar radius include (1) local metabolic changes and (2) histamine release. Local physical influences include (1) myogenic response to stretch and (2) local application of heat or cold.
- Local arteriolar vasodilatation increasing blood flow to a particular area is called "active hyperemia."
- Local chemical changes produce dilation without involving nerves or hormones.
- The single layer of specialized epithelial cells that line the lumen of all blood vessels—the endothelial cells release paracrines which act as local chemical messengers.
- Paracrines act on the underlying smooth muscle to alter its state of contraction, thus locally regulating arteriolar caliber.
- The following local chemical factors produce relaxation of arteriolar smooth muscles: decreased oxygen, increased carbon dioxide, increased acid, increased potassium, increased osmolarity, adenosine release, and prostaglandin release.
- Nitric oxide, one of the most important paracrines, causes relaxation of arteriolar smooth muscle.
- By dilating the arterioles of the penis, nitric oxide is the direct mediator of penile erection.

- Macrophages produce nitric oxide, which they use as "chemical warfare" against bacteria and cancer cells.
- Nitric oxide interferes with platelet function and blood clotting at sites of vessel damage.
- Nitric oxide serves as a neurotransmitter in the brain and elsewhere.
- Nitric oxide helps regulate peristalsis.
- Nitric oxide may play a role in relaxation of skeletal muscle.
- Nitric oxide plays a role in the changes underlying memory.
- The endothelial cells release other important chemicals, such as endothelin, which bring about vasoconstriction by causing arteriolar smooth muscle contraction.

Local histamine release pathologically dilates arterioles.
- Histamine is another local chemical mediator that influences arteriolar smooth muscle, but it is not released in response to local metabolic changes and is not derived from the endothelial cells.
- Histamine is synthesized and stored within special connective tissue cells in many tissues and in certain types of circulating white blood cells.
- When tissues are injured or during allergic reactions, histamine is released in the damaged region.
- Histamine is the major cause of vasodilatation in an injured area.

Myogenic response to stretch
- Local physical influences on arterioles include heat or cold and myogenic responses to stretch.
- Myogenic responses appear to be important in reactive hyperemia and pressure autoregulation.
- When the blood supply to a region is completely occluded, arterioles in the region dilate due to: (1) myogenic relaxation, which occurs in response to the diminished stretch accompanying no

blood flow, and (2) changes in local chemical composition.

- After the occlusion is removed, blood flow to the previously deprived tissue is transiently much higher than normal because the arterioles are widely dilated.
- This is called "reactive hyperemia."
- When mean arterial pressure falls, the driving force is reduced, so blood flow to tissues decreases.
- Widespread arteriolar dilation reduces the mean arterial pressure, which aggravates the problem.
- In the presence of sustained elevations in mean arterial pressure, local chemical and myogenic influences triggered by the initial increased flow of blood bring about an increase in arteriolar tone and resistance.
- This greater degree of vasoconstriction reduces tissue blood flow toward normal despite the elevated blood pressure.
- "Pressure autoregulation" is the term applied to these local arteriolar mechanisms, which are aimed at keeping tissue blood flow fairly constant.

Extrinsic sympathetic control of arteriolar radius is important in the regulation of blood pressure.

- Extrinsic control of arteriolar radius includes both neural and hormonal influences, with the effects of the sympathetic nervous system being the most important.
- Sympathetic nerve fibers supply arteriolar smooth muscle everywhere except in the brain.
- A certain level of ongoing sympathetic activity contributes to vascular tone.
- Increased sympathetic activity produces generalized arteriolar vasoconstriction, whereas decreased sympathetic activity leads to generalized arteriolar vasodilatation.
- The extent to which each organ actually receives blood flow is determined by local arteriolar adjustments that override the sympathetic constrictor effect.

- Cerebral vessels are almost entirely controlled by local mechanisms that maintain a constant blood flow to support a constant level of brain metabolic activity. In fact, reflex vasoconstrictor activity in the remainder of the cardiovascular system is aimed at maintaining an adequate pressure head for blood flow to the vital brain.

The medullary cardiovascular control center and several hormones regulate blood pressure.

- The main region of the brain responsible for adjusting sympathetic output to the arterioles is the cardiovascular control center in the medulla of the brain stem.
- Several hormones also extrinsically influence arteriolar radius including epinephrine, norepinephrine, vasopressin, and angiotensin II.

CAPILLARIES
Capillaries are ideally suited to serve as sites of exchange.

- Capillaries, the sites of exchange of materials between the blood and tissues, branch extensively to bring blood within the reach of every cell.
- Exchange of materials across capillary walls is accomplished primarily by the process of diffusion.
- Diffusing molecules have only a short distance to travel between the blood and surrounding cells because of the thin capillary wall and small capillary diameter, coupled with the close proximity of each and every cell to a capillary.
- Because capillaries are distributed in such incredible numbers, a tremendous total surface area is available for exchange.
- Diffusion is enhanced because blood flows more slowly in the capillaries than elsewhere in the circulatory system.

Water-filled capillary pores permit passage of small, water-soluble substances.

- Diffusion across capillary walls also depends on the walls' permeability to the materials being exchanged.
- In most capillaries, small, water filled pores exist between cells. The size of the capillary pore varies from organ to organ.
- The tightest extreme is in the brain where the cells are joined by tight junctions and this prevents passage of materials between the cells and forms part of the blood-brain-barrier.
- In most capillaries the cells are closely but not tightly joined so that narrow, water-filled clefts, or pores, are present at the junctions between the cells.
- In addition to having the narrow pores between endothelial cells, the leakier capillaries of the kidneys and intestines also have larger holes known as fenestrations, which extend through the thickness of the endothelial cells themselves. These through-the-cell passageways are important in the rapid movement of fluid across the capillaries in these organs during the formation of urine and during the absorption of a digested meal.
- Water soluble substances such as glucose, and amino acids can readily pass through the pores, but large, water-soluble materials such as plasma proteins are kept from passing through.
- Lipid-soluble substances, such as O_2 and CO_2, can readily pass through the endothelial cells themselves by dissolving in the lipid bilayer barrier of the plasma membrane surrounding the cells
- In response to appropriate signals, the endothelial cells can readjust themselves to vary the size of the pores.
- Vesicular transport also plays a limited role in the passage of materials across the capillary wall.

Interstitial fluid is a passive intermediary between blood and cells.

- Exchanges are not made directly between blood and the tissue cells.

- Cells exchange materials directly with the interstitial fluid, the type and extent of exchange being governed by the properties of the cellular plasma membranes.
- Passive diffusion down concentration gradients is the primary mechanism for exchange of individual solutes.

Bulk flow across the capillary wall is important in extracellular fluid distribution.

- Bulk flow is the process whereby a volume of protein-free plasma filters out of the capillary, mixes with the surrounding interstitial fluid, and is subsequently reabsorbed.
- The four factors that influence bulk flow are: capillary blood pressure, plasma colloid osmotic pressure, interstitial fluids hydrostatic pressure and interstitial fluid-colloid osmotic pressure.
- Bulk flow occurs because of differences in the hydrostatic and colloid osmotic pressures between the plasma and interstitial fluid.
- When pressure inside the capillary exceeds pressure on the outside, fluid is pushed out through the pores in a process known as ultrafiltration.
- Bulk flow does not play an important role in the exchange of individual solutes between blood and tissues, because the quantity of solutes moved across the capillary wall by bulk flow is extremely small compared to the much larger transfer of solutes by diffusion.
- Bulk flow plays an extremely important role in regulating the distribution of ECF between plasma and interstitial fluid.

The lymphatic system is an accessory route by which interstitial fluid can be returned to the blood.

- Even under normal circumstances, slightly more fluid is filtered out of the capillaries into the interstitial fluid than is reabsorbed from the interstitial fluid back into the plasma.

- The extra fluid filtered out as a result of this filtration-reabsorption imbalance is picked up by the lymphatic system.
- Small, blind-ended terminal lymph vessels permeate almost every tissue of the body.
- Once interstitial fluid enters a lymphatic vessel, it is called "lymph."
- Lymphatics converge to form larger and larger lymph vessels, which eventually empty into the venous system.
- The most important functions of the lymphatic system are as follows: (1) the return of excess filtered fluid to the blood, (2) the defense against disease provided by phagocytic cells in the lymph nodes, (3) the transport of absorbed fat, and (4) the return of filtered protein to the blood.

Edema occurs when too much interstitial fluid accumulates.
- Swelling of the tissues because of excess interstitial fluid is known as "edema."
- The cause of edema can be grouped into four general categories: (1) a reduced concentration of plasma proteins; (2) an increased permeability of the capillary walls;
  (3) an increased venous pressure; and (4) the blockage of lymph vessels.
- Whatever the cause of edema, an important consequence is a reduction in exchange of materials between the blood and cells.

VEINS
Veins serve as a blood reservoir as well as passageways back to the heart.
- Because the total cross-sectional area of the venous system gradually decreases as smaller veins converge into progressively fewer but larger vessels, the velocity of blood flow increases as the blood approaches the heart.
- Systemic veins serve as a blood reservoir.
- When the stored blood is needed, such as during exercise, extrinsic factors drive the extra blood from the veins to the heart.

- A delicate balance exists between the capacity of the veins, the extent of venous return, and the cardiac output.

Venous return is enhanced by a number of extrinsic factors.
- Changes in venous capacity directly influence the magnitude of venous return, which in turn is an important determinant of effective circulating blood volume.
- Sympathetic stimulation produces venous vasoconstriction, which modestly elevates venous pressure; this in turn increases the pressure gradient to drive more blood from the veins into the right atrium.
- Venous vasoconstriction enhances venous return by decreasing venous capacity.
- Increased venous return leads to increased cardiac output because of the increase in end-diastolic volume.
- Many of the large veins in the extremities lie between skeletal muscles, so when the muscles contract, the veins are compressed.
- External venous compression decreases venous capacity and increases venous pressure, in effect squeezing fluid contained in the veins forward toward the heart.
- This pumping action, known as the "skeletal muscle pump," is one way by which extra blood stored in the veins is returned to the heart during exercise.
- Increased muscular activity pushes more blood out of the veins and into the heart.
- Blood can only be driven forward because the large veins are equipped with one-way valves.
- These valves permit blood to move forward toward the heart but prevent it from moving back toward the tissues.
- Varicose veins occur when the venous valves become incompetent and can no longer support the column of blood above them.

- The most serious consequence of varicose veins is the possibility of abnormal clot formation in the sluggish, pooled blood.
- As a result of respiratory activity, the pressure within the chest cavity averages 5 mm Hg less than atmospheric pressure.
- The pressure difference between the lower veins and the chest veins squeezes blood from the lower veins to the chest veins, promoting venous return.
- This mechanism of facilitating venous return is known as the "respiratory pump."
- During ventricular contraction, the AV valves are drawn downward, enlarging the atrial cavities, which drop the atrial pressure to below 0 mm Hg, thus enhancing venous return.

BLOOD PRESSURE
Blood pressure is regulated by controlling cardiac output, total peripheral resistance, and blood volume.

- Mean arterial blood pressure is the main driving force for propelling blood to the tissues.
- The mean arterial blood pressure must be high enough to ensure sufficient driving pressure, but not so high that it creates extra work for the heart, increases the risk of vascular damage, or increases risk of possible rupture of small blood vessels.
- Let's review all the factors that have an effect on mean arterial blood pressure: (1) Mean arterial pressure depends on cardiac output and total peripheral resistance. (2) Cardiac output depends on heart rate and stroke volume. (3) Heart rate depends on the relative balance of parasympathetic and sympathetic activity. (4) Stroke volume increases in response to sympathetic activity. (5) Stroke volume increases as venous return increases. (6) Venous return is enhanced by sympathetically induced venous vasoconstriction, the skeletal pump, the respiratory pump, and cardiac suction. (7) The effective circulating blood volume influences how much blood returns to the heart. (8) The blood volume depends on the magnitude of passive bulk-flow fluid shifts between the plasma and interstitial fluid, and the salt and water balance, which are hormonally controlled by the renin-angiotensin-aldosterone system and vasopressin, respectively. (9) Total peripheral resistance depends on the radius of all arterioles as well as blood viscosity. (10) The major factor determining blood viscosity is the number of red blood cells. (11) Arteriolar radius is influenced by intrinsic metabolic controls that cause local arteriolar vasodilatation and sympathetic activity, an extrinsic mechanism that causes arteriolar vasoconstriction. (12) Arteriolar radius is also extrinsically controlled by the hormones vasopressin and angiotensin II, which are potent vasoconstrictors.
- Mean arterial pressure is constantly monitored by baroreceptors within the circulatory system.
- Short-term adjustments are accomplished by alterations in cardiac output and total peripheral resistance, mediated by means of autonomic nervous system influences on the heart, veins, and arterioles.
- Long-term control involves adjusting total blood volume by restoring normal salt and water balance through mechanisms that regulate urine output and thirst.

The baroreceptor reflex is an important short-term mechanism for regulating blood pressure through immediate effects on the heart and blood vessels.

- Any change in mean blood pressure triggers an autonomically mediated baroreceptor reflex that influences the heart and blood vessels to adjust cardiac output and total peripheral resistance in an attempt to restore blood pressure to normal.

- The carotid sinus and aortic arch baroreceptors are sensitive to changes in both mean arterial pressure and pulse pressure.
- The integrating center that receives the afferent impulses about the status of arterial pressure is the cardiovascular control center.
- The cardiovascular control center alters the ratio between sympathetic and parasympathetic activity to the effector organs.
- If arterial pressure becomes elevated above normal, the carotid sinus and aortic arch baroreceptors increase the rate of firing in their respective afferent neurons.
- The cardiovascular control center responds by decreasing sympathetic and increasing parasympathetic activity.
- These efferent signals decrease heart rate, decrease stroke volume, and produce arteriolar and vasodilatation.
- When blood pressure falls below normal, baroreceptor activity decreases, inducing the cardiovascular center to increase sympathetic cardiac and vasoconstrictor nerve activity while decreasing its parasympathetic output.
- This efferent activity leads to an increase in heart rate and stroke volume coupled with arteriolar and venous vasoconstriction.

Hypertension is a serious national public health problem, but its causes are largely unknown.

- The underlying cause is known in primary hypertension. Hypertension that occurs secondary to another known primary problem is called secondary hypertension. Only 10% of all cases are considered primary.
- There is a strong genetic tendency to develop primary hypertension, which can be hastened or worsened by contributing factors such as obesity, stress, smoking, and excessive ingestion of salt.
- Whatever the underlying defect, once initiated, hypertension appears to be self-perpetuating.

- Constant exposure to elevated blood pressure predisposes vessel walls to the development of atherosclerosis, which further elevates blood pressure.
- The baroreceptors do not respond to bring the blood pressure back to normal during hypertension because they adapt or are "reset" to operate at a higher level.
- Complications of hypertension include congestive heart failure, strokes, or heart attacks.

Orthostatic hypotension results from transient inadequate sympathetic activity.

- Orthostatic hypotension occurs either when there is a disproportion between vascular capacity and blood volume or when the heart is too weak to impart sufficient driving pressure to the blood.
- It is due to inadequate sympathetic activity.

Circulatory shock can become irreversible.

- When blood pressure falls so low that adequate blood flow to the tissues can no longer be maintained, the condition known as "circulatory shock" occurs.
- Circulatory shock is categorized into four main types: (1) hypovolemic shock, (2) cardiogenic shock, (3) vasogenic shock, and (4) neurogenic shock.
- Following severe loss of blood, the resultant reduction in circulating blood volume leads to a decrease in venous return and a subsequent fall in cardiac output and arterial pressure.
- Compensatory measures immediately attempt to maintain adequate blood flow to the brain by increasing blood pressure toward normal, followed by longer-range measures aimed at restoring plasma volume and replacing lost red blood cells
- The steps to restore blood pressure are as follows.
- The baroreceptor reflex response to the fall in blood pressure brings about increased sympathetic and decreased parasympathetic activity to the heart.

- As a result of increased sympathetic activity to the veins, generalized venous vasoconstriction occurs.
- Sympathetic stimulation of the heart increases the heart's contractility, as does increasing the stroke volume.
- Sympathetically induced generalized arteriolar vasoconstriction leads to an increase in total peripheral resistance.
- The increase in cardiac output and total peripheral resistance brings about a compensatory increase in arterial pressure.
- The original fall in arterial pressure results in fluid shifts from the interstitial fluid into the capillaries to expand the plasma volume.
- The ECF fluid shift is enhanced by plasma-protein synthesis by the liver following the hemorrhage.
- Urinary output is reduced, thereby conserving water.
- The reduced plasma volume also triggers increased secretion of the hormone vasopressin and activation of the salt-and-water-conserving renin-angiotensin-aldosterone hormonal pathway.

- Increased thirst is also stimulated by a fall in plasma volume.
- Lost red blood cells are replaced through increased red blood cell production, triggered by a reduction in oxygen delivery to the kidneys.
- Compensatory mechanisms are often insufficient in the face of substantial fluid loss.
- Fluid volume must be replaced from the outside through drinking, transfusion, or a combination of both.
- A point may be reached at which blood pressure continues to drop rapidly because of tissue damage, despite vigorous therapy.
- This condition is termed "irreversible shock," in contrast to reversible shock, which can be corrected by compensatory mechanisms and effective therapy.

Key Terms

Active hyperemia
Arteries
Arterioles
Baroreceptor reflex
Baroreceptors
Blood pressure
Bulk flow
Capacitance vessels
Capillaries
Capillary blood pressure
Cardiovascular control center
Carotid sinus
Circulatory shock
Compliance
Diastolic pressure
Distensibility
Edema
Endothelial cells
Flow rate
Hypertension
Hypotension

Initial lymphatics
Interstitial fluid-colloid osmotic pressure
Interstitial fluid hydrostatic pressure
Korotkoff sounds
Lymph
Lymphatic system
Lymph nodes
Mean arterial blood pressure
Metarteriole
Microcirculation
Myogenic response
Nitric oxide
Orthostatic (postural) hypotension
Plasma-colloid osmotic pressure
Pores
Poiseuille's law
Precapillary sphincters

Primary hypertension
Pressure autoregulation
Pressure gradient
Pulse pressure
Reabsorption
Reactive hyperemia
Resistance
Respiratory pump
Secondary hypertension
Skeletal muscle pump
Sphygmomanometer
Systolic pressure
Total peripheral resistance
Ultrafiltration
Varicose veins
Vascular tone
Vasodilatation
Veins
Venous capacity
Venous return
Venules
Viscosity

Review Exercises

Answers are in the appendix.

True/False-if an answer is false, underline the incorrect portion of the statement and rewrite it to make the statement true.

_____ 1. Blood flow to the brain is variable.

_____ 2. Diffusion across capillary walls also depends on the wall's permeability to the materials being exchanged.

_____ 3. The size of the capillary pores is consistent throughout the body.

_____ 4. Capillaries have no smooth muscle, so they cannot actively participate in the regulation of their own blood flow.

_____ 5. Solute exchanges are made directly between blood and the tissue cells.

_____ 6. Bulk flow occurs because of differences in the hydrostatic and colloid osmotic pressures between the plasma and interstitial fluid.

_____ 7. Resistance to blood flow is directly proportional to blood viscosity and vessel length.

_____ 8. Vessel radius plays a more important role in regulating resistance to blood flow than does vessel length.

_____ 9. Reconditioning organs can withstand temporary reductions in blood flow.

_____ 10. The greater the pressure gradient forcing blood through a vessel, the lesser the rate of flow through that vessel.

_____ 11. Pulse pressure is the difference between systolic pressure and heart rate.

_____ 12. Systolic pressure averages 120 mm Hg and diastolic pressure averages 80 mm Hg.

_____ 13. To determine blood pressure, a stethoscope is placed over the brachial artery.

_____ 14. Arterial pressure is essentially the same throughout the arterial tree.

_____ 15. Orthostatic hypotension caused by emotional stress can also cause dizziness or fainting.

_____ 16. The sounds heard when measuring blood pressure are different from the sounds heard when listening to the heart with a stethoscope.

_____ 17. Mean arterial blood pressure is the main driving force for propelling blood to the tissues.

_____ 18. Long-term adjustments are accomplished by alterations in cardiac output and total peripheral resistance, mediated by means of autonomic nervous system influences on the heart, veins, and arterioles.

_____ 19. The cardiovascular control center alters the ratio between sympathetic and parasympathetic activity to the effector organs.

_____ 20. The blood flow to the brain stays constant during exercise.

_____ 21. Cerebral blood vessels are not under sympathetic regulation.

_____ 22. Arterioles are the major resistance vessels in the vascular tree.

_____ 23. Local (intrinsic) controls are changes within a tissue that alter the radii of the vessels and hence adjust blood flow through the tissue by directly affecting the smooth muscle of the tissue's arterioles.

_____ 24. Local influences can only be chemical not physical in nature.

_____ 25. Capillaries in the brain have fenestrations.

_____ 26. Histamine is a local chemical mediator that influences arteriolar smooth muscle.

_____ 27. An increase in venous return leads to a decrease in cardiac output.

_____ 28. Extrinsic control of arteriolar radius includes both neural and hormonal influences.

_____ 29. Systemic veins also serve as a blood reservoir.

_____ 30. Autoregulation attempts to maintain a constant mean arterial pressure even as systemic pressure fluctuates significantly.

_____ 31. Increased muscular activity pushes more blood out of the veins and into the heart.

_____ 32. Blood can be driven forward or backward through the veins.

_____ 33. Abnormal clot formation in the sluggish, pooled blood is a serious risk to a person with varicose veins.

_____ 34. Most cases of hypertension are considered secondary hypertension.

Fill in the Blank

35. _____ is when the blood pressure is above 140/90 mm Hg.

36. _____, in its extreme form, is circulatory shock.

37. Bulk flow is offset by _____.

38. _____ is a transient hypotensive condition resulting from insufficient compensatory responses to the gravitational shifts in blood that occur when a person moves from a horizontal to a vertical position.

39. The _____ is responsible for maintaining vascular tone.

40. _____ is a measure of blood "thickness."

41. _____ shock is when a point is reached where the blood pressure continues to drop rapidly because of tissue damage, despite vigorous therapy.

42. Veins are often referred to as _____.

43. _____ refers to the volume of blood that the veins can accommodate.

44. _____ refers to the volume of blood entering each atrium per minute from the veins.

45. _____ occur when the venous valves become incompetent and can no longer support the column of blood above them.

46. Mean arterial pressure is constantly monitored by _____ within the circulatory system.

47. _____ and _____ house mechanoreceptors which are sensitive to changes in both mean arterial pressure and pulse pressure.

48. _____ carry blood from the heart.

49. Arterioles, capillaries, and venules are collectively referred to as the _____ _____.

50. _____ of blood through a vessel is directly proportional to the pressure gradient and inversely proportional to vascular resistance.

51. The maximum pressure exerted in the arteries when blood is ejected into them during systole averages _____.

52. The pressure difference between systolic and diastolic is known as the _____.

53. _____ is the average pressure responsible for driving blood forward into the tissues throughout the cardiac cycle.

54. When a small artery reaches the organ it is supplying, it branches into numerous _____.

55. Exchange of materials across capillary walls is accomplished primarily by the process of _____.

56. Capillaries containing tight junctions are found in the _____.

57. _____ is a volume of protein-free plasma that actually filters out of the capillary, mixes with the surrounding interstitial fluid, and is subsequently reabsorbed.

58. _____ is the fluid or hydrostatic pressure exerted on the inside of the capillary walls by the blood.

59. The extra fluid filtered out as a result of this ultrafiltration reabsorption imbalance is picked up by the _____.

60. Swelling of the tissues because of excess interstitial fluid is known as _____.

61. _____ is the term applied to such narrowing of a vessel.

62. Arteriolar smooth muscle normally displays a state of partial constriction known as _____.

63. _____ causes local changes in arteriolar resistance.

64. _____ is a single layer of specialized epithelial cells that line the lumen of all blood vessels.

65. The main region of the brain responsible for adjusting sympathetic output to the arterioles is the _____.

Matching
Match the statement to variable response.

a. increases
b. decreases
c. unchanged

_____ 66. Blood flow to brain during exercise
_____ 67. Cardiac output under sympathetic control
_____ 68. Venous return under sympathetic control
_____ 69. Blood flow in a single vessel when its radius decreases
_____ 70. Blood flow in a single vessel when it undergoes vasodilatation
_____ 71.
_____ 72. Vascular tone with increased sympathetic control
_____ 73. Cardiac output during exercise
_____ 74. Resistance in a single blood vessel during vasoconstriction
_____ 75. Heart rate during exercise
_____ 76. Mean arterial blood pressure when blood pressure rises slightly
_____ 77. Chance of edema when plasma colloid-osmotic pressure decreases
_____ 78. Resistance in a blood vessel when it undergoes vasoconstriction
_____ 79. Pulse pressure when blood pressure is 120/80 compared to when blood pressure is 100/60

Multiple Choice

80. Which of the following is caused by atherosclerosis?
    a. neurogenic hypertension
    b. endocrine hypertension
    c. cardiovascular hypertension
    d. pheochromocytoma
    e. hypotension

81. Which disorder is associated elevated blood pressure due to a tumor on the adrenal medulla causing elevated levels of epinephrine?
    a. neurogenic hypertension
    b. secondary hypertension
    c. cardiovascular hypertension
    d. hypotension
    e. primary hypertension

82. Which is associated with a decrease sympathetic response to standing up too quickly?
    a. neurogenic hypertension
    b. cardiovascular hypertension
    c. Conn's syndrome
    d. orthostatic hypotension
    e. none of the above

83. Which of the following does not hasten the development of primary hypertension?
    a. obesity
    b. smoking
    c. stress
    d. excessive ingestion of salt
    e. none of the above

84. Which of the following can lead to an increased risk of developing blood clot?
    a. varicose veins
    b. renal hypotension
    c. pheochromocytoma
    d. Conn's syndrome
    e. cardiovascular syndrome

Modified Multiple Choice
Indicate which of the vessels performs the function listed by writing the appropriate letter in the blank using the answer code below.

A = arteries
B = arterioles
C = capillaries
D = veins
E = lymphatics

85. _____ Site of exchange of nutrients and waste products between the blood and tissues.
86. _____ Serve as low resistance passageways from the heart to the tissues.
87. _____ Serve as blood reservoir to accommodate variations in blood volume.

88. _____ Major resistance vessels.
89. _____ Portion of the circulatory system through which the velocity of blood flow is the slowest.
90. _____ Serve as low-resistance passageways from the tissues to the heart.
91. _____ Act as a pressure reservoir to drive blood forward through the vasculature during diastole.
92. _____ Changes in the radius of this vessel type regulate the distribution of the cardiac output to various areas of the body.
93. _____ Vessels that pick up fluid that is filtered but not reabsorbed across capillaries

If the blood pressure is recorded as 118/76, indicate the correct value of the pressure in question.

> A = 118 mm Hg
> B = 42 mm Hg
> C = 97 mm Hg
> D = 76 mm Hg
> E = 90 mm Hg

94. _____ What is the systolic pressure?
95. _____ What is the diastolic pressure?
96. _____ What is the pulse pressure?
97. _____ What is the mean pressure?

Indicate the relative comparison of each of the paired items by writing the appropriate letter in the blank using the answer code below.

> A = A is greater than B
> B = B is greater than A
> C = A and B are equal

98. _____ A. Blood flow through an arteriole upon increased sympathetic activity.
      B. Blood flow through an arteriole upon decreased sympathetic activity.

99. _____ A. Blood flow through a vein upon increased sympathetic activity.
      B. Blood flow through a vein upon decreased sympathetic activity.

100. _____ A. Velocity of blood flow through the veins.
      B. Velocity of blood flow through the capillaries.

101. _____ A. Local arteriole radius in the presence of local decreased oxygen concentration and increased carbon dioxide concentration.
      B. Local arteriole radius with normal local concentration of both oxygen and carbon dioxide.

Points to Ponder

1. Having learned about active hyperemia, can you think of an example of passive hyperemia?

2. How do the palace guards in London avoid fainting?

3. Why is the cardiovascular center in the medulla as opposed to some cortical, hypothalamic, or thalamic area?

4. Since there is blood in the blood vessels, why do they need an addition external blood supply?

5. With respect to blood vessels, what causes dark circles under your eyes?

6. Based on your knowledge of capillaries, why do your cheeks turn red on a very cold day?

7. What changes occur in the circulatory system during exercise?

8. How does lifting weights cause a fall in cardiac output and blood pressure?

9. One of the consequences of the Frank-Starling law is that the outputs of the right and left ventricles are matched. Why is this important, and how is this matching accomplished?

10. Why is hypotension a result of hypovolemia?

Clinical Perspectives

1. How does hypertension affect the risk of coronary heart disease?

2. How would you explain to your friend the fact that blood pressure is higher at one time of the day than during other times?

3. Since you are taking a physiology course, how would you explain to your friend the cause of elephantitis and how it is related to edema?

4. During prolonged exercise on a very hot day, a substantial amount of water will be lost from the body through sweating. This causes hyperthermia. Why can't the body be immediately rehydrated by drinking pure water?

5. How would you explain to your friend how the class of drugs known as ASE inhibitors work?

6. Cocaine inhibits the reuptake of norepinephrine in the adrenergic axons. Based on this fact, why is myocardial ischemia a common side effect of cocaine abuse?

7. How does nitric oxide contribute to hypertension?

8. How would you explain to your friend why he becomes dizzy when he stands up too fast?

9. Preeclampsia is a toxemia of late pregnancy characterized by high blood pressure, proteinuria, and edema. Based on your knowledge of blood vessels and blood pressure, why does preeclampsia occur?

10. People with congestive heart failure are often treated with the drug digitalis. Digitalis works by inhibiting the sodium-potassium pump and causing the intracellular concentration of calcium to increase. What effect is this drug going to have on the heart and blood pressure?

Experiments of the Day

1. On a classmate's arm, try to locate the venous valves.

2. The next time you go to a grocery store or pharmacy where there is a blood pressure machine, take your blood pressure.

11

The Blood and Body Defenses

Chapter Overview

Blood plays a vital role in homeostasis through the transport of nutrients, dissolved gases, wastes and substances such as hormones.

There are two general components in blood besides water: plasma and the cellular elements. All vascular transportation relies on the fluid portion of the blood. In addition, the plasma maintains an osmotic gradient that is important in the distribution of extracellular fluid. The plasma also has a buffer system against changes in pH and has the necessary constituents to begin the clotting process.

Erythrocytes contain hemoglobin, which plays a vital role in oxygen and carbon dioxide transport and the buffering capacity of the blood. Anemia results any time the oxygen carrying capacity is reduced. There are many causes of anemia.

Leucocytes defend the body against pathogens, identify and destroy cancer cells, and clean up cellular debris. Platelets are the major elements in maintaining hemostasis. All the cellular components of blood are produced in the bone marrow, even though lymphocytes simply divide in the normal healthy body.

Humans are constantly coming into contact with external agents that can be harmful if they gain entry into the body. To help prevent disease, the immune system contributes to homeostasis by providing a sophisticated, highly effective defense system. The system is essential in defending against viruses and bacteria, removing "worn out" cell and tissue debris, and destroying mutant cells.

Chapter Outline

The hematocrit represents the packed cell volume; plasma accounts for the rest of the volume.

- Blood consists of three types of specialized cellular elements, erythrocytes, leukocytes and platelets, suspended in the complex liquid plasma.
- The hematocrit represents the percentage of total blood volume occupied by erythrocytes.
- Plasma accounts for the remaining volume.
- The white blood cells and platelets represent less than 1% of the blood volume.

Plasma water is a transport medium for many inorganic and organic substances.

- Plasma is composed of 90% water.
- A large number of inorganic and organic substances are dissolved in the plasma
- Because water has a high capacity to hold heat, plasma is able to absorb and distribute much of the heat generated metabolically within the tissues.
- Heat energy not needed to maintain body temperature is eliminated to the environment.
- The most plentiful organic constituents by weight are the plasma proteins, which compose 6–8% of plasma's total weight.
- Inorganic constituents account for approximately 1% of plasma weight.
- The most abundant electrolytes in the plasma are sodium and chloride ions.

- The remaining small percentage of plasma is occupied by nutrients, waste products, dissolved gases, and hormones.
- Most of these substances are merely being transported in the plasma.

Many functions of plasma are carried out by plasma proteins.
- The plasma proteins are the one group of plasma constituents not present just for the ride.
- Because they are the largest of the plasma constituents, plasma proteins usually do not exit through the narrow pores in the capillary walls.
- Plasma proteins exist in a colloidal dispersion.
- Plasma proteins establish an osmotic gradient between blood and interstitial fluid.
- Plasma proteins are partially responsible for the plasma's capacity to buffer changes in pH.
- Some plasma proteins are involved in the process of blood clotting.
- Inactive precursor The antibodies or immunoglobulins are important for the body's defenses
- The plasma proteins are generally synthesized by the liver, with the exception of the antibodies, which are produced by lymphocytes.

ERYTHROCYTES
The structure of erythrocytes is well suited to their main function of oxygen transport in the blood.
- Erythrocytes are flat biconcave discs.
- The biconcave shape provides a larger surface area for diffusion of oxygen.
- The thinness of the cell enables oxygen to diffuse rapidly between the exterior and the innermost regions of the cell.
- The flexibility of the erythrocytes membrane facilitates their transport function.
- Erythrocytes' most important feature, which enables them to transport oxygen, is the hemoglobin they contain.

- Hemoglobin consists of: (1) the globin portion, a protein made up of four highly folded polypeptide chains, and (2) four iron-containing, nonprotein nitrogenous groups known as heme groups.
- Each hemoglobin molecule can pick up four oxygen passengers.
- Hemoglobin can also combine with carbon dioxide, the acidic hydrogen-ion portion of ionized carbonic acid, and carbon monoxide.
- Hemoglobin plays the key role in oxygen transport while contributing significantly to carbon dioxide transport and the buffering capacity of blood.
- Erythrocytes contain no nucleus, organelles, or ribosomes.
- Only a few crucial nonrenewable enzymes remain within the mature erythrocyte especially those that are involved in g glycolysis reactions that are necessary for generating the energy in the absence of oxygen.

ERYTHROPOIESIS
Erythropoiesis is controlled by erythropoietin from the kidneys.
- The bone marrow normally generates new red blood cells, a process known as "erythropoiesis," keeping pace with the demolition of old cells.
- Red marrow not only produces red blood cells but is the ultimate source for leucocytes and platelets.

Erythropoiesis is controlled by erythropoietin from the kidneys.
- Reduced oxygen delivery to the kidneys stimulates them to secrete the hormone erythropoietin into the blood.
- This hormone stimulates erythropoiesis by the bone marrow.
- This hormone can now be produced in the laboratory.

Anemia can be caused by a variety of disorders.

- Anemia refers to a reduction below normal in the oxygen-carrying capacity of the blood and is characterized by a low hematocrit.
- Nutritional anemia is caused by a dietary deficiency of a factor needed for erythropoiesis.
- Iron-deficiency anemia occurs when insufficient iron is available for the synthesis of hemoglobin.
- A dietary deficiency of folic acid can produce anemia.
- Pernicious anemia is caused by an inability to absorb adequate amounts of vitamin B_{12} from the digestive tract.
- Aplastic anemia is caused by failure of the bone marrow to produce adequate numbers of red blood cells.
- Renal anemia is caused by inadequate erythropoietin secretion as a result of kidney disease.
- Hemorrhagic anemia is caused by the loss of substantial quantities of blood.
- Hemolytic anemia is caused by the rupture of excessive numbers of circulating erythrocytes.
- Sickle cell disease is an example among various hereditary abnormalities of erythrocytes that make these cells very fragile. Sickle cell disease is caused by a genetic mutation that changes a single amino acid in the 146-long amino acid chain that makes up a hemoglobin polypeptide

Polycythemia is an excess of circulating erythrocytes.

- Polycythemia, in contrast to anemia, is characterized by too many circulating RBCs and an elevated hematocrit

Blood types depend on surface antigens on erythrocytes

- An antigen is a large, complex molecule that triggers a specific immune response against itself when it gains entry to the body

- An antibody binds with the specific antigen against which it is produced and leads to the antigen's destruction by various means
- Within the major blood group system, the ABO system, the erythrocytes of people with type A blood contain A antigens, those with type B blood contain B antigens, those with type AB blood have both A and B antigens, and those with type O blood do not have any A or B red blood cell surface antigens.
- Accordingly, the plasma of type A blood contains anti-B antibodies, type B blood contains anti-A antibodies, and both anti-A and anti-B antibodies are present in type O blood.
- High levels of these antibodies are found in the plasma of individuals who have never been exposed to a different type of blood.
- It is now known that individuals are unknowingly exposed at an early age to small amounts of A- and B-like antigens associated with common intestinal bacteria.
- Because type O individuals do not have A or B antigens, they are considered to be universal donors.
- Type O individuals can receive only type O blood.
- Type AB individuals are called "universal recipients."
- Lacking both anti-A and anti-B antibodies, they can accept donor blood of any type.
- Individuals who possess the Rh factor (an erythrocyte antigen) are said to have Rh positive blood.
- Those lacking the Rh factor are considered to be Rh-negative.
- No naturally occurring antibodies develop against the Rh factor.
- Erythroblastosis fetalis, or hemolytic disease of the newborn occurs when an Rh- negative mother develops antibodies against the erythrocytes of an Rh-positive fetus she is carrying.

- Because the fetal bone marrow cannot put out mature erythrocytes as fast as the rate of erythrocyte destruction by the maternal antibodies, it releases immature precursors of erythrocytes known as erythroblasts

PLATELETS AND HEMOSTASIS
Platelets are cell fragments derived from megakaryocytes.

- Platelets are a third type of cellular element present in the blood.
- Platelets are small cell fragments that are shed off the outer edges of extraordinarily large bone marrow-bound cells known as "megakaryocytes."
- The hormone thrombopoietin increases the number of megakaryocytes in the bone marrow and stimulates each megakaryocyte to produce more platelets.

Hemostasis prevents blood loss from damaged small vessels.

- Hemostasis is the arrest of bleeding from a broken blood vessel.
- Hemostasis involves three major steps: (1) vascular spasm, (2) formation of a platelet plug, and (3) blood coagulation.

Vascular spasm reduces blood flow through an injured vessel.

- A cut or torn blood vessel immediately constricts.
- The opposing endothelial surfaces of the vessel are pressed together by this initial vascular spasm.
- The surfaces become sticky and adhere to each other.

Platelets aggregate to form a plug at a vessel defect.

- When the smooth endothelial lining is disrupted because of vessel injury, platelets attach to the exposed collagen. Collagen activates the bound platelets.
- Activated platelets quickly reorganize their actin cytoskeletal elements to develop spiky processes, which help them adhere to the collagen and to other platelets

- Platelets release adenosine diphosphate, which causes the surface of nearby circulating platelets to become sticky so that they adhere to the first layer of platelets.
- Normal endothelium releases prostacyclin, which inhibits platelet aggregation.
- Thus, the platelet plug is limited to the defect and does not spread to normal vascular tissue.
- The aggregated platelet plug also performs three other important roles: (1) The actin-myosin protein complex within the plug contracts to compact and strengthen the plug. (2) The platelet plug releases several powerful vasoconstrictors. (3) The platelet plug releases chemicals that enhance coagulation.

Clot formation results from a triggered chain reaction involving plasma clotting factors.

- Blood coagulation, or clotting, is the transformation of blood from a liquid into a solid gel.
- The conversion of fibrinogen into fibrin is catalyzed by the enzyme thrombin at the site of vessel injury.
- Fibrin molecules adhere to the damaged vessel surface, forming a loose, netlike meshwork that traps the cellular elements of the blood.
- Fibrin strands are only loosely interlaced.
- Chemical linkages rapidly form between adjacent strands.
- This cross-linkage process is catalyzed by a clotting factor known as "factor XIII."
- Thrombin activates factor XIII.
- Thrombin exists in the plasma in the form of an inactive precursor called "prothrombin."
- Factor X converts prothrombin into thrombin.
- Altogether, 12 plasma clotting factors participate in essential steps that lead to the final conversion of fibrinogen into a stabilized fibrin network.

- Once the first factor in the sequence is activated, it in turn activates the next factor, and so on, in a series of sequential reactions known as a "cascade," until thrombin catalyzes the final conversion of fibrinogen into fibrin.
- The clotting cascade may be triggered by the intrinsic pathway or extrinsic pathway.
- The intrinsic pathway of clotting precipitates clotting within damaged vessels as well as clotting of blood samples in test tubes. All elements necessary to bring about clotting by means of the intrinsic pathway are present in the blood samples in test tubes.
- The extrinsic pathway requires contact with tissue factors external to the blood.
- When a tissue is traumatized, it releases a protein complex known as "tissue thromboplastin," which directly activates factor X, thereby bypassing all preceding steps of the intrinsic pathway.

Fibrinolytic plasmin dissolves clots.
- Fibroblasts form a scar at the vessel defect.
- The clot is slowly dissolved by a fibrinolytic enzyme called "plasmin."
- Phagocytic white blood cells gradually remove the products of clot dissolution.

Inappropriate clotting is responsible for thromboembolism.
- An abnormal intravascular clot attached to a vessel wall is known as a "thrombus," and free floating clots are called "emboli."
- Several factors can cause thromboembolism: (1) roughened vessel surfaces associated with atherosclerosis, (2) imbalances in the clotting-anticlotting systems, (3) slow-moving blood, and (4) the release of tissue thromboplastin.
- Anticoagulant drugs are sometimes given to prevent thromboembolism in people with conditions that make them more prone to the development of clots.

Hemophilia is the primary condition responsible for excessive bleeding.
- The most common cause of excessive bleeding is hemophilia, which is caused by a deficiency in one of the factors in the clotting cascade.
- Vitamin K deficiency can cause a bleeding tendency.

LEUKOCYTES
Leukocytes primarily function as defense agents outside the blood.
- Leukocytes are the mobile units of the body's immune defense system.
- The leukocytes and their derivatives: (1) defend against invasion of pathogens by phagocytizing the foreigners or causing their destruction by more subtle means; (2) identify and destroy abnormal or mutant cells that arise within the body; and (3) function as a "cleanup crew" that removes the body's "litter" by phagocytizing debris resulting from dead or injured cells.Unlike erythrocytes, leukocytes are able to exit the blood by assuming amoeba-like behavior to wriggle through narrow capillary pores and crawl to assaulted areas

There are five types of leukocytes.
- The five types of leukocytes fall into two main categories: (1) polymorphonuclear granulocytes (neutrophils, eosinophils, and basophils), and (2) mononuclear agranulocytes (monocytes and lymphocytes).
- Each of the 5 types performs different functions. Neutrophils are phagocytic specialists. Eosinophils increase in number during allergic conditions and with parasite infection. Basophils release various chemicals including histamine. Monocytes are phagocytic. Lymphocytes provide immune defense.
- There are two types of lymphocytes, B lymphocytes and T lymphocytes.
- B lymphocytes produce antibodies that circulate in the blood.

- An antibody binds with and marks for destruction specific kinds of foreign matter.
- T lymphocytes destroy their specific target cells, a process known as "cell-mediated immune response."

Leukocytes are produced at varying rates depending on the changing needs of the body.
- Granulocytes and monocytes are produced only in the bone marrow.
- Lymphocytes are actually produced by already-existing lymphocytes residing in the lymphoid tissues.

Immune responses may be either innate and nonspecific or adaptive and specific.
- Nonspecific immune responses are inherent defense responses that nonselectively defend against foreign or abnormal material of any type.

- . The adaptive, or acquired, immune system, relies on *specific* immune responses selectively targeted against a particular foreign material to which the body has already been exposed.The adaptive immune system thus takes considerably more time to mount and takes on specific foes.

INNATE IMMUNITY
- Nonspecific defenses that come into play whether or not there has been prior experience with the offending agent include the following: inflammation, interferon, natural killer cells, and the complement system.

Inflammation is a nonspecific response to foreign invasion or tissue damage.
- The ultimate goal of inflammation is to bring to the invaded or injured area phagocytes and plasma proteins that can: (1) isolate, destroy, or inactivate the invaders; (2) remove debris; and (3) prepare for subsequent healing and repair.
- Upon bacterial invasion, macrophages already present in the area immediately begin phagocytizing the foreign microbes.
- Almost immediately upon microbial invasion arterioles within the area dilate, increasing blood flow to the site of injury.
- Histamine is released in the area of tissue damage by mast cells.
- Released histamine increases the capillaries' permeability.
- Plasma proteins that normally are prevented from leaving the blood escape into the inflamed tissue.
- As the leaked plasma proteins accumulate in the interstitial fluid, they exert a colloid osmotic pressure.
- The elevation in local osmotic pressure and the increased capillary blood pressure result in localized edema.
- Upon exposure to tissue thromboplastin in the injured tissue and to specific chemicals secreted by phagocytes on the scene, fibrinogen, the final factor in the clotting system, is converted into fibrin.
- Fibrin forms interstitial fluid clots in the spaces around the bacterial invaders and damaged cells.
- This walling off of the injured region from the surrounding tissues prevents, or at least delays, the spread of bacterial invaders and their toxic products.
- Within an hour after the injury, the area is teeming with leukocytes that have exited from the vessels.
- Neutrophils are the first to arrive.
- During the next eight to twelve hours, monocytes swell and mature into macrophages.
- Leukocyte emigration from the blood into the tissues involves the processes of margination, diapedesis, amoeboid movement, and chemotaxis.
- Outside the vessel, the leukocyte moves in an amoeboid fashion toward the site of tissue damage and bacterial invasion.
- Within a few hours after the onset of the inflammatory response, the number of neutrophils in the blood may increase up to four or five times that of normal.

- Phagocytosis involves the engulfment and intracellular degradation of foreign particles and tissue debris.
- Dead tissue and many foreign materials have surface characteristics that differ from normal body cells.
- Foreign particles are deliberately marked for phagocytic ingestion by being coated with chemical mediators (opsonins) generated by the immune system.
- Microbe-stimulated phagocytes release many chemicals, which function as mediators of the inflammatory response.
- Macrophages secrete interleukin 1, which enhances the proliferation and differentiation of both B and T lymphocytes.
- TNF stimulates release of histamine from mast cells in the vicinity. Histamine, in turn, promotes the local vasodilation and increased capillary permeability of inflammation
- The ultimate purpose of the inflammatory process is to isolate and destroy injurious agents and to clear the area for tissue repair.

NSAIDs and glucocorticoid drugs suppress the inflammatory response.
- Numerous drugs can suppress the inflammatory response; the most effective are the nonsteroidal anti-inflammatory drugs and glucocorticoids.
- NSAID interfere with the inflammatory response by decreasing histamine release, resulting in a reduction of swelling, redness, and pain Glucocorticoids suppress almost every aspect of the inflammatory response.
- In addition, they destroy lymphocytes within lymphoid tissue and reduce antibody production.

Interferon transiently inhibits multiplication of viruses in most cells.
- A nonspecific defense mechanism is the release of interferon from virus-infected cells.
- Interferon binds with receptors on the plasma membranes of neighboring cells signaling these cells to prepare for the possibility of impending viral attack.
- Interferon triggers the production of viral-blocking enzymes by potential host cells.
- Binding with interferon induces these other cells to synthesize enzymes that can break down viral messenger RNA.
- Interferon markedly enhances the actions of cell-killing cells.
- Interferon also slows cell division and suppresses tumor growth.

Natural killer cells destroy virus-infected cells and cancer cells upon first exposure to them.
- Natural killer cells are naturally occurring, lymphocyte-like cells that nonspecifically destroy virus-infected cells and cancer cells by directly lysing their membranes.

The complement system punches holes in microorganisms.
- The system derives its name from the fact that it "complements" the action of antibodies.
- The complement system consists of plasma proteins that are produced by the liver and circulate in the blood in inactive form.
- Once the first component is activated, it activates the next component, and so on.
- The final five components assemble into a large protein complex, the membrane-attack complex (MAC), which attacks the surface membrane of nearby microorganisms by embedding itself, so that a large channel is created through the microbial surface membrane.
- The osmotic flux of water into the victim cell causes it to swell and burst.
- Complement-induced lysis is the major means of directly killing microbes without phagocytizing them.
- Complement components augment the inflammatory response by: (1) serving as chemotaxins, (2) acting as opsonins; (3) promoting vasodilation and increased vascular permeability, (4) stimulating the release of histamine.

ADAPTIVE IMMUNITY: GENERAL CONCEPTS

Adaptive immune responses include antibody-mediated immunity and cell-mediated immunity.

- There are two classes of specific immune responses: antibody-mediated, or humoral, immunity involving the production of antibodies by B lymphocytes derivatives known as "plasma cells," and cell-mediated immunity, involving the production of activated T lymphocytes, which directly attack unwanted cells.
- During fetal life and early childhood, some of the immature lymphocytes migrate through blood to the thymus where they undergo further processing to become T lymphocytes.
- Lymphocytes that mature without the influence of the thymus become B lymphocytes.
- The thymus gradually atrophies and becomes less important as the individual matures.
- The thymus does, however, continue to produce thymosin, a hormone important in maintaining the T cell lineage.

An antigen induces an immune response against itself.

- Both B and T cells must be able to specifically recognize unwanted cells and other material to be destroyed or neutralized as being distinct from the body's own normal cells.
- The presence of antigens enables them to make this distinction.
- Foreign proteins are the most common antigens.

B LYMPHOCYTES: ANTIBODY-MEDIATED IMMUNITY

Antibodies stimulate B cells to convert into plasma cells that produce antibodies.

- In the case of B cells, binding with an antigen induces the cell to differentiate into a plasma cell, which produces antibodies that are able to combine with the specific antigen that stimulated the antibodies' production.

- Antibodies are secreted into the blood or lymph where they are known as "gamma globulins" or "immunoglobulins."

Antibodies are Y-shaped and classified according to properties of their tail portion.

- Antibodies are grouped into five subclasses.
- Antibody proteins in all five subclasses are composed of four interlinked polypeptide chains—arranged in the shape of a Y.
- Characteristics of the arm regions of the Y determine the specificity of the antibody.
- An antibody has two identical antigen-binding sites, one at the tip of each arm.
- These antigen-binding fragments (Fab) are unique for each different antibody.
- The tail portion of every antibody within each immunoglobulin subclass is identical.

Antibodies largely amplify innate immune responses to promote antigen destruction.

- Immunoglobulins cannot directly destroy foreign organisms or other unwanted materials upon binding with antigens on their surfaces.
- Antibodies exert their protective influence in one of two general ways: physical hindrance of antigens and amplification of nonspecific immune responses.
- Antibodies mark or identify foreign material as targets for actual destruction by the complement system, phagocytes, or killer cells.
- Occasionally, an overzealous antigen-antibody response can inadvertently cause damage to normal cells as well as to invading foreign cells causing an immune-complex disease.

Clonal selection accounts for the specificity of antibody production.

- Each B lymphocyte is preprogrammed to respond to only one of the over 100 million different antigens.

- The clonal selection theory proposes that diverse B lymphocytes are produced during fetal development, each capable of synthesizing an antibody against a particular antigen before ever being exposed to it.

Selected clones differentiate into active plasma cells and dormant memory cells.

- B cells remain dormant, not actually secreting their particular antibody product until they come in contact with the appropriate antigen.
- Antigen binding causes the activated B cell clone to multiply and differentiate into two cell types—plasma cells and memory cells.
- Plasma cells switch to the production of IgG antibodies, which are secreted rather than remaining membrane bound.
- In the blood, the secreted antibodies combine with invading free antigen, marking it for destruction by the complement system, phagocytic ingestion, or other means.
- A small proportion of the new B lymphocytes become memory cells, which do not participate in the current immune attack against the antigen, but instead remain dormant and expand the specific clone.
- During initial contact with a microbial antigen, the antibody response is delayed for several days until plasma cells are formed and does not reach its peak for a couple of weeks.
- This response is known as the "primary response."
- After reaching the peak, the antibody levels gradually decline over a period of time.
- Long-term protection against the same antigen is primarily attributable to the memory cells.
- If the same antigen ever reappears, the long-lived memory cells launch a more rapid, more potent, and longer-lasting secondary response.
- This is the basis of long-term immunity against a specific disease.

- During vaccination, the individual is deliberately exposed to a pathogen that has been stripped of its disease-inducing capability, but can still induce antibody formation against it.

T LYMPHOCYTES: CELL-MEDIATED IMMUNITY
T cells bind directly with their targets.

- Unlike B cells, T cells do not secrete antibodies.
- T cells must be in direct contact with their targets, a process known as "cell-mediated immunity."
- Like B cells, T cells are clonal and exquisitely antigen-specific.
- T cells are activated by foreign antigen only when it is present on the surface of a cell that also carries a marker of the individual's own identity.

The three types of T cells are cytotoxic T cells, helper T cells and regulatory T cells

- There are three subpopulations of T cells, depending on their roles when activated by antigens: (1) cytotoxic T cells, and (2) helper T cells (3) regulatory T cells.
- Like B cells, not all activated T cell progeny become effector T cells.
- Regulatory T cells keep the rest of the immune system under tight control.
- A small proportion of them remain dormant, serving as a pool of memory T cells that are primed and ready to respond should the same foreign antigen ever reappear within a body cell.
- The targets of cytotoxic T cells most frequently are host cells infected with viruses.

Cytotoxic T cells secrete chemicals that destroy target cells.

- One means by which cytotoxic T cells and natural killer cells destroy a targeted cell is by releasing perforin molecules, which penetrate into the target cell's surface membrane and join together to form porelike channels.

- Cytotoxic T cells also induce these virus infected cells to self-destruct, a process known as "apoptosis."
- The virus released upon destruction of the host cell is then directly destroyed in the extracellular fluid by phagocytic cells, neutralizing antibodies, and the complement system.

Helper T cells secrete chemicals that amplify the activity of other immune cells.

- In contrast to cytotoxic T cells, helper T cells are not killer cells. These cells release chemicals that influence other immune cells.
- Helper T cells secrete B-cell growth factor, which enhances the antibody-secreting ability of the activated B-cell clone.
- Helper T cells secrete T cell growth factor, which augments the activity of cytotoxic T cells, suppressor T cells, and even other helper T cells responsive to the invading antigen.
- Some chemicals secreted by T cells act as chemotaxins to lure more neutrophils and macrophages-to-be to the invaded area.
- Macrophage-migration inhibition factor, another important cytokine released from helper T cells, keeps these large phagocytic cells in the region by inhibiting their outward migration.

The major histocompatibility complex is the code for self-antigens.

- The self-antigens, which the immune system learns to recognize as markers of a person's own cells, are plasma membrane-bound glycoproteins known as "MHC molecules" because their synthesis is directed by a group of genes called the "major histocompatibility complex" or MHC.
- These are the same MHC molecules that escort engulfed foreign antigen to the cell surface for presentation by antigen-presenting cells.
- More than a hundred different MHC molecules have been identified in human tissue, but each individual has a code for only three to six of these possible antigens.
- The exact pattern of MHC molecules varies from one individual to another much like a "biochemical fingerprint."
- T cells bind with MHC self-antigens only when they are in association with a foreign antigen.
- Cytotoxic T cells can respond to foreign antigens only in association with class I MHC glycoproteins, which are found on the surface of all nucleated body cells.
- Class II MHC glycoproteins, which are recognized by helper T cells, are restricted to the surface of a few special types of immune cells.
- The class I and II markers serve as signposts to guide cytotoxic and helper T cells to precise cellular locations where their immune capabilities can be most effective..
- To minimize the rejection of transplanted tissues, the tissues of donor and recipient are matched according to MHC antigens.
- New therapeutic agents have become extremely useful in selectively depressing T cells.
- The natural function of MHC antigens lies in their ability to direct the responses of T cells, not in their artificial role in the rejection of transplanted tissue.
- Each individual has two main classes of MHC-encoded glycoproteins that are differentially recognized by cytotoxic T and helper T cells.

The immune system is normally tolerant of self-antigens.

- Suppressor T cells probably also play an important role in preventing the immune system from attacking the person's own tissues, a phenomenon known as "tolerance."
- At least four different mechanisms appear to be involved in tolerance: (1) clonal deletion, (2) clonal anergy, (3) receptor editing, (4) active suppression by regulatory T cells.

- A condition in which the immune system fails to recognize and tolerate self-antigens associated with particular tissues is known as an "autoimmune disease," of which myasthenia gravis is an example.

Immune surveillance against cancer cells involves an interplay among immune cells and interferon.

- Another important function generally attributed to the T cell system is its role in recognizing and destroying newly arisen, potentially cancerous tumor cells before they have a chance to multiply and spread—a process known as "immune surveillance."
- Any normal cell may be transformed into a cancer cell if mutations occur within its genes responsible for controlling cell division and growth.
- Such mutations frequently occur by exposure to carcinogenic factors.
- A cell that has been transformed into a tumor cell defies the normal controls on its proliferation and position.
- Unrestricted multiplication of a single tumor cell results in a tumor.
- If the mass is slow growing and does not infiltrate into the surrounding tissue, it is considered a benign tumor.
- Invasive tumors are known as "malignant tumors," commonly referred to as cancer.
- Often some of the cancer cells break away from the parent tumor and are transported through the blood to new territories.
- "Metastasis" is the term applied to this spread of cancer to other parts of the body.
- Cancer cells typically remain immature and do not become specialized, often resembling embryonic cells instead.
- Such dedifferentiated malignant cells lack the ability to perform the specialized functions of the normal cell type from which they mutated.
- Most mutations do not result in malignancy.

IMMUNE DISEASES
Immune diseases result from insufficient or inappropriate immune responses.

- Abnormal functioning of the immune system can lead to immune diseases in two general ways: deficiency diseases and inappropriate immune attacks.
- Deficiency diseases occur when the immune system fails to respond adequately to foreign invasion.
- The most recent and tragically the most common acquired immune deficiency disease is AIDS, which is caused by HIV, a virus that invades and incapacitates the critical helper T cells.

Allergies are inappropriate immune attacks against harmless environmental substances.

- The other category of immune diseases involves inappropriate specific immune attacks that cause reactions harmful to the body.
- These include: (1) autoimmune responses, in which the immune system turns against one of the body's own tissues; (2) immune-complex diseases, which involve over exuberant antibody responses that "spill over" and damage normal tissues; and (3) allergies.
- An allergy is the acquisition of an inappropriate specific immune reactivity, or hypersensitivity, to a normally harmless environmental substance.
- In immediate hypersensitivity, the allergic response appears within about 20 minutes after the sensitized individual is exposed to an allergen, whereas in delayed hypersensitivity, the reaction is not generally manifested until a day or so following exposure.
- The most common allergens that provoke immediate hypersensitivities are pollen grains, bee stings, penicillin, certain foods, molds, dust, feathers, and animal fur.
- These allergens bind to and elicit the synthesis of IgE antibodies rather than IgG antibodies.

- When an individual with an allergic tendency is first exposed to a particular allergen, compatible helper T cells secrete interleukin 4, which prods compatible B cells to synthesize IgE antibodies specific for the allergen.
- IgE antibodies do not freely circulate.
- Instead, their tail portions attach to mast cells and basophils.
- Binding of an appropriate allergen with the outreached arm regions of the IgE antibodies that are lodged tail-first in a mast cell or basophil triggers the rupture of the cell's preformed granules.
- As a result, histamine and other chemical mediators spew forth into the surrounding tissue.
- The following are among the most important chemicals released during immediate allergic reactions: (1) histamine, (2) slow-reactive substance of anaphylactic, and (3) eosinophil chemotactic factor.
- If the reaction is limited to the upper respiratory passages, the released chemicals bring about the symptoms of hay fever.
- If the reaction is concentrated primarily within the bronchioles, asthma results.
- Localized swelling in the skin because of allergy-induced histamine release causes hives.
- Treatment of localized immediate allergic reactions with antihistamines often offers only partial relief of the symptoms because some of the manifestations are invoked by other chemical mediators not blocked by these drugs.
- When large amounts of these chemical mediators gain access to the blood, the extremely serious systemic reaction known as "anaphylactic shock" occurs, which is frequently fatal.

EXTERNAL DEFENSES
The skin consists of an outer protective epidermis and an inner, connective tissue dermis.

- The epidermis consists or numerous layers of epithelial cells.
- Epidermal cells are tightly bound together by spot desmosomes, which interconnect with intracellular keratin filaments to form a strong, cohesive covering.
- The dermis is a connective tissue layer that contains many elastin fibers and collagen fibers, as well as an abundance of blood vessels and specialized nerve endings.
- Special infoldings of the epidermis into the underlying dermis form the skin's exocrine glands—the sweat glands and the sebaceous glands—as well as the hair follicles.
- Melanocytes produce the pigment melanin.
- The amount and type of melanin are responsible for the different shades of skin color of the various races.
- Keratinocytes are specialists in keratin production.
- As they die, keratinocytes form the outer protective keratinized layer and are also responsible for generating hair and nails.
- Langerhans cells present antigen to helper T cells.
- Granstein cells interact with suppressor T cells, probably serving as a "brake" or skin-activated immune responses.
- The epidermis synthesizes vitamin D in the presence of sunlight.

Protective measures within body cavities discourage pathogen invasion into the body.

- Saliva secreted into the mouth at the entrance of the digestive system contains an enzyme that lyses certain bacteria.
- Many of the surviving bacteria that are swallowed are killed by the strongly acidic gastric juice that they encounter in the stomach.
- Some bacteria do manage to survive and reach the large intestine.
- These harmless resident flora competitively suppress the growth of potential pathogens that have escaped the antimicrobial measures of earlier parts of the digestive tract.

- Within the urogenital system, would-be invaders encounter hostile conditions in the acidic urine and acidic vaginal secretions.
- The urogenital organs also produce a sticky mucus, which, like flypaper, entraps small invading particles.
- Large airborne particles are filtered out of the inspired air by hairs at the entrance of the nasal passages.
- The respiratory airways are coated with a layer of thick, sticky mucus secreted by epithelial cells within the airway lining.
- This mucus sheet, laden with any inspired particulate debris, is constantly moved upward to the throat by ciliary action.
- Also contributing to defense against respiratory infections are antibodies secreted in the mucus.
- In addition, an abundance of phagocytic specialists called the "alveolar macrophages" scavenge within the air sacs or the lungs.
- Cigarette smoking suppresses these normal respiratory defenses

Key Terms

ABO system
Acquired immune deficiency syndrome (AIDS)
Adaptive (acquired) immune system
Agglutination
Allergen
Allergy
Alveolar macrophage
Anaphylactic shock
Anemia
Antibody
Antibody-mediated (humoral immunity)
Antigen
Antigen-binding fragment (Fab)
Aplastic anemia
Asthma
Autoimmune disease
B cell growth factor
B lymphocytes
Basophils
Blood coagulation (clotting)
Bone marrow
Cancer
Cell-mediated immunity
Chemotaxis
Clonal anergy
Clonal selection theory
Clone
Clot retraction
Complement system
Constant (Fc) region
Cytokines
Cytotoxic T cells
Dendritic cells
Embolus
Eosinophils
Erythroblastosis fetalis (hemolytic disease of the newborn)
Erythrocytes (red blood cells)
Erythropoiesis
Erythropoietin
Extrinsic pathway
Factor X
Fibrin
Fibrinogen
Globulins
Granstein cells
Hay fever
Helper T cells
Hematocrit (packed cell volume)

Hemoglobin
Hemolysis
Hemolytic anemia
Hemophilia
Hemorrhagic anemia
Hemostasis
Hives
Hypersensitivity
Immune complex disease
Immune privilege
Immunity
Inflammation
Innate immune system
Interferon
Interleukin 1
Interleukin 2
Intrinsic pathway
Keratinized layer
Keratinocytes
Killer cells
Langerhans cells
Leukocytes
Lymphocytes
Lymphoid tissue
Macrophages
Major histocompatibility complex (MHC)
Mast cells
Megakaryocytes
Melanocytes
Membrane attack complex (MAC)
Memory cells
Metastasis
Monocytes
Mononuclear granulocytes
Natural killer cells
Neutralization
Neutrophils
Nutritional anemia
Pernicious anemia
Plasma
Plasma cell
Plasma proteins
Plasmin
Plasminogen
Platelet plug
Platelets (thrombocytes)
Pluripotential stem cells
Polycythemia
Polymorphonuclear granulocytes
Precipitation

Primary response
Prothrombin
Pus
Red blood cell count
Red bone marrow
Renal anemia
Rh factor
Scar tissue
Sebum
Secondary response
Serum
Severe combined immunodeficiency
Sickle cell disease
Spleen

Sweat glands
T lymphocytes
Thrombin
Thrombopoietin
Thrombus
Thymosin
Tissue plasminogen activator (tPA)
Tissue thromboplastin
Tolerance
Toll-like receptors (TLRs)
Transfusion reaction
Universal recipient
Vascular spasm
White blood cell count

Review Exercises

Answers are in the appendix.

True/False-if an answer is false, underline the incorrect portion of the statement and rewrite it to make the statement true.

_____ 1. Erythrocytes' most important feature, which enables them to transport oxygen, is the hemoglobin they contain.

_____ 2. Each of us has a total of 25–30 trillion red blood cells.

_____ 3. Estrogen, a sex hormone, increases the basal rate of erythropoiesis.

_____ 4. Secondary polycythemia is caused by a tumorlike condition of the bone marrow in which erythropoiesis proceeds at an excessive, uncontrolled rate.

_____ 5. The conversion into fibrin is catalyzed by the enzyme thromboxane at the site of vessel injury.

_____ 6. A blood clot formed by blood in a test tube was formed by the extrinsic pathway.

_____ 7. An abnormal intravascular clot attached to a vessel wall is called a "thrombus."

_____ 8. Vitamin K deficiency can cause a bleeding tendency.

_____ 9. Platelets are activated by collagen that is exposed when a vessel is damaged.

_____ 10. Plasma proteins usually do not exit through the narrow pores in the capillary walls.

_____ 11. Plasma proteins contribute to blood viscosity.

_____ 12. Plasma proteins do not affect the plasma's capacity to buffer changes in pH.

_____ 13. Leukocytes lack hemoglobin.

_____ 14. Monocytes release chemical messengers including histamine.

_____ 15. The total number of leukocytes normally ranges from 5–10 million cells per milliliter of blood.

_____ 16. Heparin interferes with Vitamin K activity to decrease the blood's ability to clot.

_____ 17. Lymphocytes that mature without benefit of "thymic education" become T lymphocytes.

_____ 18. There are four classes of specific immune responses.

_____ 19. The thymus becomes less important as the individual matures.

_____ 20. Antibodies are grouped into six subclasses.

_____ 21. Immunoglobulins cannot directly destroy foreign organisms.

_____ 22. An individual can acquire antibodies by the direct transfer of antibodies actively formed by another person or animal.

_____ 23. The most common allergens that provoke immediate hypersensitivities are pollen grains, bee stings, penicillin, certain foods, molds, dust, feathers, and animal fur.

_____ 24. Immune complex disease is when the immune system turns against one of the body's own tissues.

_____ 25. A single mast cell may be coated with a number of different IgC antibodies.

_____ 26. Antibodies not only target viruses for destruction in the extracellular fluid but can also eliminate viruses inside neurons.

_____ 27. The AIDS virus selectively invades cytotoxic T cells.

_____ 28. Fair skinned people have the same number of pigment producing cells as dark skinned people.

_____ 29. Epidermal cells are tightly bound together by spot desmosomes.

_____ 30. The keratinized layer is airtight, fairly waterproof, and impervious to most substances.

_____ 31. The dermal blood vessels have no control in temperature regulation.

_____ 32. The epidermis synthesizes vitamin A in the presence of sunlight.

_____ 33. Natural killer cells are a family of proteins that nonspecifically defend against viral infections.

_____ 34. Neutrophils are the first to arrive at the site of an injury.

_____ 35. NSAIDs reduce fever by inhibiting the production of prostaglandins.

Fill in the Blank

36. _____ is the arrest of bleeding from a broken blood vessel.

37. When a tissue is traumatized, it releases a protein complex known as _____.

38. The clot is slowly dissolved by a fibrinolytic enzyme called _____.

39. The most common cause of excessive bleeding is _____.

40. Most old red blood cells meet their final demise in the _____.

41. _____ anemia is caused by failure of the bone marrow to produce adequate numbers of red blood cells.

42. _____ anemia is caused by the rupture of excessive numbers of circulating erythrocytes.

43. The remaining small percentage of plasma is occupied by: (1)_____, (2) _____, (3) _____, and (4) _____.

44. Class _____ MHC glycoproteins are restricted to the surface of a few special types of immune cells.

45. Alpha, beta, and gamma are the three subclasses of _____.

46. _____ are the mobile units of the body's immune defense system.

47. _____ are the smallest of the leukocytes.

48. _____ are phagocytic specialists.

49. T lymphocytes destroy their specific target cells, a process known as a(n) _____ _____.

50. _____ antibodies are associated with the onset of allergic reactions.

51. Antibodies mark or identify foreign material as targets for actual destruction by the _____, _____, or _____.

52. The production of antibodies as a result of exposure to an antigen is referred to as _____ against that antigen.

53. T-cells must be in direct contact with their targets, a process known as _____.

54. Natural killer cells release _____ molecules, which penetrate into the target cell's surface membrane and join together to form pore-like channels.

55. _____ are by far the most numerous of the T cells.

56. The most recent and tragically the most common acquired immune deficiency disease is _____.

57. When large amounts of chemical mediators or allergens gain access to the blood, the extremely serious systemic reaction known as _____ occurs.

58. _____ refer collectively to the tissues that store, produce, or process lymphocytes.

59. Upon exposure to tissue thromboplastin in the injured tissue and to specific chemicals secreted by phagocytes on the scene, _____, the final factor in the clotting system, is converted into _____.

60. _____ T cells probably play an important role in preventing the immune system from attacking the person's own tissues

61. There are two classes of specific immune responses: _____ and _____.

62. _____ are low-molecular-weight organic substances that are not antigenic by themselves but can become antigenic if they attach to body proteins.

63. The _____ weighs over 9 lbs and covers 21 square feet.

64. _____ release a dilute salt solution through small openings, _____ onto the surface of the body.

65. The most abundant epidermal cells are the _____.

66. _____ interact with suppressor T cells, probably serving as a "brake" on skin-activated immune responses.

67. Phagocytic specialists called _____ scavenge within the air sacs of the lungs.

Matching
Match the blood constituent to their function.

_____	68.	Carries heat	a. platelets
_____	69.	Hemostasis	b. erythrocyte
_____	70.	Become macrophages	c. fibrinogen
_____	71.	Transport oxygen	d. basophils
_____	72.	Precursor for fibrin	e. neutrophils
_____	73.	Cell-mediated immune response	f. eosinophils
_____	74.	Release histamine	g. B lymphocytes
_____	75.	Attacks parasitic worms	h. water
_____	76.	Antibody production	i. T lymphocytes
_____	77.	Phagocytizes bacteria	j. monocytes

Match the lymphoid tissue to the function.

a. thymus
b. spleen
c. bone marrow
d. lymph nodes, tonsils, adenoids, appendix, gut associated lymphoid tissue

_____ 78. Exchanges lymphocytes with blood.

_____ 79. Origin of all blood cells.

_____ 80. Stores a small percentage of red blood cells.

_____ 81. Secretes the hormone thymosin.

_____ 82. Exchanges lymphocytes with lymph.

_____ 83. Site of maturational processing for T lymphocytes.

_____ 84. Resident lymphocytes produce antibodies and sensitized T cells, which are released into the blood.

_____ 85. Site of maturational processing for B lymphocytes.

_____ 86. Resident macrophages remove microbes and other particulate debris from lymph.

_____ 87. Resident lymphocytes produce antibodies and sensitized cells, which are released into the lymph.

_____ 88. Resident macrophages remove microbes and other particulate debris, most notably worn-out red blood cells, from the blood.

Multiple Choice

89. Which cellular element has the largest percentage of total blood volume?
    a. eosinophils
    b. erythrocytes
    c. basophils
    d. neutrophils
    e. platelets

90. A condition in which the immune system fails to recognize and tolerate self-antigens associated with particular tissues is known as   a. leukemia
    b. autoimmune disease   c. anemia
    d. embolus
    e. none of the above

91. Which organ secretes erythropoietin?
    a. bone marrow
    b. heart
    c. lungs
    d. spleen
    e. kidney

92. Which disorder is characterized by a tumorlike condition of the bone marrow?
    a. hemorrhagic anemia
    b. sickle-cell anemia
    c. secondary polycythemia
    d. leukemia
    e. primary polycythemia

93. Which type of anemia is associated with Vitamin B 12 deficiency?
    a. factor X anemia
    b. sickle cell anemia
    c. aplastic anemia
    d. pernicious anemia
    e. iron deficiency anemia

94. Hemophilia results from:
    a. a deficiency of platelets
    b. inadequate hemoglobin production
    c. vitamin B_{12} deficiency
    d. a genetic inability to produce one of the factors in the clotting cascade
    e. excess production of heparin

95. Which of the following describes the major function of tissue thromboplastin?
    a. It is released from traumatized tissue and triggers the extrinsic clotting pathway.
    b. It converts fibrinogen to fibrin.
    c. It forms the network of the clot.
    d. It activates Hageman factor.
    e. More than one of the above are correct.

96. Which of the following is true with respect to the extrinsic clotting pathway?
    a. It is set off by factor XII.
    b. It clots blood in an injured blood vessel.
    c. It has more steps than the intrinsic pathway.
    d. More than one of the above are correct.
    e. None of the above are correct.

97. Platelets:
    a. convert prothrombin to thrombin
    b. form the meshwork upon which the erythrocytes become entrapped
    c. adhere to collagen
    d. release fibrinogen one a platelet plug is formed
    e. produce antibodies

98. What forms the meshwork of a blood clot?
    a. red blood cells
    b. fibrinogen
    c. platelets
    d. thrombin
    e. Hageman factor

99. What does prostacyclin do in the body?
    a. It activates the clotting cascade.
    b. It induces profound vasoconstriction of an injured blood vessel.
    c. It profoundly inhibits platelet aggregation.
    d. It is released by aggregated platelets.
    e. It dissolves the blood clot.

100. Platelets:
    a. are small cell fragments derived from large megakaryocytes in the bone marrow
    b. lack nuclei
    c. contain high concentrations of actin and myosin
    d. both a and b
    e. a, b, and c

101. Basophils:
    a. contain granules that preferentially take up basic blue dyes
    b. leave the blood to become macrophages
    c. synthesize and store histamine and heparin
    d. both a and b
    e. a, b, and c

102. Polymorphonuclear granulocytes include:
    a. neutrophils
    b. eosinophils
    c. basophils
    d. both a and b
    e. a, b, and c

103. Hemoglobin
    a. is found in the nuclei of red blood cells
    b. contains carbonic anhydrase
    c. can combine with oxygen, carbon dioxide, hydrogen ions, and carbon monoxide
    d. both a and b
    e. a, b, and c

104. Which of the following substances causes an inappropriate immune response?
    a. perforins
    b. opsonins
    c. endogenous pyrogens
    d. haptens
    e. allergens

105. Which of the following substances is a secretory product of helper T cells?
    a. kallikrein
    b. interleukin 1
    c. haptens
    d. interleukin 2
    e. allergens

106. Which of the following substances are not antigenic by themselves but can become antigenic if they attach to body proteins?
    a. opsonins
    b. allergens
    c. perforins
    d. histamines
    e. haptens

107. Which of the following substances is a secretory product of macrophages?
    a. interleukin 1
    b. interferon
    c. kallikrein
    d. interleukin 2
    e. perforin

108. Which of the following substances triggers the production of viral-blocking enzymes by potential host cells?
    a. perforin
    b. interleukin 2
    c. interleukin 1
    d. ferrokinogen
    e. interferon

Modified Multiple Choice
Indicate whether the characteristic applies to active immunity or passive immunity by writing the appropriate letter in the blank using the following answer code.

A = applies to active immunity
B = applies to passive immunity
C = applies to both active and passive immunity

109. _____ exposure to antigen required for immunity to develop
110. _____ antibodies produced by a source other than ones own body
111. _____ associated with vaccine function
112. _____ resistance to antigen exposure immediate upon injection of antibodies
113. _____ natural immunity is actually a special case of this type of immunity

Indicate which type of T cell is being described by writing the appropriate letter in the blank using the answer code below.

> A = applies to cytotoxic T cells
> B = applies to helper T cells
> C = applies to suppressor T cells
> D = applies to both helper T cells and suppressor T cells
> E = applies to all three types of T cells

114. _____ destroys host cells bearing foreign antigen
115. _____ suppresses both T-cell and B-cell activity
116. _____ secretes B cell growth factor
117. _____ enhances the development of antigen-stimulated B cells into antibody-secreting cells
118. _____ called "regulatory T cells"
119. _____ secretes interleukin 2
120. _____ releases perforin
121. _____ secretes macrophage-migration inhibition factor
122. _____ attacked by the AIDS virus
123. _____ most numerous of the T cells
124. _____ serves to limit immune reactions
125. _____ believed to play a role in tolerance
126. _____ recognizes class I MHC glycoproteins
127. _____ recognizes class II MHC glycoproteins

Points to Ponder

1. Why do you suppose that the clotting cascade is so long and complicated?

2. Over the years we have gained much knowledge about the composition of blood. Why then is it so difficult to make artificial blood?

3. What is the advantage in having erythrocytes with no nuclei, organelles, or ribosomes that are short-lived as compared to longer-lived erythrocytes that are complete cells?

4. Does drinking coffee after binge drinking help a person "sober up"? Explain your answer.

5. In your body, where is most of your blood located?

6. Where does the term "blue blooded" mean?

7. Why doesn't blood clot in blood vessels?

8. When a blood panel is done, blood is taken from a vein and not an artery. Explain why.

9. How is hemoglobin recycled?

10. What is the difference between hematocrit and reticulocyte count?

11. With the current social views in this country, do you feel that the eradication of AIDS is possible?

12. In that part of the body defenses involving B cells and T cells, cellular memory plays a major role. When and how is this memory lost?

13. Do animal rights activists hinder our fight against viral diseases?

14. How does exercise strengthen the immune system?

15. If you have an allergy, will your children inherit it? Explain why or why not.

16. A drug has been discovered that destroys all mast cells. How might this drug help prevent allergies? Would there be any negative side effects by taking this drug?

17. How does an elevated body temperature help an infection caused by a virus?

18. What is the difference between a nonspecific host defense mechanism and a specific defense mechanism? Give examples.

19. How does the clonal selection theory explain the ability of the immune response to distinguish between self and nonself?

20. What is the basis of immunological memory?

Clinical Perspectives

1. When a person cuts their wrists to commit suicide, does it make a difference whether or not they completely sever the vessels?

2. Why is 20% of a cigarette smoker's hemoglobin nonfunctional?

3. In what type of clinical situation would you receive just blood platelets during a blood transfusion?

4. Your friend tells you she has iron deficiency anemia. She knows you are taking a physiology course and asks you to explain to her what this means. What would your answer be?

5. Your mother tells you that after her yearly physical the doctor recommends that she takes one baby aspirin each day. She knows that you are taking a physiology course and asks you to explain why the doctor recommended aspirin therapy. How would you explain this therapy to her?

6. What are the reasons you give a patient tissue plasminogen activator (tPA) or streptokinase?

7. Your athlete friend asks you if she should try induced erythrocythemia (blood doping). What would you tell her?

8. In a blood panel, what does a differential white blood cell count tell the physician about homeostasis in your body?

9. How would you explain to your friend why blood is typed and cross-matched before a blood transfusion?

10. Your college roommate has infectious mononucleosis. She asks you why she is so tired all the time. What would you tell her?

11. What is the significance of a "booster shot"?

12. This summer you will direct a biological field trip to the remote areas of the Chihuahuan Desert. You and the students will spend a month living in tents as you study the desert. With regard to this chapter, what precautions should you take before you begin this trip?

13. Why don't we become immune to the cold virus?

14. Why do most children outgrow common food allergies?

15. Why is cortisone and its analogues used clinically in patients who have a transplanted organ?

16. While painting the house, you get stung by a bee. This is the first time you have ever been stung by a bee. You develop an itchy rash, which is not relieved by antihistamines prescribed by your physician. The physician then tries cortisone, which alleviates the rash. Several weeks later, you get stung again by a bee; however, unlike the first time, you respond well to the antihistamines. Why did this occur?

17. The tuberculin skin test and poison ivy are both type IV hypersensitivities. What does this mean? Explain the common mechanism involved.

18. How does hemolytic disease of the newborn develop?

19. What is the immunological basis behind blood transfusions?

20. In desensitization procedures, the allergist injects more of the same allergen to which the person is allergic. How can this be beneficial?

The Respiratory System

Chapter Overview

Respiration involves the sum of the processes that accomplish ongoing passive movement of oxygen from the atmosphere to the tissues to support cellular metabolism, as well as the continual passive movement of metabolically produced carbon dioxide from the tissues to the atmosphere. The homeostatic role of the respiratory system overlaps that of the circulatory system: supplying oxygen to the cells of the body and removing carbon dioxide, a byproduct of cellular metabolism. While these are not the only functions of the respiratory system, they are surely the most important. The entire ventilation-transport process, from air outside the body to the gaseous exchange at the systemic cellular level, involves diffusion down gradients. For the most part, they are pressure gradients.

The mechanics of ventilation rely heavily on the physical properties of the alveolar wall and surface tension. Ventilation is controlled by the nervous system at both conscious and subconscious levels.

The bonding of oxygen and carbon dioxide to hemoglobin is important for transport by the blood. Oxygen bonds to the heme portion and carbon dioxide attaches to the globin fraction of the hemoglobin molecule. This is, however, only one step in the process of respiration.

From alveolar air to the mitochondria of the various cells involves many barriers through which oxygen and carbon dioxide must traverse. Crossing the numerous barriers, oxygen and carbon dioxide diffuse down pressure gradients involving differences in their respective partial pressures. Using quite sophisticated control mechanisms, the respiratory system plays a major role in maintaining homeostasis.

Chapter Outline

INTRODUCTION
The Respiratory system does not participate in all steps of respiration.

- The respiratory system does not participate in all steps of respiration.
- The primary function of the respiratory system is to obtain oxygen for use by the body's cells and to eliminate carbon dioxide the cells produce.
- Internal or cellular respiration refers to the intracellular metabolic processes carried out within the mitochondria, which use oxygen and produce carbon dioxide during the derivation of energy from nutrient molecules.
- External respiration refers to the entire sequence or events in the exchange of oxygen and carbon dioxide between the external environment and the cells of the body.
- External respiration, the topic or this chapter, encompasses four steps: (1) Air is alternately moved in and out of the lungs so that exchange of air can occur between the atmosphere and the alveoli of the lungs. (2) Oxygen and carbon dioxide are exchanged between air in the alveoli and blood within the pulmonary capillaries by the process of

diffusion. (3) Oxygen and carbon dioxide are transported by the blood between the lungs and the tissues. (4) The exchange of oxygen and carbon dioxide takes place between the tissues and the blood by the process of diffusion across the systemic capillaries.

- The respiratory system additionally performs the following nonrespiratory functions: (1) It provides a route for water loss and heat elimination. (2) It enhances venous return. (3) It enables speech, singing, and other vocalization. (4) It defends against inhaled foreign matter. (5) It removes, modifies, activates, or inactivates various materials passing through the pulmonary circulation. (6) The nose serves as the organ of smell. (7) It contributes to the maintenance of normal acid-base balance.

The respiratory airways conduct air between the atmosphere and alveoli.

- The respiratory system includes the respiratory airways leading into the lungs, the lungs themselves, and the structures of the thorax involved in producing movement of air through the airways into and out of the lungs.
- The airways include: (1) nasal passages, (2) pharynx, (3) larynx, (4) trachea, (5) bronchi, and (6) bronchioles. The term "vocal chords" has been replaced by the more descriptive term "vocal folds."
- Air passes into the larynx through the space between the vocal folds. This laryngeal opening is known as the glottis. As air moves through the opening the vocal folds vibrate and produce different sounds.

The gas-exchanging alveoli are thin-walled, inflatable air sacs encircled by pulmonary capillaries.

- The alveoli are clusters of thin walled, inflatable, grape like sacs at the terminal branches of the conducting airways.
- The lungs contain about 500 million alveoli with a total surface area of 75 sq meters.
- The alveolar walls consist of a single layer of flattened Type 1 alveolar cells.
- The interstitial space between an alveolus and the surrounding capillary network forms an extremely thin barrier.
- In addition to the Type I cells, the alveolar epithelium also contains Type II alveolar cells, which secrete pulmonary surfactant.

The lungs occupy much of the thoracic cavity.

- There is no muscle within the alveolar walls to cause them to inflate or deflate during the breathing process.
- The lungs, heart and associated vessels, esophagus, thymus, and some nerves occupy the thoracic cavity.
- The outer chest wall is formed by 12 pairs of curved ribs, which join the sternum anteriorly and the thoracic vertebrae posteriorly.
- The diaphragm forms the floor of the thoracic cavity with the ribs and skeletal muscles forming the walls of the cavity

A pleural sac separates each lung from the thoracic wall.

- Separating each lung from the thoracic wall and other surrounding structures is a doubled-walled, closed sac called the "pleural sac."
- The surfaces of the pleura secrete a thin intrapleural fluid, which lubricates the pleural surfaces.

RESPIRATORY MECHANICS
Interrelationships among pressures inside and outside the lungs are important in ventilation.

- Air tends to move from a region of higher pressure to a region of lower pressure down a pressure gradient.

- Three different pressure considerations are important in ventilation: (l) atmospheric pressure, (2) intra-alveolar pressure, and (3) intrapleural pressure.

The lungs are normally stretched to fill the larger thorax.
- The transmural pressure gradient hold the lungs and thoracic wall in tight apposition, even though the lungs are smaller than the thorax.
- As the thoracic cavity enlarges, the lungs enlarge; that is, the lungs follow the movements of the chest wall.
- Because of the lungs' elasticity, they try to pull inward away from the thoracic wall as they are stretched to fill the larger thoracic cavity. The transmural pressure gradient, however, prevents the lungs from pulling away except to the slightest degree. The resultant ever-so-slight expansion of the pleural cavity is sufficient to drop the pressure in this cavity by 4 mm Hg.
- Thus means the intra-alveolar pressure is greater than the intrapleural pressure, so a greater pressure is pushing outward than is pushing inward across the lung wall.
- A similar transmural pressure gradient exists across the thoracic wall.
- The atmosphere pressure pushing inward on the thoracic wall is greater than the intrapleural pressure pushing outward on the same wall.
- The stretched lungs have a tendency to pull inward away from the thoracic wall, whereas the compressed thoracic wall tends to move away from the lungs.

Flow of air into and out of the lungs occurs because of cyclical changes in intra-alveolar pressure.
- Because air flows down a pressure gradient, the intra-alveolar pressure must be less than atmospheric pressure for air to flow into the lungs during inspiration.
- Intra-alveolar pressure can be changed by altering the volume of the lungs, in accordance with Boyle's law.
- At the onset of inspiration, the inspiratory muscles—the relaxed diaphragm and external intercostal muscles—are stimulated to contract, resulting in enlargement of the thoracic cavity.
- The intra-alveolar pressure is now less than atmospheric pressure; air flows into the lungs down the pressure gradient from higher to lower pressure.
- During inspiration, the intrapleural pressure falls.
- At the end of inspiration, the inspiratory muscles relax.
- The chest wall and the stretched lungs recoil to their preinspiratory size because of their elastic properties.
- In a resting expiration, the intra-alveolar pressure increases.
- Air now leaves the lungs down a pressure gradient from high intra-alveolar pressure to lower atmospheric pressure.
- Expiration is normally a passive process.
- Inspiration is always an active process.
- To produce active or forced expiration, the expiratory muscles must contract to further reduce the volume of the thoracic cavity and lungs.

Airway resistance influences airflow rates.
- The primary determinant to resistance to airflow is the radius of the conducting airway.
- Normally, modest adjustments in airway size can be accomplished by autonomic nervous system regulation to suit the body's needs.
- Parasympathetic stimulation promotes bronchiolar smooth muscle contraction, which increases airway resistance by producing bronchoconstriction.

- Sympathetic stimulation and its associated hormone, epinephrine, bring about bronchodilation and decreased airway resistance by promoting bronchiolar smooth muscle relaxation.
- Resistance becomes an extremely important impediment to airflow when airway lumens become narrowed as a result of disease.

Airway resistance is abnormally increased with chronic obstructive pulmonary disease.

- Chronic obstructive pulmonary disease is a group of lung diseases characterized by increased airway resistance resulting from the narrowing of the lumen of the lower airways.
- Chronic obstructive pulmonary disease encompasses three chronic diseases: chronic bronchitis asthma, , and emphysema.
- Chronic bronchitis is a long-term inflammation of the lower respiratory airways, generally triggered by irritating cigarette smoke, polluted air, or allergens.
- In asthma, airway obstruction is due to: (1) profound constriction of the smaller airways caused by allergy-induced spasm of the smooth muscle in the walls of these airways; (2) plugging of airways by excess secretion of very thick mucus; and (3) thickening of the walls of the airways due to inflammation and histamine-induced edema.
- Emphysema is characterized by collapse of the smaller airways and a breakdown of alveolar walls.
- When airway resistance is increased as a result of chronic obstructive lung disease of any type, expiration is more difficult to accomplish than inspiration.
- When COPD of any type increases airway resistance,

expiration is more difficult than inspiration. The smaller airways tend to collapse upon expiration making the process more difficult.
A person with asthma has more difficulty expiring than inspiring, giving rise to the characteristic "wheeze" as air is forced out through the narrowed airways.

Elastic behavior of the lungs is a result of elastic connective tissue and alveolar surface tension.

- Two interrelated concepts are involved in pulmonary elasticity: elastic recoil and compliance.
- Elastic recoil refers to how readily the lungs rebound after having been stretched.
- Compliance refers to how much effort is required to stretch or distend the lungs.
- A highly compliant lung stretches further for a given increase in the pressure difference than does a less compliant lung.
- Pulmonary elastic behavior depends mainly on two factors: highly elastic connective tissue in the lungs and alveolar surface tension.
- Pulmonary connective tissue contains large quantities of elastin fibers arranged into a meshwork that amplifies their elastic behavior.
- Surface tension is responsible for a twofold effect.
- First, the liquid layer resists any force that increases its surface area.
- Second, the liquid surface area tends to become as small as possible because the surface water molecules try to get as close together as possible.

Pulmonary surfactant decreases surface tension and contributes to lung stability.

- The tremendous surface tension of pure water is normally counteracted by secretion of pulmonary surfactant by the Type II alveolar cells.
- By lowering the alveolar surface tension, pulmonary surfactant provides two important benefits: (1) it increases pulmonary compliance, thus reducing the work of inflating the lungs; and (2) it reduces the lungs' tendency to recoil, so that they do not collapse readily.

- Pulmonary surfactant's role in reducing the alveoli's tendency to recoil is important in helping maintain lung stability.
- A deficiency of pulmonary surfactant is responsible for newborn respiratory distress syndrome.
- In an infant born prematurely, pulmonary surfactant may be insufficient to reduce the alveolar surface tension to manageable levels.
- The resultant collection of symptoms that develop is referred to as "newborn respiratory distress syndrome."
- It is more difficult to expand a collapsed alveolus by a given volume than to increase an already partially expanded alveolus by the same volume.
- With newborn respiratory distress syndrome, lung expansion may require considerably larger transmural pressure gradients.
- The problem is compounded by the fact that the newborn's muscles are still weak

The lungs normally operate at about "half full."
- On average, the maximum amount of air that the lungs can hold is about 5.7 liters in males and 4.2 liters in females.
- Anatomical build, age, the distensibility of the lungs, and the presence or absence of respiratory disease affect this total lung capacity.
- At the end of a normal quiet expiration, the lungs still contain about 2,200 ml of air.
- About 500 ml of air are inspired, and the same quantity expired during quiet breathing.
- Using a spirometer, the following lung volumes and lung capacities can be determined: (1) tidal volume, (2) inspiratory reserve volume, (3) inspiratory capacity, (4) expiratory reserve volume, (5) residual volume, (6) functional residual capacity, (7) vital capacity, (8) total lung capacity, and (9) forced expiratory volume in one second.
- Two general categories of respiratory dysfunction yield abnormal results during spirometry: obstructive and restrictive lung disease.
- Additional respiratory dysfunction includes diseases impairing gas exchange; reduced ventilation due to muscle and nerve issues; inadequate blood flow to lungs; ventilation and perfusion imbalances
- To determine what abnormalities are present, the diagnostician relies on a variety or respiratory function tests in addition to spirometry, including X-ray examination, blood gas determinations, and tests to measure the diffusion capacity of the alveolar capillary membrane.

Alveolar ventilation is less than pulmonary ventilation because of the presence of dead space.
- Not all of the inspired air gets down to the site of gas exchange in the alveoli.
- Part of the air remains in the conducting airways, where it is not available for gas exchange.
- This volume is considered to be anatomic dead space.
- Alveolar ventilation—the volume of air exchanged between the atmosphere and the alveoli per minute—is more important than pulmonary ventilation.

GAS EXCHANGE
Gases move down partial pressure gradients.
- The ultimate purpose of breathing is to provide a continual supply of fresh oxygen for pickup by the blood and to constantly remove carbon dioxide unloaded from the blood.
- Gas exchange at both the pulmonary-capillary and tissue-capillary level involves simple passive diffusion of oxygen and carbon dioxide down partial pressure gradients.
- If, as in the case with oxygen, the alveolar partial pressure of a gas is higher than the

partial pressure of that gas in the blood entering the pulmonary capillaries, the higher alveolar partial pressure drives more oxygen into the blood.

- Conversely, if the alveolar partial pressure of a gas is lower than its partial pressure in the entering blood—the situation that exists for carbon dioxide—the lower alveolar partial pressure permits some of the carbon dioxide to escape from solution in the blood.
- A gas always diffuses down its partial pressure gradient from the area of higher partial pressure to the area of lower partial pressure.

Oxygen enters and carbon dioxide leaves the blood in the lungs passively down partial pressure gradients.

- The alveolar partial pressure of oxygen remains relatively constant throughout the respiratory cycle.
- The partial pressure of oxygen in the blood likewise remains fairly constant at the same value.
- A similar situation in reverse exists for carbon dioxide.
- The appropriate partial pressure gradients between the alveoli and blood are maintained to ensure that oxygen enters the blood and carbon dioxide leaves the blood.
- The amount of oxygen picked up in the lungs matches the amount extracted and used by the tissues.
- The amount of carbon dioxide given up to the alveoli from the blood matches the amount of carbon dioxide picked up at the tissues.

Factors other than the partial pressure gradient influence the rate of gas transfer.

- According to Fick's law of diffusion, the rate of diffusion of a gas depends on the surface area and thickness of the membrane, and on the diffusion coefficient of the gas.
- Surface area is reduced in emphysema.
- Inadequate gas exchange can also occur when the thickness of the barrier separating the air and blood is pathologically increased.
- The thickness increases in: (1) pulmonary edema, (2) pulmonary fibrosis, and (3) pneumonia.

Gas exchange across the systemic capillaries also occurs down partial pressure gradient.

- Just as they do at the pulmonary capillaries, oxygen and carbon dioxide move between the systemic capillary blood and the tissue cells by simple passive diffusion down partial pressure gradients.
- The amount of oxygen transferred to the cells and the amount of carbon dioxide carried away from the cells depend on the rate of cellular metabolism.

GAS TRANSPORT
Most oxygen in the blood is transported bound to hemoglobin.

- Oxygen is present in the blood in two forms: physically dissolved and chemically bound to hemoglobin.
- Very little oxygen is physically dissolved in the plasma water because oxygen is poorly soluble in body fluids.
- Of the oxygen transported by the blood, 98.5% is transported in combination with hemoglobin.
- Hemoglobin, an iron-bearing protein molecule contained within the red blood cells, has the ability to form a loose, easily reversible combination with oxygen.

The partial pressure of oxygen is the primary factor determining the percent of hemoglobin saturation.

- Each hemoglobin molecule can carry up to four molecules of oxygen.
- The most important factor determining the percent of hemoglobin saturation is the partial pressure of oxygen in the blood, which in turn is related to the concentration of oxygen physically dissolved in the blood.

- When the partial pressure or oxygen in the blood is increased, as it is in the pulmonary capillaries, the reaction is driven toward an increase in the formation of oxyhemoglobin.
- When the partial pressure of oxygen is decreased, as it is in the systemic capillaries, the reaction is driven toward an increase in the formation of reduced hemoglobin.

Oxygen-HB dissociation curve
- The relationship between partial pressure of oxygen and Hb saturation can be seen in the oxygen-Hb dissociation curve.
- Important portions of the curve include the plateau portion where oxygen is being loaded onto Hb; the steep portion where oxygen is unloaded from Hb.
- Depending on physiological conditions, the curve can shift to the right or left to alter oxygen delivery to tissue.

Hemoglobin promotes the net transfer of oxygen at both the alveolar and tissue levels.
- Net diffusion of oxygen from alveoli to blood occurs continuously until the hemoglobin becomes saturated with oxygen as completely as it can be at that particular partial pressure.
- At a normal partial pressure for oxygen, hemoglobin is 97.5% saturated.
- Not until hemoglobin can store no more oxygen does all of the oxygen transferred into the blood remain dissolved and directly contribute to the partial pressure.
- Once the partial pressure of oxygen in the blood equilibrates with the alveolar partial pressure, no further oxygen transfer can take place.

- The reverse situation occurs at the tissue level.
- When the partial pressure of oxygen in the blood falls, hemoglobin is forced to unload some of its stored oxygen because the percent of hemoglobin saturation is reduced.
- Only when hemoglobin is no longer able to release any more oxygen into solution can the partial pressure of oxygen in the blood become as low as in the surrounding tissue.
- At this time, further transfer of oxygen ceases.
- Hemoglobin plays an important role in the total quantity of oxygen that the blood can pick up in the lungs and drop off in the tissues.

Factors at the tissue level promote the unloading of oxygen from hemoglobin.
- An increase in the partial pressure of carbon dioxide reduces the amount or oxygen and hemoglobin that can be combined.
- An increase in acidity also reduces the amount of oxygen and hemoglobin that can be combined.
- The influence of carbon dioxide and acid on the release of oxygen is known as the "Bohr effect."
- Local elevation in temperature enhances oxygen release from hemoglobin for use by the more active tissues.
- Thus, increases in carbon dioxide, acidity, and temperature at the tissue level enhance the effect of a drop in the partial pressure of oxygen in facilitating the release of oxygen and hemoglobin.
- These effects are largely reversed at the pulmonary level, where the extra acid-forming carbon dioxide is blown off and the local environment is cooler.
- Inside the red blood cell, 2,3-diphosphoglycerate is produced during cellular metabolism.
- This erythrocyte constituent can bind reversibly with hemoglobin and reduce its affinity for oxygen.

Hemoglobin has a higher affinity for carbon monoxide than for oxygen.

- Carbon monoxide and oxygen compete for the same binding sites on hemoglobin, but the affinity of hemoglobin for carbon monoxide is 240 times that of the bond strength between hemoglobin and oxygen.
- Fortunately, carbon monoxide is not a normal constituent of inspired air.
- If carbon monoxide is being produced in a closed environment, it can reach lethal levels without the victim ever being aware of the danger.

Most carbon dioxide is transported in the blood as bicarbonate.

- Carbon dioxide is transported in the blood in three ways: (1) physically dissolved, (2) bound to hemoglobin, and (3) as bicarbonate.
- The amount of carbon dioxide physically dissolved in the blood is only 10% of the blood's total carbon dioxide content at normal systemic venous partial pressure levels.
- Another 30% of the carbon dioxide combines with hemoglobin to form carbamino hemoglobin.
- Carbon dioxide binds with the globin portion of hemoglobin.
- By far the most important means of carbon dioxide transport is as bicarbonate.
- Carbon dioxide combines with water to form carbonic acid.
- This reaction proceeds swiftly within red blood cells because of the presence of the erythrocyte enzyme carbonic anhydrase, which catalyzes the reaction.
- The one carbon and two oxygen atoms of the original carbon dioxide molecule are thus present in the blood as an integral part of the bicarbonate ion.

- Bicarbonate and hydrogen start to accumulate within the red blood cells in the systemic capillaries.
- Bicarbonate diffuses down its concentration gradient out of erythrocytes into the plasma.
- Chloride ions, the dominant plasma anions, diffuse into the red blood cells down the electrical gradient to restore neutrality.
- This inward shift of chloride ions in exchange for the outflux of carbon dioxide-generated bicarbonate ions is known as the "chloride shift."
- Reduced hemoglobin has a greater affinity for hydrogen ions than does oxyhemoglobin.

Various respiratory states are characterized by abnormal blood gas levels.

- Hypoxia refers to insufficient oxygen at the cellular level.
- There are four general categories of hypoxia: (1) hypoxic hypoxia, (2) anemic hypoxia, (3) circulatory hypoxia, and (4) histotoxic hypoxia.
- Hypercapnia refers to excess carbon dioxide in the arterial blood.
- Hypercapnia is caused by hypoventilation.

CONTROL OF RESPIRATION
Respiratory centers in the brain stem establish a rhythmic breathing pattern.

- Respiratory control centers housed in the brain stem are responsible for generating the rhythmic pattern of breathing.
- The generation of respiratory rhythm is now widely believed to lie in the preBotzinger complex.
- The primary respiratory control center is the medullary respiratory center.
- There are two other respiratory centers higher in the brain stem in the pons—the apneustic center and the pneumotaxic center.
- The medullary respiratory center consists of two neuronal clusters known as the "dorsal respiratory group" and the "ventral respiratory group."
- The dorsal respiratory group consists mostly of inspiratory neurons whose descending fibers terminate on the motor neurons that supply the inspiratory muscles.

- This ventral respiratory group is composed of inspiratory neurons and expiratory neurons.
- The ventral respiratory group is called into play by the dorsal respiratory group as an "overdrive" mechanism during periods when demands for ventilation are increased.
- The pontine centers exert "fine-tuning" influences over the medullary center to help produce normal, smooth inspirations and expirations.
- The pneumotaxic center sends impulses to the dorsal respiratory group that help "switch off" the inspiratory neurons, thereby limiting the duration of inspiration.
- The apneustic center prevents the inspiratory neurons from being switched off.
- The pneumotaxic center is dominant over the apneustic center.
- When the tidal volume is large, as during exercise, the Hering-Breuer reflex is triggered to prevent over inflation of the lungs.
- Action potentials from the pulmonary stretch receptors travel through afferent nerve fibers to the medullary center and inhibit the inspiratory neurons.

The magnitude of ventilation is adjusted in response to three chemical factors: the partial pressure of oxygen, the partial pressure of carbon dioxide, and the hydrogen ion concentration.

- Arterial blood gases are maintained within a normal range by varying the magnitude of ventilation to match the body's need for oxygen uptake and carbon dioxide removal.
- The medullary respiratory center receives inputs that provide information about the body's needs for gas exchange.

Decreased arterial partial pressure of oxygen increases ventilation only as an emergency mechanism.

- The medullary respiratory center receives inputs that provide information about the body's needs for gas exchange.
- The arterial partial pressure of oxygen is monitored by peripheral chemoreceptors known as the "carotid bodies" and "aortic bodies," which are located at the bifurcation of the common carotid arteries and in the arch of the aorta, respectively.
- The peripheral chemoreceptors are not sensitive to modest reductions in the arterial partial pressure of oxygen.
- Reflex stimulation of respiration by the peripheral chemoreceptors serves as an important emergency mechanism in dangerously low arterial partial pressures of oxygen conditions.

Carbon dioxide-generated hydrogen ions in the brain are normally the primary regulator of ventilation.

- The arterial partial pressure of carbon dioxide is the most important input regulating the magnitude of ventilation under resting conditions.
- An increase in the arterial partial pressure of carbon dioxide reflexly stimulates the respiratory center, with the resultant increase in ventilation promoting elimination of excess carbon dioxide.
- Conversely, a fall in arterial partial pressure of carbon dioxide reflexly reduces respiratory drive.
- There are no important receptors that monitor the arterial partial pressure of carbon dioxide, per se.
- Located in the medulla in the vicinity of the respiratory center, central chemoreceptors are sensitive to changes in carbon dioxide-induced hydrogen ion concentration in the brain extracellular fluid.
- Increased partial pressure of carbon dioxide within the brain extracellular fluid causes a corresponding increase in the concentration of hydrogen ions.

Adjustments in ventilation in response to changes in arterial hydrogen ion are important in acid-base balance.

- An elevation in the hydrogen ion concentration in the brain extracellular fluid directly stimulates the central chemoreceptors, which in turn increase ventilation by stimulating the respiratory center.
- A decline in arterial partial pressure of carbon dioxide below normal is paralleled by a fall in partial pressure of carbon dioxide and hydrogen ions in the brain extracellular fluid, the result of which is a central chemoreceptor-mediated decrease in ventilation.
- The powerful influence of the central chemoreceptors on the respiratory center is responsible for your inability to deliberately hold your breath for more than about a minute.
- Carbon dioxide induced changes in the hydrogen ion concentration in the arterial blood are detected by the peripheral chemoreceptors.
- The result is reflex stimulation of ventilation in response to an increase in arterial hydrogen ions and depression of ventilation is association with a decrease in arterial hydrogen ion concentration.

During apnea, a person "forgets to breathe"; during dyspnea, a person feels "short of breath."

- Apnea is the transient cessation of ventilation with the expectation that breathing will resume spontaneously.
- The condition is called "respiratory arrest."
- In exaggerated cases of sleep apnea, the victim may be unable to recover from an apneic period, and death results.
- This is the case of sudden infant death syndrome.
- Most evidence suggests that the baby "forgets to breathe" as a result of the immaturity of the respiratory control mechanisms.
- Certain risk factors make babies more vulnerable to sudden infant death syndrome (SIDS).
- Among them are sleeping position (a higher incidence of SIDS is associated with sleeping on the abdomen rather than on the back or side) and exposure to nicotine during fetal life or after birth.
- People who have dyspnea have the subjective sensation that they are not getting enough air.

Key Terms

2,3-bisphosphoglycerate
Alveolar dead space
Alveolar surface tension
Alveolar ventilation
Alveoli
Anatomical dead space
Apnea
Apneusis
Apneustic center
Asthma
Atmospheric barometric pressure
Boyle's law
Breathing (ventilation)
Bronchi
Bronchioles
Bronchoconstriction
Bronchodilation
Carbamino hemoglobin
Carbon monoxide
Carbonic anhydrase
Carboxyhemoglobin (HbCO)
Carotid bodies
Central chemoreceptors
Chronic bronchitis
Chronic obstructive pulmonary disease (COPD)
Compliance
Deoxyhemoglobin
Dyspnea
Elastic recoil
Emphysema
Expiration
Expiratory reserve volume
External respiration
Forced expiratory volume in one second (FEV1)
Functional residual capacity
Hering-Breuer reflex
Hypercapnia
Hyperoxia
Hyperventilation

Hypocapnia
Hypoventilation
Hypoxia
Inspiration
Inspiratory capacity
Inspiratory reserve volume
Internal (cellular) respiration
Larynx (voice box)
Law of mass action
Lungs
Newborn respiratory distress syndrome
Oxygen-hemoglobin dissociation curve
Oxygen toxicity
Oxyhemoglobin
Partial pressure
Partial pressure gradient
Percent (%Hb) hemoglobin saturation
Pneumotaxic center
Pneumothorax
Pre-Botzinger complex
Pulmonary surfactant
Pulmonary ventilation
Residual volume
Respiration
Respiratory airways
Respiratory arrest
Respiratory rate
Respiratory system
Sleep apnea
Spirogram
Spirometer
Sudden infant death syndrome (SIDS)
Thorax
Tidal volume
Total lung capacity
Transmural pressure gradient
Type I alveolar cells
Type II alveolar cells
Ventilation (breathing)
Vital capacity
Vocal folds

Review Exercises

Answers are in the appendix

True/False-if an answer is false, underline the incorrect portion of the statement and rewrite it to make the statement true.

_____ 1. Pulmonary connective tissue contains large quantities of elastin fibers.

_____ 2. The work of breathing may be increased when pulmonary compliance is increased.

_____ 3. In obstructive lung disease emptying the lungs is made more difficult.

_____ 4. The maximum amount of air that the lungs can hold is about 4.2 liters in males.

_____ 5. The lungs still contain about 2,200 ml of air at the end of a normal quiet expiration.

_____ 6. Alveolar ventilation is more important than pulmonary ventilation.

_____ 7. A gas always diffuses down its partial pressure gradient from the area of higher to the area of lower partial pressure.

_____ 8. The peripheral chemoreceptors are activated when the partial pressure of oxygen in tbe blood drops below 150 mm Hg.

_____ 9. Surface area is reduced in emphysema.

_____ 10. The rate of gas transfer is directly proportional to the diffusion coefficient.

_____ 11. The amount of oxygen transferred to the cells and the amount of carbon dioxide carried away from the cells depend on the rate of cellular metabolism.

_____ 12. The net diffusion of oxygen occurs first between the blood and the tissues.

_____ 13. Respiratory control centers housed in the brain stem are responsible for generating the rhythmic pattern of breathing.

_____ 14. VRG is generally regarded as being responsible for the basic rhythm of ventilation.

_____ 15. The pneumotaxic center is dominant over the apneustic center.

_____ 16. Arterial partial pressure of carbon dioxide is the most important input regulating the magnitude of ventilation under resting conditions.

_____ 17. A fall in arterial partial pressure of carbon dioxide reflexly reduces the respiratory drive.

_____ 18. A decrease in hydrogen ion concentrations in the brain ECF directly stimulates the central chemoreceptors.

_____ 19. The cause of increased ventilation during exercise is still largely speculative.

_____ 20. Hyperventilation affects carbon dioxide levels more significantly than oxygen levels.

_____ 21. 98.5 % of the oxygen in the blood is bound to Hb.

_____ 22. Each hemoglobin molecule can carry up to four molecules of oxygen.

_____ 23. In the systemic capillaries the blood partial pressure of carbon dioxide is increased.

_____ 24. At a normal partial pressure of oxygen of 100 mm Hg, hemoglobin is 98.5% saturated.

_____ 25. Local elevation in temperature enhances oxygen release from hemoglobin for use by the more active tissues.

_____ 26. Carbon dioxide binds with the globin portion of hemoglobin.

_____ 27. Histotoxic hypoxia arises when too little oxygenated blood is delivered to the tissues.

_____ 28. There are no muscles within the alveolar walls.

_____ 29. Air tends to move from a region of lower pressure to a region of higher pressure.

_____ 30. The intra-pleural pressure is greater than the intra-alveolar pressure.

_____ 31. The stretched lungs have a tendency to pull inward.

_____ 32. During inspiration, the intrapleural pressure increases.

_____ 33. At resting expiration, the intra-alveolar pressure decreases.

_____ 34. Expiration is normally a passive process.

_____ 35. If the blood flow is greater than the airflow to a given alveolus, the oxygen level in the alveolus and surrounding tissues will fall below normal.

Fill in the Blank

36. _____ oxygen delivery to the tissues is normal, but the cells are unable to use the oxygen available to them.

37. _____ refers to excess carbon dioxide in the arterial blood that is caused by hypoventilation.

38. The relationship between partial pressure of oxygen and percent hemoglobin saturation is depicted by an S-shaped curve known as the _____.

39. The laryngeal opening is known as the _____

40. The combination of carbon monoxide and hemoglobin is known as _____.

41. The _____ moves downward during inspiration and upward during expiration.

42. _____ is characterized by a low arterial blood partial pressure of oxygen accompanied by inadequate hemoglobin saturation.

43. _____ is characterized by inflammatory fluid accumulation within or around the alveoli.

44. The individual pressure exerted independently by a particular gas within a mixture of gases is known as its _____.

45. A difference in partial pressure between pulmonary blood and alveolar air is known as _____.

46. Across systemic capillaries: oxygen partial pressure gradient from blood to tissue cell is _____ and carbon dioxide partial pressure gradient from tissue cell to blood is _____.

47. _____ is the volume of air in the lungs at the end of a normal passive expiration.

48. Two general categories of respiratory dysfunction yield abnormal results during spirometry: _____ and _____ lung disease.

49. Not all of the inspired air gets down to the site of gas exchange in the alveoli. Part of it remains in the conducting airways, where it is not available for gas exchange. The volume is considered to be _____.

50. _____ is the pressure exerted by the weight of the air in the atmosphere on objects on the earth's surface.

51. _____ refers to decreased airway resistance by promoting bronchiolar smooth muscle relaxation.

52. _____ is characterized by collapse of the smaller airways and a breakdown of alveolar walls.

53. _____ refers to how much effort is required to stretch or distend the lungs.

54. A second factor that contributes to alveolar stability is the _____ of neighboring alveoli.

55. _____ reduces surface tension within the alveoli.

56. _____ is the volume of air entering or leaving the lung during a single breath.

57. The primary respiratory control center is the _____.

58. There are two other respiratory centers higher in the brain stem in the pons:
    (1) _____ and (2) _____.

59. The _____ consists mostly of inspiratory neurons.

60. _____ is the enzyme found inside red blood cells that acts on water and carbon dioxide.

61. Arterial partial pressure of oxygen is monitored by peripheral chemoreceptors known as the _____ and _____.

62. _____ is the transient cessation of ventilation with the expectation that breathing will resume spontaneously.

63. People who have _____ have the subjective sensation that they are not getting enough air.

64. _____ involves the sum of the processes that accomplish ongoing passive movement of oxygen from the atmosphere to the tissues to support cellular metabolism.

65. Two tubes lead from the pharynx: the _____, through which air is conducted to the lungs, and the _____, the tube through which food passes to the stomach.

66. _____ are two bands of elastic tissue that lie across the opening of the larynx, which can be stretched and positioned in different shapes by laryngeal muscles.

67. The alveolar walls consist of a single layer of flattened _____.

68. _____ secrete pulmonary surfactant.

69. _____ chemoreceptors are found in the brain and are sensitive to changes in pH of the CSF. are present in the alveolar walls to permit airflow between adjacent alveoli.

Matching
Match the description of the lung volume or lung capacity to the name of lung volume or capacity.

a. total lung capacity
b. tidal volume
c. vital capacity
d. functional residual capacity
e. inspiratory reserve volume
f. residual volume
g. inspiratory capacity
h. expiratory reserve volume

_____ 70. The vital capacity plus the residual volume

_____ 71. The minimum volume of air remaining in the lungs even after maximal expiration

_____ 72. The extra volume of air that can be maximally inspired over and above the typically resting tidal volume

_____ 73. The maximum volume of air that the lungs can hold

_____ 74. The tidal volume plus the inspiratory reserve volume

_____ 75. The volume of air in the lungs at the end of a normal passive expiration

_____ 76. The volume of air entering or leaving the lungs during a single breath

_____ 77. The extra volume of air that can be actively expired by maximal contraction of the expiratory muscles beyond that which is normally passively expired at the end of a typical resting tidal volume

_____ 78. The tidal volume, plus the inspiratory reserve volume, plus the expiratory reserve volume

_____ 79. The expiratory reserve volume plus the residual volume

_____ 80. The maximum volume of air that can be inspired at the end of a normal quiet expiration

_____ 81. The maximum volume of air that can be moved out during a single breath following a maximal inspiration

Multiple Choice

82. Which condition is characterized by the person "forgetting to breathe"?
    a. newborn respiratory distress syndrome
    b. atelectasis
    c. sudden infant death syndrome
    d. dyspnea
    e. hypercapnia

83. Which of the following conditions is characterized by air entering the pleural cavity?
    a. apnea
    b. pleurisy
    c. dyspnea
    d. asthma
    e. pneumothorax

84. Which of the following conditions refers to insufficient oxygen at the cellular level?
    a. hypoxia
    b. hypercapnia
    c. hyperpnea
    d. emphysema
    e. atelectasis

85. Which of the following conditions indicates that the partial pressure of carbon dioxide is above normal in arterial blood?
    a. hyperoxia
    b. hypercapnia
    c. hyperpnea
    d. atelectasis
    e. hypercarbonium

86. If the intra-pulmonary pressure is 756 mm Hg what is the intrapleural pressure (in mm Hg)?
    a. 756
    b. 760
    c. 764
    d. 0
    e. 752

87. Which of the following conditions is characterized by the lungs collapsing to their unstretched size?
    a. pleurisy
    b. atelectasis
    c. emphysema
    d. dyspnea
    e. pneumothorax

88. Which of the following refers to an inflammation of the pleural sac?
    a. pleuritis
    b. pneumothorax
    c. pleurisy
    d. emphysema
    e. hyperpnea

89. As the thoracic cavity enlarges, the lungs?
    a. flatten
    b. collapse
    c. do not change
    d. expel air
    e. enlarge

90. Which of the following conditions refers to a situation in alveolar surface tension due to insufficient pulmonary surfactant?
    a. sudden infant death syndrome
    b. newborn respiratory distress syndrome
    c. pleurisy
    d. pleuritis
    e. pulmonary distress syndrome

91. Intra-alveolar pressure can be changed by altering the volume of the lungs, in accordance with: ?
    a. Boyle's Law
    b. Law of Mass Action
    c. Bohr effect
    d. Herring-Breuer reflex
    e. none of the above

92. Which of the following conditions refers to increased ventilation to meet increased oxygen needs?
    a. hypopnea
    b. apnea
    c. dyspnea
    d. hyperpnea
    e. hypoxia

93. Which of the following conditions is characterized by inflammation of the lower respiratory airways?
    a. dyspnea
    b. chronic bronchitis
    c. asthma
    d. pleuritis
    e. pleurisy

94. Which of the following conditions is a subjective sensation of not getting enough air?
    a. dyspnea
    b. emphysema
    c. apnea
    d. hyperpnea
    e. pleurisy

95. At high altitudes:
    a. The alveolar P_{O2} is higher than normal.
    b. The alveolar P_{O2} is lower than normal.
    c. The alveolar P_{CO2} is higher than normal.
    d. Both a and c above are correct.
    e. Both b and c above are correct.

96. The receptors that are stimulated by a large drop in the blood partial pressure of oxygen are located where?
    a. In the respiratory center of the brain.
    b. In the carotid and aortic bodies.
    c. In the tissue capillaries.
    d. Two of the above are correct.
    e. All of the above (a-c) are correct.

97. The apneustic center:
    a. is located in the medulla
    b. stimulates the inspiratory neurons
    c. inhibits inspiratory activity
    d. both a and b above
    e. both a and c above

98. Hypercapnia:
    a. refers to excess carbon dioxide in the arterial blood
    b. occurs when carbon dioxide is blown off to the
       atmosphere at a rate faster than it is being produced by
       the tissues
    c. always accompanies hypoxia
    d. two of the above
    e. all of the above (a-c)

99. A person with asthma:?
    a. can expire normally
    b. has trouble with expiration because the small airways
       tend to collapse
    c. has trouble only with inspiration
    d. would show hypocapnia
    e. none of the above

100. The normal percent saturation of hemoglobin in venous
     blood is:
     a. 97%
     b. 75%
     c. 50%
     d. 40%
     e. 10%

*Indicate what each of the following conditions would have on
oxygen delivery to tissue. by writing the appropriate letter in the
blank using the following answer code.*

> A = increases
> B = decreases
> C = has no effect on

101. _____ breathing air with 20% oxygen
102. _____ elevated temperature around tissue
103. _____ carbon monoxide poisoning
104. _____ moderate increase in carbon dioxide production by
       tissue
105. _____ hemoglobin deficiency
106. _____ elevation in, 2,3-diphosphoglycerate levels

Points to Ponder

1. The lung of a frog is quite different when compared to that
   of a human. The frog lung is a thin cylindrical tube much
   like a finger on a rubber glove. How does this very simple
   structure adequately serve the frog, whereas humans require
   millions of alveoli to survive?

2. How would you explain the position of the trachea being ventral to the esophagus: the food must pass over the trachea to be swallowed?

3. What should you do when a small child declares that he will hold his breath until he dies if his demands are not met?

4. Why do some athletes breathe pure oxygen on the sidelines?

5. Why do we yawn?

6. Why is it important for the respiratory system to have a dual blood supply?

7. Why is it important that the capillaries surrounding the alveoli have a small diameter?

8. Under normal conditions, why is the first breath of life difficult?

9. After driving from sea level to a trail high in the Sierras, you get out of your SUV and feel dizzy.
What do you suppose is causing your dizziness?

10. Nicotine from cigarette smoke causes the buildup of mucus and paralyzes the ciliated epithelial cells that line the bronchioles. How might these conditions affect pulmonary function tests?

Clinical Perspectives

1. How do you avoid carbon monoxide in a closed environment?

2. How would you relate drowning to this chapter?

3. How does taking a dive in a hyperbaric chamber enhance treatment against certain anaerobic bacteria?

4. You have just come upon an accident victim. After summoning help you see that the victim has a pneumothorax. How could you possibly help this person until medical help arrives?

5. Your college roommate has just returned from a football game and is very hoarse from all of the cheering he did. He knows you are taking a physiology course and wants to know why he is hoarse.   How would you explain to him that yelling makes us hoarse?

6. Are common colds spread mostly through sneezes? Explain why or why not.

7. Why is the disease asthma often treated with glucocorticoid drugs?

8. When or under what conditions would the use of hyperbaric oxygen therapy be useful?

9. People who hyperventilate during psychological stress are sometimes told to breathe into a paper bag? What is the physiological basis for this act?

10. Your friend asks you to explain to her what "mountain sickness" is. How would you explain this to her?

Experiments of the Day

1. Exhale normally, then exhale a little more and a little more until you can no longer exhale. From which of the lung volumes or capacities are you exhaling? Repeat using inhalation until you can no longer inhale. Into which lung volumes or capacities are you inhaling?

2. Do some exercise until you are ventilating rapidly. Place a bag over your mouth and nose. What are the results and why?

The Urinary System

Chapter Overview

The survival and proper functioning of cells depend on the maintenance of stable concentrations of salt, acids, and other electrolytes in the internal environment. Cell survival also depends on the continual removal of toxic metabolic wastes produced by the cells as they perform life-sustaining chemical reactions. The urinary system, more specifically the kidneys, is a major contributor to the phenomenon of multicellular organisms. The existence of multicellular organisms depends on the presence of the extracellular fluid. Since virtually all cells interface with the extracellular fluid, the homeostasis of this fluid is of utmost importance. For the most part, it is the role of the kidney to provide this homeostasis.

To fulfill this role the kidneys regulate most extracellular fluid ions; maintain pH, plasma volume, water balance, and osmolarity; eliminate waste products of metabolism and many foreign compounds; and in a secondary capacity, produce erythropoietin and convert vitamin D into its active form.

While the functions of the kidneys can be listed quite simply in a short paragraph the physiology of the kidney is extremely complex. The kidneys are true guardians of homeostasis in the multicellular organism.

Chapter Outline

INTRODUCTION
The kidneys perform a variety of functions aimed at maintaining homeostasis.

- The individual cells of complex multicellular organisms are able to function and survive only in a fluid environment of essentially constant composition similar to the sea.
- This salty internal fluid environment is the extracellular fluid that bathes all the cells of the body and must be homeostatically maintained.
- Terrestrial animals are able to live on dry land independent of the sea because of their kidneys, the organs that, in concert with the hormonal and neural inputs that control their function, are primarily responsible for maintaining the stability of extracellular fluid volume and electrolyte composition.
- The kidneys are the primary route for elimination of potentially toxic metabolic wastes and foreign compounds from the body.
- The kidneys make adjustments in urinary output of water, salt, and other electrolytes to compensate for their abnormal losses through heavy sweating, vomiting, diarrhea, or hemorrhage.
- The following are specific functions performed by the kidneys, (1) maintain water balance in the body, (2) regulate the quality and concentration of most ECF ions, (3) maintain proper plasma volume, (4) assist in maintaining the proper acid-base balance, (5) maintain proper osmolarity, (6) excrete end products of bodily metabolism, (7) excrete many foreign compounds, (8) secrete erythropoietin, (9) secrete renin, and (10) convert vitamin D into its active form.

The kidneys form urine; the remainder of the urinary system carries the urine to the outside.

- The urinary system consists of the urine-forming organs—the kidneys—and the structures that carry the urine from the kidneys to the outside for elimination from the body.
- The kidney acts on the plasma flowing through it to produce urine, conserving materials to be retained in the body and eliminating unwanted materials into the urine.
- The renal artery carries blood into the kidney, the renal vein carries blood out of the kidney.

The nephron is the functional unit of the kidney.

- Each kidney is composed of about one million microscopic functional units known as "nephrons."
- A nephron is the smallest unit capable of urine formation.
- The arrangement of nephrons within the kidneys gives rise to two distinct regions: an outer granular-appearing renal cortex, and an inner region, the renal medulla.
- Each nephron consists of a vascular component and a tubular component.
- The dominant portion of the vascular component is the glomerulus.
- On entering the kidney, the renal artery systematically subdivides to ultimately form many small vessels known as "afferent arterioles," one of which supplies each nephron.
- The afferent arteriole delivers blood to the glomerular capillaries, which rejoin to form another arteriole, the efferent arteriole.
- No oxygen or nutrients are extracted from the blood for use by the kidney tissues nor are waste products picked up from the surrounding tissue.
- The efferent arteriole quickly divides into a second set of capillaries, the peritubular capillaries, which supply the renal tissue with blood and are important in exchanges between the tubular system and blood during conversion of the filtered fluid into urine.
- The peritubular capillaries rejoin to form venules that ultimately drain into the renal vein, by which the blood leaves the kidney.
- The tubular component begins with Bowman's capsule.
- From Bowman's capsule, the filtered fluid passes into the proximal tubule, which lies entirely in the cortex.
- The next segment, the loop of Henle, dips into the renal medulla.
- The descending limb of Henle's loop plunges from the cortex into the medulla; the ascending limb traverses back up into the cortex.
- The ascending limb passes through the fork formed by the afferent and efferent arterioles.
- Both the tubular and vascular cells at this point are specialized to form the juxtaglomerular apparatus.
- Beyond the juxtaglomerular apparatus, the tubule once again becomes highly coiled to form the distal tubule.
- The distal tubule empties into a collecting tubule.
- Each collecting duct or tubule plunges down through the medulla to empty its fluid contents into the renal pelvis.

The three basic renal processes are glomerular filtration, tubular reabsorption, and tubular secretion.

- Glomerular filtration is the first step in urine formation.
- The filtrate flows through the tubules.
- Selective movement of substances from inside the tubule into the blood is referred to as "tubular reabsorption."
- Tubular secretion refers to the selective transfer of substances from the peritubular capillary blood into the tubular lumen.
- Urine excretion refers to the elimination of substances from the body in the urine and is the result of the first three processes.

GLOMERULAR FILTRATION

The glomerular membrane is considerably more permeable than capillaries elsewhere.

- Fluid filtered from the glomerulus into Bowman's capsule must pass through three layers that make up the glomerular membrane: (1) the wall of the glomerular capillaries, (2) an acellular gelatinous layer known as the "basement membrane," and (3) the inner layer of Bowman's capsule
- The glomerular capillary wall is perforated by many large pores or fenestrae.
- A basement membrane composed of collagen and glycoproteins is sandwiched between the glomerulus and Bowman's capsule.
- The final layer consists of podocytes, octopuslike cells that encircle the glomerular tuft.
- The narrow slits between adjacent foot processes, known as filtration slits, provide a pathway through which fluid leaving the glomerular capillaries can enter the lumen of Bowman's capsule..Collectively, these layers function as a fine molecular sieve that retains the blood cells and plasma proteins but permits H_2O and solutes to filter through

Glomerular capillary blood pressure is the major force that causes glomerular filtration.

- No active transport mechanisms or local energy expenditures are involved in moving fluid from the plasma across the glomerular membrane.
- Passive physical forces are responsible for glomerular filtration.
- Three physical forces are involved in glomerular filtration: (1) glomerular capillary blood pressure, (2) plasma-colloid osmotic pressure, and (3) Bowman's capsule hydrostatic pressure.
- The glomerular capillary blood pressure is higher than capillary blood pressure elsewhere.
- Plasma-colloid osmotic pressure is caused by the unequal distribution of plasma proteins across the glomerular membrane.
- The resultant tendency for water to move by osmosis down its own concentration gradient from Bowman's capsule into the glomerulus opposes glomerular filtration.
- The fluid in Bowman's capsule exerts a hydrostatic pressure.
- This pressure, which tends to push fluid out of Bowman's capsule, opposes glomerular filtration.
- The total force favoring filtration is attributable to the glomerular capillary blood pressure.
- The total of the two forces opposing filtration is less.
- The net difference favoring filtration is referred to as the "net filtration pressure."

Changes in GFR result primarily from changes in glomerular capillary blood pressure.

The glomerular filtration rate (GFR) is controlled by two mechanisms: (1) autoregulation, which is aimed at preventing spontaneous changes in the GFRI; and (2) extrinsic sympathetic control, which is aimed at long-termed regulation of arterial blood pressure.

- The GFR can be changed on purpose by extrinsic-control mechanisms that override the autoregulatory responses.
- Extrinsic control of GFR, which is mediated by sympathetic nervous system input to the afferent arterioles, is aimed at the regulation of arterial blood pressure.
- The parasympathetic nervous system does not exert any influence on the kidneys.
- No new mechanism is required to decrease the GFR.
- It is reduced as a result of the baroreceptor reflex response to a fall in blood pressure.
- The afferent arterioles constrict as a result of increased sympathetic activity.

- The resultant decrease in GFR in turn leads to a reduction in urine volume.
- When the baroreceptors detect a rise in blood pressure, sympathetic vasoconstrictor activity to the arterioles, including the renal afferent arterioles, is reflexly reduced, allowing afferent arteriolar vasodilation to occur.
- As more blood enters the glomeruli through the dilated afferent arterioles, glomerular capillary blood pressure rises, increasing the GFR.

The kidneys normally receive 20% to 25% of the cardiac output.

- Most of the blood goes to the kidneys not to supply the renal tissue but to be adjusted and purified by the kidneys.
- Only by continuously processing such a large proportion of the blood are the kidneys able to precisely regulate the volume and electrolyte composition of the internal environment and adequately eliminate large quantities of metabolic waste products that are constantly produced.

TUBULAR REABSORPTION
Tubular reabsorption is tremendous, highly selective, and variable.

- Tubular reabsorption is a highly selective process.
- The tubules have a high reabsorptive capacity for substances needed by the body and a poor or no reabsorptive capacity for substances of no value.
- Only a small percentage of filtered plasma constituents that are useful to the body are present in the urine, most having been reabsorbed and returned to the blood.
- Only excess amounts of essential materials such as electrolytes are excreted in the urine.
- As water and other valuable constituents are reabsorbed, the waste products remaining in the tubular fluid become highly concentrated..Of the 125 ml/min filtered, typically 124 ml/min are reabsorbed.

Tubular reabsorption involves transepithelial transport.

- To be reabsorbed, a substance must traverse five distinct barriers: (1) It must leave the tubular fluid by crossing the luminal membrane of the tubular cell. (2) It must pass through the cytosol from one side of the tubular cell to the other. (3) It must traverse the basolateral membrane of the tubular cell to enter the interstitial fluid. (4) It must diffuse through the interstitial fluid. (5) It must penetrate the capillary wall to enter the blood plasma.
- This entire sequence is known as "transepithelial transport."
- There are two types of tubular reabsorption: passive reabsorption and active reabsorption.

An active sodium-potassium ATPase pump in the basolateral membrane is essential for sodium reabsorption.

- Of the Na^+ filtered, 99.5% is normally reabsorbed
- Sodium reabsorption in the proximal tubule plays a pivotal role in the reabsorption of glucose, amino acids, water, chloride ions, and urea.
- Sodium reabsorption in the loop of Henle, along with chloride ion reabsorption, plays a critical role in the kidneys' ability to produce urine of varying concentrations and volumes, depending on the body's need to conserve or eliminate water.
- Sodium reabsorption in the distal portions of the nephron is variable and subject to hormonal control, being important in the regulation of ECF volume.
- Sodium reabsorption is also linked in part with potassium and hydrogen ion secretion.

Aldosterone stimulates sodium reabsorption in the distal and collecting tubules.

- The most important and best known hormonal system involved in the regulation of sodium is the renin-angiotensin-aldosterone system, which

stimulates sodium reabsorption in the distal and collecting tubules.
- Cells of the juxtaglomerular apparatus secrete a hormone, renin, into the blood.
- Renin acts as an enzyme to activate angiotensinogen into angiotensin I.
- On passing through the lungs, angiotensin I is converted to angiotensin II.
- Angiotensin II is the primary stimulus for the secretion of the hormone aldosterone.
- Aldosterone increases sodium reabsorption by the distal and collecting tubules.
- The renin-angiotensin-aldosterone system thus promotes salt retention and a resultant water retention, and elevation of arterial blood pressure.
- Diuretics are therapeutic agents that cause diuresis.
- ACE inhibitor drugs block the action of angiotensin-converting enzyme.

Atrial natriuretic peptide inhibits sodium reabsorption.
- Atrial natriuretic peptide is released from the cardiac atria when the ECF volume is expanded.
- The primary action of atrial natriuretic peptide is to inhibit sodium reabsorption in the distal parts of the nephron.
- This increases Na^+ excretion thereby promoting accompanying osmotic H_2O excretion in the urine

Glucose and amino acids are reabsorbed by sodium-dependent secondary active transport.
- Glucose and amino acids are transferred by means of secondary active transport by symport carriers.
- These symport carriers, such as the sodium and glucose cotransporter (SGLT) simultaneously transfer both Na^+ and the specific organic molecule from the lumen into the cell Once transported into the tubular cells, glucose and amino acids passively diffuse down their concentration gradients across the basolateral membrane into the plasma, facilitated by a carrier, such as the *glucose transporter (GLUT)*.
- Glucose and amino acids get a "free ride" at the expense of energy already used in the reabsorption of sodium.

In general, actively reabsorbed substances exhibit a tubular maximum.
- The tubular maximum in the kidney is the maximum amount of a substance that tubular cells can actively transport within a given time period.
- The quantity of any substance filtered per minute is known as the "filtered load."

Glucose is an example of an actively reabsorbed substance that is not regulated by the kidneys.
- At a constant GFR, the glucose filtered load is directly proportional to the plasma glucose concentration.
- The plasma glucose concentration can become extremely high in diabetes mellitus.
- Consequently, although glucose does not normally appear in the urine, it is found in the urine of persons with diabetes when the plasma glucose concentration exceeds the renal threshold.

Phosphate is an example of an actively reabsorbed substance that is regulated by the kidneys.
- The kidneys do directly contribute to the regulation of many electrolytes such as calcium and phosphate.
- Our diets are generally rich in phosphates, but because the tubules can reabsorb up to the normal plasma concentration's worth of phosphate and no more, the excess ingested phosphate is quickly spilled into the urine.
- Unlike the reabsorption or organic nutrients, the reabsorption of phosphate and calcium is also subject to hormonal control.

- Parathyroid hormone can alter the renal thresholds for phosphate and calcium, thus adjusting the quantity of these electrolytes conserved depending on the body's momentary needs.

Active sodium reabsorption is responsible for the passive reabsorption of chloride ions, water, and urea.

- The negatively charged chloride ions are passively reabsorbed down the electrical gradient created by the active reabsorption of positively charged sodium ions.
- For the most part, chloride ions pass between, not through, the tubular cells. Water is passively reabsorbed by osmosis throughout the length of the tubule.
- Water passes through recently discovered water channels, or aquaporins, formed by specific plasma membrane proteins in the tubular cells.
- No part of the tubule directly requires energy for this tremendous reabsorption of H_2O.
- The passive reabsorption of urea is also indirectly linked to active sodium reabsorption. About 50% of the filtered urea is passively reabsorbed

In general, unwanted waste products are not reabsorbed.

- Urea molecules, being the smallest of waste products, are the only wastes able to be passively reabsorbed as a result of the concentrating effect of the kidneys.
- The waste products, failing to be reabsorbed, generally remain in the tubules and are excreted in the urine in a highly concentrated form.

TUBULAR SECRETION

Hydrogen ion secretion is important in acid-base balance.

- The most important substances secreted by the tubules are hydrogen ions, potassium ions, and organic anions and cations, many of which are compounds foreign to the body.
- Renal hydrogen ion secretion is extremely important in the regulation of acid-base balance in the body.

- Hydrogen ions can be added to the filtered fluid by being secreted by the proximal, distal, and collecting tubules.

Potassium ion secretion is controlled by aldosterone.

- Potassium concentrations are tightly controlled because small changes in K+ levels in the ECF can have significant influence on membrane excitability.
- Potassium is actively reabsorbed in the proximal tubule and actively secreted in the distal and collecting tubules.
- Potassium-ion secretion in the distal and collecting tubules is coupled to sodium ion reabsorption by means of the energy-dependent basolateral sodium-potassium pump.
- Increased aldosterone secretion always promotes simultaneous sodium reabsorption and potassium secretion.

Organic anion and cation secretion helps efficiently eliminate foreign compounds from the body.

- The proximal tubule contains two distinct types of secretory carriers, one for the secretion or organic anions and a separate system for the secretion of organic cations.
- The ability of the organic-ion secretory systems to eliminate many foreign compounds from the body is most important.
- Nonselectivity permits the organic-ion secretory systems to hasten the removal of many foreign organic chemicals, including food additives, environmental pollutants, drugs, and other non-nutritive organic substances that have gained entrance to the body.

URINE EXCRETION AND PLASMA CLEARANCE

- On the average, one milliliter of urine is excreted per minute.
- Typically, of the 125 ml of plasma filtered per minute, 124 ml/min are reabsorbed, so the final quantity of urine formed averages 1 ml/min.

- Thus, 1.5 liter of urine per day are excreted, out of the 180 liters per day filtered.
- A relatively small change in the quantity of filtrate reabsorbed can bring about a large change in the volume of urine formed.

Plasma clearance refers to the volume of plasma cleared of a particular substance per minute.

- Clearance refers not to the amount of the substance removed but to the volume of plasma from which that amount was removed
- If a substance is filtered but not reabsorbed or secreted, its plasma clearance rate equals the GFR.
- If a substance is filtered and reabsorbed but not secreted, its plasma clearance rate is always less than the GFR.
- If a substance is filtered and secreted but not reabsorbed, its plasma clearance rate is always greater than the GFR.
- If a substance is filtered but not reabsorbed or secreted, its plasma clearance rate equals the GFR.
- If a substance is filtered and reabsorbed but not secreted, its plasma clearance rate is always less than the GFR.
- If a substance is filtered and secreted but not reabsorbed, its plasma clearance is always greater than the GFR.

The kidneys can excrete urine of varying concentrations depending on the body's state of hydration.

- The ECF osmolarity depends on the relative amount of water compared to solute.
- At normal fluid balance and solute concentration, the body fluids are said to be isotonic.
- If there is too much water relative to the solute load, the body fluids are hypotonic.
- If a water deficit exists relative to the solute load, the body fluids are too concentrated, or are hypertonic.

The medullary vertical osmotic gradient is established by means of countercurrent multiplication.

- A large vertical osmotic gradient is uniquely maintained in the interstitial fluid of the medulla of each kidney.
- The concentration of the interstitial fluid progressively increases from the cortical boundary down through the depth of the renal medulla until it reaches a maximum of 1,200 mosm/liter in humans, at the junction with the renal pelvis.
- The presence of this gradient enables the kidneys to produce urine that ranges in concentration from 100 to 1,200 mosm/liter.
- When the body is in ideal fluid balance, isotonic urine is formed.
- When the body is overhydrated, the kidneys are able to produce a large volume of dilute urine thus eliminating the excess water into the urine.
- Conversely, the kidneys are able to put out a small volume of concentrated urine when the body is dehydrated.
- The juxtamedullary nephrons' long loops of Henle establish the vertical gradient. Their vasa recta prevent the dissolution of this gradient while providing blood to the renal medulla.

Vasopressin-controlled, variable water reabsorption occurs in the final tubular segments.

- The collecting tubules of all nephrons use the gradient, in conjunction with the hormone vasopressin, to produce urine of varying concentrations.
- Vasopressin causes cells in the distal tubules and collecting ducts to insert more water channels into their membranes to increase water permeability.
- Collectively, this entire functional organization is known as the medullary countercurrent system.
- The long loops of Henle of juxtamedullary nephrons establish the medullary vertical osmotic gradient by means of countercurrent multiplication.

- The descending limb is highly permeable to water but does not actively extrude sodium.
- The ascending limb actively transports NaCl out of the tubular lumen into the surrounding interstitial fluid and is always impermeable to water, so salt leaves the tubular fluid without water osmotically following along.
- The vertical osmotic gradient in the medullary interstitial fluid is used by the collecting ducts to concentrate the tubular fluid so that urine more concentrated than normal body fluids can be excreted.
- The fact that the fluid is hypotonic as it enters the distal portions of the tubule enables the kidneys to excrete urine more dilute than normal body fluids.
- The active salt pump in the ascending limb is able to transport sodium chloride out of the lumen until the surrounding interstitial fluid is 200 mosm/liter more concentrated than the tubular fluid in this limb.
- When the ascending limb pump starts actively extruding salt, the medullary interstitial fluid becomes hypertonic.
- Water cannot follow osmotically from the ascending limb because this limb is impermeable to water.
- Net diffusion of water does occur from the descending limb into the interstitial fluid.
- The tubular fluid entering the descending limb from the proximal tubule is isotonic.
- Because the descending limb is highly permeable to water, net diffusion of water occurs by osmosis out of the descending limb into the more concentrated interstitial fluid.
- The tubular fluid entering the loop of Henle immediately starts to become more concentrated as it loses water.
- The concentration of tubular fluid is progressively increasing in the descending limb and progressively descending in the ascending limb.
- The fluid in the descending limb becomes progressively more hypertonic until it reaches a maximum

concentration of 1200 mosm/liter at the bottom of the loop.
- The tubular fluid becomes hypotonic as it leaves the ascending limb to enter the distal tubule at a concentration of 100 mosm/liter.

Renal failure has wide-ranging consequences.
- When the functions of both kidneys are disrupted to the point that they are unable to perform their regulatory and excretory functions sufficiently to maintain homeostasis, renal failure is said to exist.
- Among the causes of renal failure are: (1) infectious organisms, (2) toxic agents, (3) inappropriate immune responses, (4) obstruction of urine flow, and (5) an insufficient renal blood supply.
- Acute renal failure is reversible; chronic renal failure is not.

Urine is temporarily stored in the bladder, from which it is emptied by micturition.
- Peristaltic contractions of the smooth muscles within the ureteral wall propel the urine forward from the kidneys to the bladder.
- As the bladder fills, the ureteral ends within its walls are compressed closed.
- Urine can still enter because ureteral contractions generate sufficient pressure to overcome the resistance.
- As is characteristic of smooth muscle, bladder muscle is able to stretch tremendously without a build up in bladder wall tension.
- The bladder smooth muscle is richly supplied by parasympathetic fibers, stimulation of which causes bladder contraction.
- The exit from the bladder is guarded by two sphincters: the internal urethral sphincter and the external urethral sphincter.
- The internal urethral sphincter is under involuntary control.
- The external urethral sphincter and pelvic diaphragm are under voluntary control.

- Micturition, the process of bladder emptying, is governed by two mechanisms: the micturition reflex and voluntary control.
- Urinary incontinence, or inability to prevent discharge of urine, occurs when descending pathways in the spinal cord that mediate voluntary control of the external sphincter are disrupted

Key Terms

ACE inhibitor drugs
Active reabsorption
Afferent arterioles
Angiotensinogen
Angiotensinogen I
Angiotensinogen II
Antidiuretic hormone
Aquaporins (water channels)
Atrial natriuretic peptide (ANP)
Blood urea nitrogen (BUN)
Bowman's capsule
Collecting tubule (duct)
Distal tubule
Diuretics
Efferent arterioles
Filtration fraction
Filtration slits
Glomerular filtration
Glomerular filtration rate (GFR)
Glomerular membrane
Glomerulus
Hypertonic
Hypotonic
Inulin
Juxtaglomerular apparatus
Kidneys
Lateral spaces
Loop of Henle
Medullary countercurrent system
Micturition (urination)
Micturition reflex
Nephrons
Net filtration pressure
Para-aminohippuric acid (PAH)
Passive reabsorption
Peritubular capillaries
Plasma clearance
Podocytes

Proximal tubule
Renal cortex
Renal failure
Renal medulla
Renal pelvis
Renal pyramids
Renal threshold
Renin-angiotensin-aldosterone system (RAAS)
Sodium and glucose cotransporter (SGLT)
Sodium load
Transepithelial transport
Tubular maximum (T_m)
Tubular reabsorption
Tubular secretion
Ureter
Urethra
Urinary bladder
Urinary incontinence
Urinary system
Urine excretion
Vasopressin (antidiuretic hormone)
Water channels (aquaporins)

Review Exercises

Answers are in the appendix.

True/False-if an answer is false, underline the incorrect portion of the statement and rewrite it to make the statement true.

_____   1.   The afferent arteriole delivers blood to the glomerular capillaries.

_____   2.   The efferent arteriole carries venous blood to the renal vein to exit the kidney.

_____   3.   The peritubular capillaries supply the renal tissue with oxygen.

_____   4.   Glomerular capillaries contain filtration slits to increase permeability.

_____   5.   ANP increases sodium reabsorption.

_____   6.   The GFR is dependent upon the net filtration pressure only.

_____   7.   Under normal conditions, 0.5% off all sodium filtered is excreted in the urine.

_____   8.   Aldosterone stimulates sodium reabsorption in the collecting tubules.

_____   9.   50% of the urea that is filtered is reabsorbed.

_____   10.   The filtered load is known as the level at which the tubular maximum has been reached and the particular substance begins to appear in the urine.

_____   11.   The negatively charged chloride ions are passively reabsorbed in the proximal tubule.

_____   12.   Urea is passively reabsorbed in the distal and collecting tubule.

_____   13.   Organic cations are secreted by the proximal tubules.

_____   14.   Water channels are stored in vesicles and inserted into the membrane of distal tubule cells and collecting ducts when stimulated by vasopressin.

_____   15.   The most important substances secreted by the distal and collecting tubules are sodium and potassium.

_____   16.   Potassium is actively reabsorbed in the proximal tubule.

_____   17.   Potassium is actively secreted in the distal and collecting tubules under the control of vasopressin.

_____   18.   The Na-K pumps are located on the basolateral surface of tubules cells.

_____   19.   If a substance is filtered and reabsorbed but not secreted, its plasma clearance is always   greater than the GFR.

_____   20.   If a substance is filtered and secreted but not reabsorbed, its plasma clearance is always   less than the GFR.

_____ 21. If a substance is filtered but not secreted, its plasma clearance rate equals the GFR.

_____ 22. Vasopressin is also known as aldosterone.

Fill in the Blank

23. _____ is a waste product resulting from the breakdown of proteins.

24. The passive reabsorption of urea is also directly linked to active _____ reabsorption.

25. The parathyroid hormone can alter renal thresholds for _____.

26. Major metabolic wastes excreted by the kidney include _____, _____, _____, and _____

27. Fluid retention and edema that accompany congestive heart failure are brought about by _____ activity.

28. Renin acts as an enzyme to activate angiotensinogen into _____.

29. _____ is the primary stimulus for the secretion of aldosterone.

30. _____ increases sodium reabsorption by the distal and collecting tubules.

31. Tubular reabsorption involves _____ transport.

32. _____ channels in the distal tubules and collecting ducts are always open.

33. _____ renal failure is reversible whereas _____ renal failure is not.

34. From Bowman's capsule, the filtered fluid passes into the _____ tubule.

35. Solute excretion is always accompanied by comparable water excretion because of _____ considerations.

36. The early portion of the collecting tubule is _____ to urea.

37. Causes of renal failure are: a. _____, b. _____, c. _____, d. _____, and e. _____.

38. _____ contractions of the smooth muscle within the ureteral wall propel urine forward from the kidneys to the bladder.

39. The exit from the bladder is guarded by two sphincters: _____ and _____.

40. The passive exchange of solutes and water between the two limbs of the vasa recta and the interstitial fluid is known as _____.

41. Body fluids are said to be isotonic at an osmolarity of _____ mosm/l.

42. Potassium-ion secretion in the distal and collecting tubules is coupled to sodium reabsorption by means of the energy-dependent basolateral _____.

43. Increased aldosterone secretion always promotes simultaneous sodium reabsorption and

    _____.

Matching
Using the code below match the reabsorption or secretion function to all the tubules that perform the function. Some questions might have more than one correct answer.

a. proximal tubule
b. distal tubule
c. collecting tubule

_____ 44.   Glucose might normally be found in the filtrate here.

_____ 45.   Variable water reabsorption, controlled by vasopressin

_____ 46.   Variable hydrogen ion secretion, depending on acid-base status of the body

_____ 47.   All filtered glucose and amino acids reabsorbed by secondary active transport; not subject to control

_____ 48.   Variable sodium reabsorption, controlled by aldosterone; chloride follows passively

_____ 49.   All filtered potassium reabsorbed; not subject to control

_____ 50.   50% of filtered urea passively reabsorbed; not subject to control

_____ 51.   Variable amounts of phosphate ions and other electrolytes reabsorbed; subject to control

_____ 52.   Variable potassium secretion, depending on aldosterone

_____ 53.   67% of filtered sodium actively reabsorbed; not subject to control; chloride follows passively

_____ 54.   65% of filtered water osmotically reabsorbed; not subject to control

Multiple Choice

55. Which of the following substances is secreted by the kidneys and stimulates red blood cell production?
    a. aldosterone
    b. vasopressin
    c. para-aminohippuric acid
    d. renin
    e. erythropoietin

56. Which substance is secreted by the hypothalamus, stored in the posterior pituitary, and stimulates the reabsorption of water in the collecting tubules?
    a. aldosterone
    b. angiotensin I
    c. vasopressin
    d. para-aminohippuric acid
    e. inulin

57. Which of the following substances causes sodium reabsorption and potassium secretion?
    a. aldosterone
    b. inulin
    c. renin
    d. atrial natriuretic peptide
    e. vasopressin

58. Which of the following substances is the primary stimulus for the secretion of the hormone aldosterone?
    a. angiotensin I
    b. renin
    c. inulin
    d. angiotensin II
    e. atrial natriuretic peptide

59. Which substance inhibits sodium reabsorption in the distal parts of the nephron?
    a. atrial natriuretic peptide
    b. para-aminohippuric acid
    c. vasopressin
    d. aldosterone
    e. renin

60. Which of the following substances is a hormone of the kidney?
    a. renin
    b. aldosterone
    c. inulin
    d. vasopressin
    e. erythropoietin

61. Which substance functions as an enzyme to activate angiostensinogen into angiotensin I?
    a. aldosterone
    b. vasopressin
    c. inulin
    d. renin
    e. para-hippuric acid

62. Which of the following is the substrate for angiotensin converting enzyme?
    a. angiotensin I
    b. para-aminohippuric acid
    c. angiotensin II
    d. aldosterone
    e. vasopressin

63. Which of these substances must pass through the lungs before they can become a stimulus for hormone release?
    a. angiotensin I
    b. angiotensin II
    c. angiotensinogen
    d. inulin
    e. aldosterone

64. Which of the following substances increases sodium reabsorption by the distal tubule?
    a. renin
    b. angiotensin I
    c. para-aminohippuric acid
    d. atrial natriuretic peptide
    e. aldosterone

65. What is the concentration of the tubular fluid as it leaves the ascending limb and enters the distal tubule?
    a. 200 mosm/liter
    b. 600 mosm/liter
    c. 100 mosm/liter
    d. 1,200 mosm/liter
    e. 800 mosm/liter

66. With maximum levels of vasopressin, what is the concentration of the tubular fluid at the end of the collecting tubules?
    a. 200 mosm/liter
    b. 600 mosm/liter
    c. 100 mosm/liter
    d. 1,200 mosm/liter
    e. 800 mosm/liter

67. What is the minimum volume of urine required to excrete wastes each day?
    a. 1 ml
    b. 50 ml
    c. 500 ml
    d. 600 ml
    e. 1,200 ml

68. In which tubules or renal structures does vasopressin influence water permeability?
    a. proximal tubules and loop of Henle
    b. descending and ascending limbs of loop of Henle
    c. descending limb or loop of Henle and distal tubule
    d. loop of Henle only
    e. distal and collecting tubules

69. Which of the following is not a function of the kidneys?
    a. excrete the end products of bodily metabolism
    b. maintain proper plasma volume
    c. regulate blood glucose levels
    d. maintain proper osmolarity of body fluids
    e. assist in maintaining the proper acid-base balance of the body

70. The functional unit of the kidney is:
    a. the renal medulla
    b. the nephron
    c. the countercurrent system
    d. the loop of Henle
    e. the glomerulus

71. Which of the following is not part of the nephron?
    a. the glomerulus
    b. the proximal tubule
    c. the renal artery
    d. the collecting duct
    e. Bowman's capsule

72. The peritubular capillaries:
    a. supply nutrients and oxygen to the tubular cells
    b. take up the substances that are reabsorbed by the tubules
    c. supply substances that are secreted by the tubules
    d. all of the above
    e. none of the above

73. The blood that flows through the kidneys:
    a. is normally about 20–25% of the total cardiac output
    b. is all filtered through the glomeruli
    c. is all used to supply the renal tissue with oxygen and nutrients
    d. both a and b
    e. all of the above (a-c)

74. The glomerular filtration rate:
    a. averages 125 ml/min
    b. averages 180 L/day
    c. represents about 20–25% of the total cardiac output
    d. both a and b above
    e. all of the above (a-c)

75. The glomerular filtrate as it enters Bowman's capsule:
    a. contains among other things proteins
    b. contains among other things glucose
    c. contains only substances that are not needed by the body
    d. is formed as a result of active forces
    e. is formed at a constant rate under all circumstances

76. Which of the following does not normally appear in the glomerular filtrate?
    a. plasma proteins
    b. glucose
    c. sodium
    d. urea
    e. calcium

77. Renal clearance
    a. is a measure of the surface area and permeability of the glomerular membrane
    b. is a constant value
    c. can be varied by contraction of the podocytes and mesangial cells
    d. is the amount of a substance that is removed from the blood
    e. is the volume of blood cleared of a substance in a specific period of time

78. Filtered substances do not pass through which of the following as they move across the glomerular membrane?
    a. podocytes
    b. filtration slits
    c. neither of the above

79. Which of the following forces oppose glomerular filtration?
    a. blood-colloid osmotic pressure
    b. bowman's capsule hydrostatic pressure
    c. glomerular capillary blood pressure
    d. both a and b above
    e. both a and c above

80. Bowman's capsule:
    a. filters water and solute from the blood
    b. exerts a hydrostatic pressure that opposes filtration
    c. exerts a hydrostatic pressure that favors filtration
    d. both a and b above
    e. both a and c above

81. Which of the following statements concerning the process of glomerular filtration is correct?
    a. Bowman's capsule hydrostatic pressure opposes filtration.
    b. The glomerular filtration rate is limited by a tubular max.
    c. All the plasma that enters the glomerulus is filtered.
    d. Two of the above are correct.
    e. All of the above (a-c) are correct.

82. Renin:
    a. is synthesized and released by cells in the juxtaglomerular apparatus.
    b. targets cells in the juxtaglomerular apparatus.
    c. causes the kidney to secrete EPO.
    d. both a and c above
    e. both b and c above

83. Tubular reabsorption:
    a. refers to the movement of a substance from the peritubular capillary blood into the tubular fluid
    b. occurs by either active or passive transport
    c. involves the process of transepithelial transport
    d. both a and c above
    e. all of the above (a-c)

84. The vessels that substances enter during tubular reabsorption are the:
    a. afferent arterioles
    b. efferent arterioles
    c. peritubular capillaries
    d. glomerular capillaries
    e. collecting tubules

85. The proximal tubule:
    a. reabsorbs about 65% of the filtered water.
    b. is the site of action of aldosterone.
    c. is the location where glucose is reabsorbed.
    d. both a and c above
    e. all of the above (a-c)

86. Which of the following plasma constituents is not regulated by the kidneys?
    a. glucose
    b. sodium
    c. hydrogen
    d. phosphate
    e. water

87. Which of the following substances exhibits a tubular max?
    a. sodium
    b. chloride
    c. glucose
    d. urea
    e. water

88. The juxtaglomerular apparatus:
    a. is a combination of specialized tubular and vascular cells
    b. secretes aldosterone
    c. secretes renin
    d. both a and b above
    e. both a and c above

89. Aldosterone:
    a. stimulates sodium reabsorption in the distal and collecting tubules
    b. is secreted by the juxtaglomerular apparatus
    c. stimulates potassium secretion in the distal tubules
    d. both a and b above are correct
    e. both a and c above

90. Aldosterone secretion:
    a. occurs in the kidneys
    b. is stimulated by angiotensin II
    c. is controlled by the plasma concentration of chloride ion
    d. all of the above (a-c)
    e. none of the above

91. Sodium reabsorption in the distal portions of the nephron is stimulated by:
    a. atrial natritic peptide
    b. vasopressin
    c. aldosterone
    d. renin
    e. angiotensin II

92. What percentage of urea is normally reabsorbed?
    a. 0
    b. 50
    c. 75
    d. 100

93. The distal and collecting tubules:
    a. are the site of the co-transport carriers for glucose and amino acid reabsorption
    b. are the site of the organic ion secretory systems
    c. are the site of aldosterone and vasopressin action
    d. both a and c above
    e. both b and c above

94. Water reabsorption is under the control of vasopressin:
    a. along the entire length of the nephron
    b. only in the loop of Henle
    c. only in the distal and collecting tubules
    d. only in the proximal tubule
    e. only in the glomerulus

95. Water reabsorption:
    a. occurs passively by osmosis in the proximal tubule
    b. is under control of vasopressin in the distal and collecting tubules
    c. occurs by active transport in the distal and collecting tubules
    d. both a and b above
    e. all of the above (a-c)

96. Water reabsorption:
    a. cannot occur from any portion of the nephron in the absence of vasopressin
    b. occurs to the greatest extent in the proximal convoluted tubule
    c. is under vasopressin control in the proximal tubule
    d. is under vasopressin control in the distal and collecting tubules
    e. both b and c above

97. Urea:
    a. is the waste product with the smallest molecular size in the glomerular filtrate
    b. is in greater concentration at the end of the proximal tubule than in other body fluids
    c. clearance rate is less than the GFR
    d. both a and b above
    e. all of the above (a-c)

98. Urea:
    a. reabsorption occurs by active transport
    b. is only 50% reabsorbed in the proximal tubule because the carrier has a low tubular max
    c. is a waste product of protein metabolism
    d. both b and c above
    e. all of the above (a-c)

99. Secretion of a foreign substance such as drugs generally occurs in the:
    a. distal tubule
    b. loop of Henle
    c. proximal tubule
    d. collecting duct
    e. glomerulus

100. Potassium:
    a. is actively reabsorbed in the proximal tubule
    b. is actively secreted in the distal and collecting tubules
    c. secretion is controlled by aldosterone
    d. both b and c above
    e. all of the above

Modified Multiple Choice

Indicate whether the first item in the statement increases, decreases, or has no effect on the second item by filling in the appropriate letter using the following code.

A = increases
B = decreases
C = has no effect on

101. Increased osmolarity of body fluids _____ vasopressin secretion.
102. Decreased vasopressin secretion _____ sodium reabsorption.
103. Decreased sodium in body fluids (sodium depletion) _____ renin secretion.
104. Increased renin secretion _____ angiotensin I activation.
105. Increased vasopressin secretion _____ urinary output.
106. Increased angiotensin II activation _____ aldosterone secretion.
107. Increased aldosterone secretion _____ potassium secretion.
108. Increased vasopressin secretion _____ sodium secretion

Indicate which substance in the right column undergoes the process in the left column by writing the appropriate letter in the blank.

109. _____ Filtered and actively reabsorbed.
110. _____ Filtered and passively reabsorbed.
111. _____ Filtered and both actively reabsorbed and actively secreted.
112. _____ Filtered but not reabsorbed or secreted.
113. _____ Not filtered.

A. potassium
B. glucose
C. insulin
D. plasma protein
E. urea
F. hydrogen ion

Points to Ponder

1. What is the advantage of sweat containing salt and other ions? Would water work just as well?

2. What makes the urine yellow?

3. Mammals excrete urea, uric acid, and creatinine as liquids while other vertebrates excrete these waste products as solids. Is there an advantage to this?

4. What would happen if we cut one ureter and inserted the proximal end into a vein so that the urine flowed directly back into the blood? The other kidney functions normally.

5. How often does the average adult urinate each day? A baby? How can you explain this difference?

6. Why does drinking beer stimulate the need to urinate?

7. How and when does the autonomic nervous system regulate kidney function?

8. Why does the glomerular ultrafiltrate have a low protein concentration?

9. What is the functional significance of the vasa recta and the countercurrent exchange mechanism?

10. What happens to urinary bicarbonate excretion when a person hyperventilates?

Clinical Perspective

1. A favorite fraternity pledge prank is to put methyl blue in the coffee of your big brother. Explain why the urine turns blue.

2. Why does urinary incontinence occur more often in the elderly?

3. How would you explain to your fraternity brother how kidney stones form?

4. How would you explain to your classmate what diabetes insipidus is and why these patients drink a lot of fluids?

5. Why is creatinine used clinically as an index of kidney function?

6. Based on what the kidney does to penicillin, why must large amounts of penicillin be given to an individual in order to be effective?

7. Why does removal of both adrenal glands cause hyperkalemia?

8. One of the complications of diuretics is hypokalemia. Why does this occur?

9. The drug acetazolamide is used to treat acute mountain sickness. Its pharmacology involves inhibiting renal carbonic anhydrase. How does this drug work in the kidney?

10. Suppose a student has a history of polycystic kidney disease. This student will also have proteinuria and an elevated blood creatinine and reduced inulin clearance. What do these physiological parameters indicate?

11. Mammals excrete urea, uric acid, and creatinine as liquids while other vertebrates excrete these waste products as solids. Is there an advantage to this?

12. What would happen if we cut one ureter and inserted the proximal end into a vein so that the urine flowed directly back into the blood? The other kidney functions normally.

Experiment of the Day

1. Eat some asparagus. Note the time of ingestion. Drink plenty of fluids. During the next 24 hours check your urine for any noticeable changes. Note the time that the changes are first observed. Explain the changes should you become aware of any. If you observe no changes, ask a classmate for their changes.

14

Fluid and Acid-Base Balance

Chapter Overview

Homeostasis depends on maintaining a balance between the input and output of all constituents in the internal fluid environment. As changes occur, homeostasis is maintained by body systems responding to the changes. This chapter deals with two very important areas of change: water (control of solute concentration and salt balance) and pH.

Water is located primarily in two major compartments: extracellular fluid (plasma and interstitial fluid) and intracellular fluid. All of these have different amounts of water and ions. To regulate and maintain these fluid balances, adjustments are made in the extracellular volume and the extracellular osmolarity.

Acid-base balance is very critical in the survival of cells. The somewhat narrow pH range is maintained by four buffer systems: carbonic acid/carbonate buffer system, protein buffer system, hemoglobin buffer system, and the phosphate buffer system.

The buffers are the first line of defense against pH changes. The second line is the lungs. By removing carbon dioxide less carbonic acid is added to body fluids. The last line of defense in the prevention of change in pH is the kidney. The big role of the kidney is to excrete the necessary amounts of hydrogen ions, bicarbonate ions, and ammonium ions. Maintaining a relatively constant pH is truly remarkable considering the great variability of substances brought into and produced by the body.

Chapter Outline

BALANCE CONCEPT
The internal pool of a substance is the amount of that substance in the ECF.
- The quantity of any particular substance in the ECF is considered to be a readily available internal pool.
- The ECF pool can be further altered by transferring a particular constituent into storage within the cells or bones.

To maintain stable balance of an ECF constituent, its input must equal its output.
- If the quantity of a substance is to remain stable within the body, its input by means of ingestion or metabolic production must be balanced by an equal output, by means of excretion or metabolic consumption.

FLUID BALANCE
- Water is the most abundant component of the human body.
- The H_2O content of an individual remains fairly constant, largely because the kidneys efficiently regulate H_2O balance. Known as the "balance concept," this relationship is extremely important in the maintenance of homeostasis.
- ₁f the body as a whole has a surplus or a deficit of a particular stored substance, the storage site can be expanded or partially depleted to maintain the extracellular fluid concentration of the substance within homeostatically prescribed limits.
- When total body input of a particular substance equals its total body output, a stable balance exists.

- When the gains via input exceed its losses via output, a positive balance exists.
- When the losses of a substance exceed its gains, a negative balance exists.

Body water is distributed between the ICF and ECF compartments.
- The intracellular fluid compartment comprises about two-thirds of the total body water.
- The remaining one-third of the body water found in the extracellular fluid is further subdivided into plasma and interstitial fluid.

The plasma and interstitial fluid are similar in composition, but the ECF and ICF differ markedly.
- Plasma and interstitial fluid are nearly identical in composition, except that interstitial fluid lacks plasma proteins.
- Any change in one of these ECF compartments is quickly reflected in the other compartment because they are constantly mixing.
- The composition of the ECF differs considerably from that of the ICF.
- Among the major differences are: (1) the presence of cellular proteins in the ICF that are unable to permeate the enveloping membranes to leave the cells, and (2) the unequal distribution of sodium and potassium and their attendant anions as a result of the action of the membrane-bound sodium-potassium ATPase pump, which is present in all cells.
- Sodium is the primary ECF cation, and potassium is primarily found in the ICF.
- In the extracellular fluid, sodium is accompanied primarily by the anion chloride and to a lesser extent by bicarbonate.
- The major intracellular anions are phosphate and the negatively charged proteins trapped within the cell.
- The movement of water between the plasma and the interstitial fluid across capillary walls is governed by relative imbalances between capillary blood pressure and colloid osmotic pressure.
- The net transfer of water between the interstitial fluid and the intracellular fluid across the cellular plasma membranes occurs as a result of osmotic effects alone.

Fluid balance is maintained by regulating ECF volume and osmolarity.
- Any control mechanism that operates on the plasma in effect regulates the entire extracellular fluid.
- The ICF, in turn, is influenced by changes in the ECF to the extent permitted by the permeability of the membrane barriers surrounding the cells.
- The factors regulated to maintain fluid balance in the body are ECF volume and ECF osmolarity.
- Extracellular fluid volume must be closely regulated to help maintain blood pressure.
- Extracellular fluid osmolarity must be closely regulated to prevent swelling or shrinking of the cells.

Control of ECF volume is important in the long-term regulation of blood pressure.
- Two short-term compensatory measures come into play to transiently adjust the blood pressure until the ECF volume can be restored to normal: (1) Baroreceptor reflex mechanisms alter both cardiac output and total peripheral resistance through autonomic nervous system effects on the heart and blood vessels. (2) Fluid shifts occur temporarily and automatically between the plasma and interstitial fluid.
- A reduction in plasma volume is partially compensated for by a shift of fluid out of the interstitial compartment into the blood vessels, thereby expanding the circulating plasma volume at the expense of the interstitial compartment.
- Conversely, when the plasma volume is too large, much of the excess fluid is shifted into the interstitial compartment.

- Responsibility for long-term regulation of blood pressure rests with the kidneys and the thirst mechanism, which control urinary output and fluid intake.

Control of salt balance is primarily important in regulating ECF volume.

- The total mass of sodium salts in the ECF determines the ECF's volume, and appropriately, regulation of the ECF volume depends primarily on controlling salt balance.
- To maintain salt balance at a set level, salt input must equal salt output.
- Since we typically consume salt in excess of our needs, it is obvious that salt intake in humans is not well controlled.
- The three avenues for salt output are obligatory loss of salt in sweat and feces and controlled excretion of salt in the urine.
- Precise control of salt output in the urine. By regulating the rate of urinary salt excretion, the kidneys keep the total sodium mass in the ECF constant despite any notable changes in dietary intake of salt or unusual losses through sweating, diarrhea, or other means.
- The amount of sodium excreted in the urine represents the amount of sodium that is filtered but is not subsequently reabsorbed.
- The kidneys accordingly adjust the amount of salt excreted by controlling two processes: (1) the glomerular filtration rate, and (2) the tubular reabsorption of sodium controlled by the rennin-angiotensin-aldosterone system.
- A fall in arterial blood pressure brings about a twofold effect in the renal handling of sodium: (1) a reflex reduction in the glomerular filtration rate to decrease the amount of sodium filtered, and (2) a hormonally adjusted increase in the amount of sodium reabsorbed.
- A rise in arterial blood pressure brings about: (1) increases in the amount of sodium filtered, and (2) a reduction in renin-angiotensin-aldosterone activity,

which decreases salt and water reabsorption.

Control of ECF osmolarity prevents changes in ICF volume.

- The osmotic activity across the capillary wall is due to the unequal distribution of plasma proteins.
- Plasma proteins are present only in the plasma.
- Osmotic activity across the cellular plasma membranes is directly related to any differences in ionic concentration between the ECF and ICF.
- Plasma proteins play no role in osmosis of water across the cell membranes because they are absent in both the interstitial fluid and the ICF, which are separated by cell membranes.
- Sodium and its attendant anions, being by far the most abundant solutes in the ECF in terms of numbers of particles, account for the vast majority of the ECF's osmotic activity.
- Potassium and its accompanying intracellular anions are responsible for the ICF's osmotic activity.
- Normally, the osmolarities of the ECF and ICF are the same.

During ECF hypertonicity, the cells shrink as water leaves them.

- Hypertonicity of the ECF, or the excessive concentration of ECF solutes, is usually associated with dehydration, or a negative free water balance.
- Dehydration with accompanying hypertonicity can be brought about in three major ways: (1) insufficient water intake, (2) excessive water loss, and (3) diabetes insipidus.
- Whenever the ECF compartment becomes hypertonic, water moves out of the cells by osmosis into the more concentrated ECF until the ICF osmolarity equilibrates with the ECF.
- The cells shrink as water leaves them.
- Shrinking of brain neurons causes disturbances in brain function.

- Circulatory problems may range from slight reduction in blood pressure to circulatory shock and death.

During ECH hypotonicity, the cells swell as water enters them.
- Hypotonicity of the ECF is usually associated with over hydration; that is, excess of free water is present.
- Hypotonicity can arise in three ways: (1) Patients with renal failure who are unable to excrete a dilute urine become hypotonic when they consume relatively more water than solutes. (2) Hypotonicity can occur transiently in healthy people if water is rapidly ingested to such an excess that the kidneys are unable to respond quickly enough to eliminate the extra water. (3) Hypotonicity can occur when excess water without solute is retained in the body as a result of inappropriate secretion of vasopressin.
- The resultant difference in osmotic activity between ECF and ICF induces water to move by osmosis from the more dilute ECF into the cells, with the cells swelling.
- Pronounced swelling in brain cells also leads to brain dysfunction and circulatory disturbances, including hypertension and edema.
- When an isotonic fluid is injected into the ECF compartment, the ECF volume increases but the concentration of ECF solutes remains unchanged; the ECF and ICF are still in osmotic equilibrium.
- In the case of an isotonic fluid loss such as occurs in hemorrhage, the loss is confined to the ECF with no corresponding loss of fluid from the ICF.

Control of water output in the urine by vasopressin and thirst
- Of the many sources of water input and output only two can be regulated to maintain water balance: (1) control of water input by thirst, and (2) control of water output in the urine by vasopressin.

- A thirst center is located in the hypothalamus in close proximity to the vasopressin-secreting cells.
- Thirst increases water input, whereas vasopressin, by reducing urine production, decreases water output.
- Vasopressin and thirst are both stimulated by a free water deficit and suppressed by a free water excess.
- The predominant excitatory input for both vasopressin secretion and thirst come from the hypothalamus osmoreceptors located near the vasopressin-secreting cells and thirst center.
- Left atrial volume receptors monitor the blood pressure, which is a reflection of the ECF volume.

Vasopressin secretion and thirst are largely triggered simultaneously.
- In response to a major reduction in ECF volume and arterial pressure, as during hemorrhage, the left atrial volume receptors reflexly stimulate both vasopressin secretion and thirst.
- The outpouring of vasopressin and increased thirst lead to decreased urine output and increased fluid intake.
- Vasopressin exerts a potent vasoconstrictor effect on arterioles in addition to having an effect on the kidney tubules.
- Vasopressin and thirst are both inhibited when the ECF/plasma volume and arterial blood pressure are elevated.
- Aldosterone controlled sodium reabsorption is the most important factor in regulating ECF volume, with the vasopressin and thirst mechanisms playing only a supportive role.

ACID-BASE BALANCE
Acids liberate free hydrogen ions, whereas bases accept them.
- Acids are a special group of hydrogen-containing substances that dissociate when in solution to liberate free hydrogen ions.

- A base is a substance that can combine with a free hydrogen ion thus removing it from solution.

The pH designation is used to express [H].
- The pH equals the logarithm to the base 10 of the reciprocal of the hydrogen ion concentration.
- The greater the hydrogen ion concentration the lower the pH.
- Every unit change in pH actually represents a tenfold change in hydrogen ion concentration
- Solutions having a pH less than 7.0 are considered to be acidic.
- Solutions having a pH greater than 7.0 are considered to be basic or alkaline.
- The pH of arterial blood is normally 7.45, and the pH of venous blood is 7.35, for an average blood pH of 7.4. The pH of venous blood is slightly lower (more acidic) than that of arterial blood because H^+ is generated by the formation of H_2CO_3 from CO_2 picked up at the tissue capillaries.
- Acidosis exists whenever blood pH falls below 7.35, whereas alkalosis occurs when blood pH is above 7.45

Fluctuations in hydrogen-ion concentration alter nerve, enzyme, and potassium activity.
- Changes in excitability of nerve and muscle cells are among the major clinical manifestations of pH abnormalities.
- Hydrogen-ion concentration exerts a marked influence on enzyme activity.
- Changes in hydrogen-ion concentration influence potassium levels in the body. When H+ secretion increases to compensate for acidosis, less K+ can be secreted and vice versa when H+ secretion is reduced in alkalosis.

Hydrogen ions are continually being added to the body fluids as a result of metabolic activities.
- Normally hydrogen is continually being added to the body fluids from the three following sources: (1) carbonic acid formation, (2) inorganic acids produced during the breakdown of nutrients, and (3) organic acids resulting from intermediary metabolism.
- Three lines of defense against changes in hydrogen ion concentration operate to maintain the hydrogen ion concentration of body fluids at a nearly constant level despite unregulated input: (1) the chemical buffer systems, (2) the respiratory mechanism of pH control, and (3) the renal mechanism of pH control.

Chemical buffer systems minimize changes in pH by binding with or yielding free hydrogen ions.
- A chemical buffer system is a mixture in a solution of two chemical compounds that minimizes pH changes when either an acid or a base is added or removed from the solution.
- There are four buffer systems in the body: (1) the carbonic acid/bicarbonate buffer system, (2) the protein buffer system, (3) the hemoglobin buffer system, and (4) the phosphate buffer system.

The carbonic/bicarbonate buffer pair is the primary ECF buffer for non-carbonic acids.
- The carbonic acid/bicarbonate buffer pair is the most important buffer system in the ECF for buffering pH changes brought about by causes other than fluctuations in carbon dioxide-generated carbonic acid.

The protein buffer system is primarily important intracellularly.
- The protein buffer system is primarily important intracellularly.

The hemoglobin buffer system buffers hydrogen generated from carbonic acid.
- The hemoglobin buffer system buffers hydrogen ion generated from carbonic acid.

The phosphate buffer system is an important urinary buffer.
- The phosphate buffer system is an important urinary buffer.

- The phosphate buffer system is composed of an acid phosphate salt and a basic phosphate salt.
- This system contributes significantly to intracellular buffering.
- The phosphate system serves as an excellent urinary buffer.

Chemical buffer systems act as the first line of defense against changes in hydrogen ion concentration.

- All chemical buffer systems act immediately, within fractions of a second, to minimize changes in pH.

The respiratory system regulates hydrogen-ion concentration by controlling the rate of carbon dioxide removal.

- When arterial hydrogen ion concentration increases, the respiratory center in the brain stem is reflexly stimulated to increase pulmonary ventilation.
- When arterial hydrogen ion concentration falls, pulmonary ventilation is reduced.
- Carbon dioxide diffuses from the cells into the blood faster than it is removed from the blood by the lungs, so higher-than-usual amounts of acid-forming carbon dioxide accumulate in the blood, thus restoring hydrogen ion concentration toward normal.

The respiratory system serves as the second line of defense against changes in [H].

- If a deviation in hydrogen ion concentration is not swiftly and completely corrected by the buffer systems, the respiratory system comes into action a few minutes later, thus serving as the second line of defense against changes in hydrogen ion concentration.

The kidneys help maintain acid-base balance by adjusting their rate of [H] excretion, HCO_3 excretion, and NH_3 secretion.

- The kidneys control the pH of the body fluids by adjusting three interrelated factors: (1) hydrogen excretion, (2) bicarbonate excretion, and (3) ammonia secretion.
- The magnitude of hydrogen secretion depends on a direct effect of the plasma acid-base status on the kidneys' tubular cells.
- When the hydrogen ion concentration of the plasma passing through the peritubular capillaries is elevated above normal, the tubular cells respond by secreting greater-than-usual amounts of hydrogen from the plasma into the tubular fluid to be excreted in the urine.
- When the plasma hydrogen ion concentration is lower than normal, the kidneys conserve hydrogen by reducing their secretion and subsequent excretion in the urine.
- When the plasma hydrogen ion concentration is increased above normal during acidosis, renal compensation includes the following: (1) increased secretion and subsequent increased excretion of hydrogen in the urine, thereby eliminating the excess hydrogen and decreasing plasma hydrogen ion concentration; and (2) reabsorption of all filtered bicarbonate,
- When plasma hydrogen ion concentration is reduced below normal during alkalosis, renal responses include the following: (1) decreased secretion and subsequent reduced excretion of hydrogen in the urine, resulting in conservation of hydrogen and increased plasma hydrogen ion concentration; and (2) incomplete reabsorption of filtered bicarbonate and subsequent increased excretion of bicarbonate, resulting in reduction of plasma bicarbonate ion concentration.
- There are two important urinary buffers: (1) filtered phosphate buffers, and (2) secreted ammonia.
- Normally, secreted hydrogen is first buffered by the phosphate buffer system.

- When acidosis exists, the tubular cells secrete ammonia into the tubular fluid once the normal urinary phosphate buffers are saturated.

- Ammonia enables the kidneys to continue secreting additional hydrogen ions, because ammonia combines with free hydrogen in the tubular fluid to form ammonium ions.
- The tubular membranes are not very permeable to ammonium, so the ammonium ions remain in the tubular fluid and are lost in the urine, each one taking a hydrogen ion with it.

The kidneys are a powerful third line of defense against changes in [H].

- By simultaneously removing acid from and adding base to the body fluids, the kidneys are able to restore the pH toward normal more efficiently than the lungs.
- Also contributing to the kidneys' acid-base regulatory potency is their ability to return the pH almost exactly to normal.

Acid-base imbalances can arise from either respiratory dysfunction or metabolic disturbances.

- Respiratory acidosis is the result of abnormal carbon dioxide retention arising from hypoventilation.
- The primary defect in respiratory alkalosis is excessive loss of carbon dioxide from the body as a result of hyperventilation.
- The following are the most common causes of metabolic acidosis: (1) severe diarrhea, (2) diabetes mellitus, (3) strenuous exercise, and (4) uremic acidosis.
- Metabolic alkalosis arises most commonly from the following: (1) vomiting, and (2) ingestion of alkaline drugs.

Key Terms

Acid
Acid-base balance
Acidic
Acidosis
Alkalosis
Ammonia (NH_3)
Ammonium ion (NH_4^+)
Balance concept
Bases
Basic (alkaline)
Chemical buffer system
Diabetes insipidus
Extracellular fluid (ECF)
Fluid balance
Hypothalamic osmoreceptors
Hypotonicity
Insensible loss
Interstitial fluid

Intracellular fluid (ICF)
Left atrial volume receptors
Metabolic acidosis
Metabolic alkalosis
Metabolic water
Negative balance
Osmolarity
pH
Plasma
Positive balance
Respiratory acidosis
Respiratory alkalosis
Salt balance
Sensible water loss
Stable balance
Thirst center
Transcellular fluid

Review Exercises
Answers are in the appendix.

True/False-if an answer is false, underline the incorrect portion of the statement and rewrite it to make the statement true.

_____   1. When the gains via input exceed losses via output, a negative balance exists.

_____   2. The ICF compartment comprises about one-half of the total body water.

_____   3. Water loss during sweating is classified as insensible.

_____   4. The major intracellular anions are phosphate ions.

_____   5. The phosphate buffer system is not important in the urine.

_____   6. Vasopressin decreases water output by reducing urine production.

_____   7. When the kidney secretes excess H+ it decreases secretion of K+.

_____   8. Bases are a special group of hydrogen-containing substances that dissociate when in solution to liberate free hydrogen.

_____   9. A base is a substance that can combine with free hydrogen and thus remove it from solution.

_____ 10. The greater the hydrogen ion concentration, the lower the pH.

_____ 11. The first line of defense against change in pH is the respiratory control of pH.

_____ 12. Tubular cells secrete ammonium ions that can combine with free hydrogen to form ammonia.

_____ 13. Metabolic acidosis can be caused by diabetes insipidus.

_____ 14. Metabolic alkalosis can be caused by vomiting.

Fill in the Blank

15. When total body input of a particular substance equals its total body output, a(n) _____ exists.

16. Two prominent organic acids are _____ which is produced during fat metabolism and _____ which is produced during exercise.

17. The _____ buffer system is an important urinary buffer because it allows the kidney to excrete H+ in the urine without significantly altering urine pH.

18. The three avenues for salt output are _____, _____, and controlled excretion of salt in the _____.

19. The main factor controlling the extent of sodium reabsorption in the distal tubule and collecting tubule is the powerful _____ system.

20. Dehydration with accompanying hypertonicity can be brought about in three major ways:
(1) _____, (2) _____ and
(3)_____.

21. The predominant excitatory input for both vasopressin secretion and thirst comes from _____.

22. The _____ equals the logarithm to the base 10 of the reciprocal of the hydrogen ion concentration.

23. The _____ serves as an excellent urinary buffer.

24. Respiratory acidosis is the result of abnormal carbon dioxide retention arising from _____.

25. Hyperventilating because of an anxiety attack causes _____.

26. Vomiting causes _____.

27. Bicarbonate cannot buffer urinary hydrogen as it does in the ECF because bicarbonate is not excreted in the urine simultaneously with _____.

28. The kidneys control the pH of the body fluids by adjusting three interrelated factors:
(1) _____, (2) _____, and
(3)_____.

Matching
Match the cause on the left to the acid-base condition on the right.

_____ 29. diabetes mellitus   a. metabolic alkalosis

_____ 30. hyperventilation   b. respiratory acidosis

_____ 31. vomiting   c. metabolic acidosis

_____ 32. excess ingestion of antacids d. respiratory alkalosis

_____ 33. strenuous exercise

_____ 34. uremic acidosis

_____ 35. hypoventilation

_____ 36. severe diarrhea

Multiple Choice

37. Which of the following disorders is characterized by a deficiency of vasopressin?
    a. diabetes mellitus
    b. hemorrhage
    c. hypotonicity
    d. colloidal isotonicity
    e. diabetes insipidus

38. Which of the following disorders or symptoms can be caused by hypotonicity?
    a. diabetes insipidus
    b. hypertension and edema
    c. hemorrhage
    d. diarrhea
    e. diabetes mellitus

39. Which of the following disorders or symptoms causes changes in the ECF but not in the ICF?
    a. diabetes insipidus
    b. hypertension and edema
    c. hemorrhage
    d. dehydration
    e. overhydration

40. Which of the avenues for salt output is linked to aldosterone?
    a. tears
    b. feces
    c. hemorrhage
    d. sweat
    e. controlled excretion in urine

41. Which is the primary cation found in the ECF?
    a. bicarbonate
    b. phosphate
    c. sodium
    d. chloride
    e. potassium

42. Which is the primary cation found in the ICF?
    a. bicarbonate
    b. sodium
    c. chloride
    d. phosphate
    e. potassium

43. Which of the following organs can respond to acid-base disturbances within a few minutes of their onset?
    a. lungs
    b. heart
    c. spleen
    d. kidneys
    e. intestines

44. Which fluid compartment has the greatest percentage of body fluid?
    a. lymph
    b. plasma
    c. ICF
    d. ECF
    e. transcellular fluid

45. Which fluid compartment represents the greatest percentage of body fluid within the ECF?
    a. lymph
    b. interstitial fluid
    c. synovial fluid
    d. plasma
    e. cerebrospinal fluid

46. When the plasma volume is too large, to which fluid compartment is the excess fluid shifted?
    a. lymph
    b. interstitial fluid
    c. intracellular
    d. cerebrospinal fluid
    e. none of the above

Modified Multiple Choice
Indicate which fluid imbalance is being described by writing the appropriate letter in the blank using the answer code below.

> A = overhydration
> B = dehydration
> C = both overhydration and dehydration
> D = neither overhydration nor dehydration

47. _____ Symptoms include dry skin, parched tongue, and sunken eyeballs.
48. _____ Water enters the cells by osmosis.
49. _____ The body fluids are hypotonic.
50. _____ ECF and ICF become hypertonic.
51. _____ Cells become swollen.
52. _____ No fluid shift occurs between the ECF and ICF.
53. _____ Occurs as a consequence of water deprivation.
54. _____ Cells lose water and shrink
55. _____ ECF and ICF become hypertonic.
56. _____ Convulsions and coma may occur.
57. _____ Occurs as a consequence of diabetes insipidus.
58. _____ Occurs as a consequence of heavy vomiting.
59. _____ Osmolarity of the body fluids is decreased.
60. _____ Occurs as a consequence of excessive fluid intake.
61. _____ Occurs as a consequence of excessive vasopressin secretion.
62. _____ Vasopressin secretion is stimulated as a compensatory mechanism.
63. _____ Increased urinary output as a compensatory mechanism.

Indicate which type of acid-base imbalance might occur in each of the following situations by writing the appropriate letter in the blank.

A = respiratory acidosis
B = respiratory alkalosis
C = metabolic acidosis
D = metabolic alkalosis

64. _____ Fever.
65. _____ Excessive ingestion of antacids
66. _____ Aspirin poisoning.
67. _____ Anxiety.
68. _____ Severe exercise.
69. _____ Uremia.
70. _____ Damage to the respiratory centers of the brain.
71. _____ Severe diarrhea.
72. _____ Pneumonia.
73. _____ Excessive vomiting
74. _____ Diabetes mellitus.

When a person has diarrhea, s/he loses excessive salt and water from the body. This fluid loss results in sodium depletion, dehydration, a decreased extracellular fluid volume, a reduction in plasma volume, and a decreased systemic arterial blood pressure. The following refers to a sequence of events that occur to compensate for this fluid loss. Indicate whether each factor listed:

A = exhibits no change
B = is increased to compensate for fluid loss
C = is decreased to compensate for fluid loss

75. _____ Parasympathetic activity to the afferent arterioles of the nephrons.
76. _____ Diameter of the afferent arterioles.
77. _____ Glomerular capillary blood pressure.
78. _____ Net filtration pressure.
79. _____ GFR.
80. _____ Amount of sodium and water filtered.
81. _____ Renin secretion.
82. _____ Angiotensin I and II production.
83. _____ Aldosterone secretion.
84. _____ Amount of sodium reabsorbed.
85. _____ Amount of sodium secreted.
86. _____ Vasopressin secretion.
87. _____ Permeability of distal and collecting tubules to water.
88. _____ Amount of water reabsorbed.
89. _____ Amount of water excreted.
90. _____ Urinary volume.
91. _____ Thirst.
92. _____ Amount of potassium lost.

Points to Ponder

1. As you shop in your favorite grocery store you may select a meat product that has been "sugar cured." How does this process protect the meat and make it safe for you to eat?

2. At present one of the makers of a common antacid boasts that their product uses calcium instead of sodium, and that this is better. Do you find any truth in this advertisement? Explain.

3. Most Americans are aware of the caffeine in these popular drinks: coffee, tea, and most soft drinks. How are the other ingredients related to this chapter? Are these drinks diuretics? Are these beverages acidic or basic? How does your body handle these fluids?

4. What forces control the movement of water between the plasma and the interstitial fluid?

5. What are the two main forces that move water between compartments?

6. How is water intake regulated?

7. Why can't humans drink sea water to quench our thirst?

8. What is the role of thirst in regulating fluid intake?

9. What is meant by fluid balance?

10. Define the term "body fluid."

Clinical Perspective

1. The patient has lost a lot of blood due to the traumatic injuries received in an auto accident. Intravenous fluids are administered. Start with the hemorrhage and explain the sequence of events with reference to this chapter.

2. What causes a blister?

3. Why do high sodium diets tend to cause a person to gain weight?

4. For the sake of convenience, long-distance truck drivers often limit their intake of fluids, such as coffee. During this time, the urine becomes very dark in color. What makes it dark?

5. With respect to the above question (#4), if the truck driver drinks several glasses of water, the urine will be much lighter in color. Why does increased fluid intake lighten the urine?

6. How could a person go into hypochloremia? Hyponatremia?

7. How are alkalosis and acidosis compensated and treated?

8. What are the main physiological effects of acidosis and alkalosis?

9. What are some common causes of respiratory acidosis?

10. What are some characteristics of hypocalcemia?

Experiment of the Day

Equipment:
3 glasses
3 dried prunes
sugar
salt

1. Place water in all three glasses. Place several spoons of salt into one glass and several spoons of sugar into still another. The third glass has plain water. Drop a prune into each glass. Note time. Explain the results.

The Digestive System

Chapter Overview

To maintain homeostasis, nutrient molecules used for energy production must continually be replaced by the acquisition of new, energy-rich nutrients. Cells need these nutrients, vitamins, and minerals to survive. The digestive system meets these cellular needs by removing the substances from the external environment and presenting them to the cells in a form that the cells can utilize. The food systematically passes through the mouth, pharynx, esophagus, stomach, small intestine, large intestine, and anus. The digestive tract receives enzymes and other secretions from the salivary glands, the exocrine pancreas, the liver, and the exocrine cells in the digestive tract's epithelial lining. During this trip, carbohydrates, proteins, and fats are broken down by various enzymes and other substances to yield monosaccharides, amino acids, and monoglycerides respectively. These smaller nutrient molecules are then absorbed into the small intestine and pass to the liver or lymphatic system. The liver performs the final modification on the nutrient molecules. Undigested or unutilized material, along with bacteria, are passed to the outside through the anus. Nutrition prepared by the digestive system is another important step in homeostasis. The digestive system is under the control of the central nervous system and several hormones.

Chapter Outline

INTRODUCTION
- The primary function of the digestive system is to transfer nutrients, water, and electrolytes from the food we eat into the body's internal environment.

The digestive system performs four basic digestive processes.
- There are four basic digestive processes: (1) motility, (2) secretion, (3) digestion, and (4) absorption.
- Propulsive movements push the contents forward through the digestive tract at varying speeds.
- Mixing movements serve a twofold function.
- First, by mixing food with the digestive juices, these movements promote digestion of the food.
- Second, they facilitate absorption by exposing all portions of the intestinal contents to the absorbing surfaces of the digestive tract.
- Each digestive secretion consists of water, electrolytes, and specific organic constituents that are important in the digestive process, such as enzymes, bile salts, or mucus.
- Digestion refers to the breaking-down process whereby the structurally complex foodstuffs of the diet are converted to smaller absorbable units by the enzymes produced within the digestive system.
- Digestion is accomplished by enzymatic hydrolysis.

- Through the process of absorption, the small absorbable units that result from digestion, along with water, vitamins, and electrolytes, are transferred from the digestive tract lumen into the blood or lymph.

The digestive tract and accessory digestive organs make up the digestive system.
- The accessory digestive organs include the salivary glands, the exocrine pancreas, and the biliary system (liver and gallbladder).
- The digestive tract includes the following organs: mouth, pharynx, esophagus, stomach, small intestine, large intestine, and anus.

The digestive tract wall has four layers.
- A cross section of the digestive tube reveals four tissue layers: mucosa, submucosa, muscularis externa, and serosa.
- The mucosa lines the luminal surface of the digestive tract; the submucosa is connective tissue that provided distensibility and elasticity; the mucularis externa is the major smooth muscle layer; the serosa is the outermost layer. .

Regulation of digestive function is complex and synergistic.
- Four factors are involved in the regulation of digestive system function: (1) autonomous smooth muscle function, (2) intrinsic nerve plexuses, (3) extrinsic nerves, and (4) gastrointestinal hormones.
- The prominent type of self-induced electrical activity in digestive smooth muscle is slow-wave potentials, alternately referred to as the digestive tracts basic electrical rhythm or "pacesetter potential."
- The interstitial cells of Cajal are the pacesetter cells responsible for instigating cyclic slow-wave activity. These cells are connected with smooth muscle cells via gap junctions.

- The second factor involved in the regulation of digestive tract function is the intrinsic nerve plexuses.
- Two major networks of nerves form the plexus of the digestive tract: the myenteric plexus and the submucous plexus.
- The two plexuses are often termed the "enteric nervous system."
- The intrinsic plexuses influence all facets of digestive tract activity
- The intrinsic plexuses can be influenced by endocrine, paracrine and nerve signals.
- The extrinsic nerves are from both branches of the autonomic system.
- Neurons that release acetylcholine as a neurotransmitter promote contraction of digestive tract smooth muscle.
- The sympathetic system tends to inhibit or slow down digestive tract contraction and secretion.
- The parasympathetic nervous system dominates in quiet, relaxed situations when general maintenance types of activities such as digestion can proceed optimally.
- Parasympathetic nerve fibers, which arrive primarily by way of the vagus nerve, tend to increase smooth muscle motility and promote secretion of digestive enzymes and hormones.
- Tucked within the mucosa of certain regions of the digestive tract are endocrine gland cells that release hormones into the blood upon appropriate stimulation.
- Gastrointestinal hormones are carried to other areas of the digestive tract where they exert either excitatory or inhibitory influences on smooth muscle and exocrine gland cells.

Receptor activation alters digestive activity through neural reflexes and hormonal pathways.
- The wall of the digestive tract contains three different types of sensory receptors that respond to local chemical or mechanical changes in the digestive

tract: (1) chemoreceptors, (2) mechanoreceptors, and (3) osmoreceptors.
- The effector cells include smooth muscle, exocrine gland, and endocrine gland cells.

MOUTH

The oral cavity is the entrance to the digestive tract.
- Entry to the digestive tract is through the mouth or oral cavity.
- The palate separates the mouth from the nasal passages.
- Toward the front of the mouth, the palate is made of bone, forming what is known as the "hard palate."
- The palate toward the rear of the mouth is called the "soft palate."
- The tongue is composed of voluntarily controlled skeletal muscle.
- The pharynx acts as a common passageway for both the digestive system and the respiratory system.

The teeth are responsible for chewing.
- The first step in the digestive process is mastication.
- Occlusion allows food to be ground and crushed between the tooth surfaces.
- The purposes of chewing are: (1) to grind and break food up into smaller pieces to facilitate swallowing and to increase the surface area on which salivary enzymes will act (2) to mix food with saliva, and (3) to stimulate taste buds.

Saliva begins carbohydrate digestion, is important in oral hygiene, and facilitates speech.
- Saliva is produced by three major pairs of salivary glands: the sublingual, submandibular and parotid glands.
- Saliva is composed of water, proteins, and electrolytes.
- The process of salivation occurs in two stages.
- The most important salivary proteins—amylase, mucus, and lysozyme—

contribute to the functions of saliva as follows: (1) Salivary amylase breaks polysaccharides down into disaccharides. (2) Saliva facilitates swallowing by providing mucus. (3) Saliva exerts some antibacterial action by means of lysozyme, an enzyme that lyses or destroys certain bacteria. (4) Saliva serves as a solvent for molecules that stimulate the taste buds. (5) Saliva aids speech by facilitating movements of the lips and tongue. (6) Saliva plays an important role in oral hygiene. (7) Saliva buffers the acids in food as well as acids produced by bacteria in the mouth, thereby helping to prevent dental caries.

Salivary secretion is continuous and can be reflexly increased.
- The simple, or unconditioned, salivary reflex occurs when chemoreceptors and pressure receptors within the oral cavity respond to the presence of food.
- These receptors initiate impulses in afferent nerve fibers that carry the information to the salivary center located in the medulla of the brain stem.
- The salivary center sends impulses via the extrinsic autonomic nerves to the salivary glands to promote increased salivation.
- With the acquired, or conditioned, salivary reflex, salivation occurs without oral stimulation.
- Just thinking about, seeing, smelling, or hearing the preparations of pleasant food initiates salivation through this reflex.
- Inputs that arise outside the mouth and are mentally associated with the pleasure of eating act through the cerebral cortex to stimulate the medullary salivary center.
- Parasympathetic stimulation, which exerts the dominant role in salivary secretion, produces a prompt and abundant flow of watery saliva that is rich in enzymes.
- Sympathetic stimulation produces a much smaller volume of thick saliva that is rich in mucus.

Digestion in the mouth is minimal; no absorption of nutrients occurs.

- Digestion in the mouth involves the hydrolysis of polysaccharides into disaccharides by amylase
- Importantly, some therapeutic agents can be absorbed by the oral mucosa, a prime example being a vasodilator drug, nitroglycerin, which is used by certain cardiac patients to relieve anginal attacks associated with myocardial ischemia.

PHARYNX AND ESOPHAGUS
Swallowing is a sequentially programmed all-or-none reflex.

- The pressure of the bolus in the pharynx stimulates pharyngeal pressure receptors, which send afferent impulses to the swallowing center located in the medulla.
- Swallowing is an example of a sequentially programmed all-or-none reflex in which multiple responses are triggered in a specific timed sequence.
- Swallowing is initiated voluntarily, but once it is initiated, it cannot be stopped.

During the oropharyngeal stage of swallowing, food is prevented from entering the wrong passageways.

- Food is prevented from reentering the mouth during swallowing by the position of the tongue against the hard palate.
- The uvula is elevated and lodges against the back of the throat, sealing off the nasal passage.
- Contraction of laryngeal muscles aligns the vocal folds in tight apposition to each other, thus sealing the glottis entrance.
- The bolus tilts the epiglottis backward down over the closed glottis.
- Because the respiratory passages are temporarily sealed off during swallowing, respiration is briefly inhibited.

- Pharyngeal muscles contract to force the bolus into the esophagus.

The pharyngoesophageal sphincter prevents air from entering the digestive tract during breathing.

- The upper esophageal sphincter is the pharyngoesophageal sphincter and the lower sphincter is the gastroesophageal sphincter.

Peristaltic waves push food through the esophagus.

- The swallowing center initiates a primary peristaltic wave that sweeps from the beginning to the end of the esophagus, forcing the bolus ahead of if through the esophagus to the stomach.
- Peristalsis refers to the ringlike contractions of the circular smooth muscle that move progressively forward with a stripping motion.

The gastroesophageal sphincter prevents reflux of gastric contents.

- If gastric contents do flow back into the esophagus, the acidity of these contents irritates the esophagus, causing the esophageal discomfort known as heartburn.

Esophageal secretion is entirely protective.

- Esophageal secretion is entirely mucus.
- Mucus provides lubrication for passage of food and protects the esophageal wall from acid and enzymes in the gastric juice if gastric reflux should occur.

STOMACH
The stomach stores food and begins protein digestion.

- The stomach performs 3 main functions: stores ingested food until it can be emptied into the small intestines; secretes HCl and enzymes that begin protein digestion; by motility coverts food into a thick liquid mixture chyme.

- It is important that the stomach store food and meter it into the duodenum at a rate that does not exceed the small intestine's capacities.

Gastric filling involves receptive relaxation.
- There are four aspects to gastric motility: (1) gastric filling, (2) gastric storage, (3) gastric mixing, and (4) gastric emptying.
- Gastric filling relies on receptive relaxation
- The stomach can expand to 20 times in volume with little change in tension or rise in pressure.

Gastric storage takes place in the body of the stomach.
- Food emptied into the stomach from the esophagus is stored in the relatively quiet body without being mixed.
- Food is gradually fed from the body into the antrum, where mixing does take place.
- Pacemaker cells at the upper parts of the stomach generate slow-wave potentials that sweep down the stomach.

Gastric mixing takes place in the antrum of the stomach.
- The strong antral peristaltic contractions are responsible for the mixing of food with gastric secretions to produce chyme.
- The antral peristaltic contractions provide the driving force for gastric emptying.
- The main gastric factor that influences the strength of contraction is the amount of chyme in the stomach.

Gastric emptying is controlled by factors in the duodenum.
- Factors in the duodenum are of primary importance in controlling the rate of gastric emptying.
- The four most important duodenal factors that influence gastric emptying are: (1) fat,

(2) acid, (3) hypertonicity, and (4) distention.
- The presence of one or more of these stimuli in the duodenum activates duodenal receptors, thereby triggering either a neural or hormonal response.
- The subsequent reduction in antral peristaltic activity slows down the rate of gastric emptying.
- The neural response is mediated through the intrinsic nerve plexuses and the autonomic nerves.
- The hormonal response involves the release from the duodenal mucosa of several hormones collectively known as enterogastrones: secretin, cholecystokinin, (CCK) and gastric inhibitory peptide.
- Collectively, these reflexes are called the enterogastric reflex.
- Peristaltic contractions occur in the empty stomach before the next meal.

Emotions can influence gastric motility.
- Sadness and fear generally tend to decrease motility, whereas anger and aggression tend to increase it.
- Intense pain from any part of the body tends to inhibit motility.

The stomach does not actively participate in vomiting.
- Vomiting, or emesis, is the forceful expulsion of gastric contents out through the mouth.
- The major force for expulsion comes from contraction of the respiratory muscles: namely, the diaphragm and the abdominal muscles.
- The complex act of vomiting is coordinated by a vomiting center in the medulla.
- Nausea, retching, and vomiting can be initiated by afferent input to the vomiting center from a number of receptors throughout the body.
- The causes of vomiting include the following: (1) tactile stimulation of the back of the throat; (2) irritation or

distension of the stomach or duodenum; (3) elevated intracranial pressure; (4) rotation or acceleration of the head, producing dizziness; (5) chemical agents, emetics; and (6) psychogenic vomiting induced by emotional factors.

Gastric digestive juice is secreted by glands located at the base of gastric pits.
- The cells responsible for gastric secretion are located in the lining of the stomach, the gastric mucosa, which is divided into two distinct areas: (1) the oxyntic mucosa, which lines the body and fundus, and (2) the pyloric gland area, which lines the antrum.
- The mucosal gland cells are found in gastric pits.
- Three types of secretory cells are found in the walls of the pits in the oxyntic mucosa.
- The mucous neck cells secrete a thin, watery mucus.
- Chief cells secrete the enzyme precursor pepsinogen, and parietal cells secrete HCl and intrinsic factor.
- The gastric pits of the pyloric gland area secrete mucus and pepsinogen.
- No acid is secreted in the pyloric gland area.
- More importantly, endocrine cells in the pyloric gland area secrete the hormone gastrin into the blood.

Hydrochloric acid activates pepsinogen.
- Parietal cells actively secrete HCl.
- Hydrochloric acid: (1) activates the enzyme precursor pepsinogen to an active enzyme, pepsin, and provides an acid medium that is optimal for pepsin activity; (2) denatures proteins; and (3) along with salivary lysozyme, kills most of the microorganisms ingested with the food.

Pepsinogen, once activated, begins protein digestion.
- The major digestive constituent of gastric secretion is pepsinogen.

- Hydrochloric acid cleaves off a small fragment of the pepsinogen molecule, converting it to the active form of the enzyme pepsin.
- Pepsin initiates protein digestion by splitting proteins to yield peptide fragments.

Mucus is protective.
- The surface of the gastric mucosa is covered by a layer of mucus.
- Mucus protects the gastric mucosa against mechanical injury.
- Mucus helps protect the stomach wall from self-digestion.
- Being alkaline, mucus helps protect against acid injury by neutralizing HCl in the vicinity of the gastric lining.

Intrinsic factor is essential for absorption of vitamin B_{12}.
- Intrinsic factor is important in the absorption of certain vitamins.
- Binding of intrinsic factor with vitamin B_{12} triggers the receptor-mediated endocytosis of this complex
- In the absence of intrinsic factor pernicious, anemia results.

Multiple regulatory pathways influence the parietal and chief cells.
- Four chemical messengers primarily influence the secretion of gastric juices. These include acetylcholine, histamine, gastrin, and somatostatin.
- Of the three somatostatin is inhibitory to digestive function, the others are stimulatory.

Control of gastric secretion involves three phases.
- Gastric secretion is divided into three phases: (1) the cephalic, (2) the gastric, and (3) the intestinal phase.
- The cephalic phase of gastric secretion refers to the increased secretion of HCl and pepsinogen even before the food reaches the stomach.

- Thinking about, tasting, smelling, chewing, and swallowing food increases gastric secretion by means of vagal nerve activity.
- Vagal stimulation of the intrinsic plexuses promotes increased secretion of HCl and pepsinogen by the secretory cells.
- Vagal stimulation of the pyloric gland cells causes a release of gastrin.
- The gastric phase of gastric secretion occurs when food actually reaches the stomach.
- Stimuli acting in the stomach—namely, proteins, distention, caffeine, or alcohol— increase gastric secretion by means of overlapping efferent pathways.
- Protein initiates short local reflexes in the intrinsic nerve plexuses to stimulate the secretory cells.
- Protein directly stimulates the release of gastrin.
- The intestinal phase of gastric secretion encompasses the factors originating in the small intestine that influence gastric secretion.
- The presence of the products of protein digestion in the duodenum stimulates further gastric secretion by triggering the release of an intestinal gastrin that is carried by the blood to the stomach.
- This is known as the "excitatory component."
- The inhibitory component of the intestinal phase of gastric secretion is dominant over the excitatory component.
- The inhibitory component is important in helping to shut off the flow of gastric juices as chyme begins to be emptied into the small intestine.

Gastric secretion gradually decreases as food empties from the stomach into the intestine.
- As the meal is gradually emptied into the duodenum, the major stimulus for enhanced gastric secretion is withdrawn.

- After food leaves the stomach the pH falls very low.
- Gastric secretion is inhibited.
- The same stimuli that inhibit gastric motility inhibit gastric secretion.

Carbohydrate digestion continues in the body of the stomach; protein digestion begins in the antrum.
- Carbohydrate digestion continues under the influence of salivary amylase.
- Because food is not mixed with gastric secretions in the body of the stomach, very little protein digestion occurs here.

The stomach absorbs alcohol and aspirin but no food.
- No food or water is absorbed into the blood from the stomach mucosa.
- Even though none of the ingested food is absorbed from the stomach, two noteworthy non-nutrient substances are absorbed directly by the stomach: ethyl alcohol and aspirin.

PANCREATIC AND BILIARY SECRETIONS
The pancreas is a mixture of exocrine and endocrine tissue.
- The pancreas is a mixed gland that contains both exocrine and endocrine tissue.
- The predominant exocrine portion consists of grapelike clusters of secretory cells that form sacs known as "acini."
- The smaller endocrine portion consists of isolated islands of endocrine tissue, the islets of Langerhans.

The exocrine pancreas secretes digestive enzymes and an aqueous alkaline fluid.
- The exocrine pancreas secretes a pancreatic juice consisting of two components: a potent enzymatic secretion and an aqueous alkaline secretion that is rich in sodium bicarbonate.

- The three types of pancreatic enzymes are: (1) proteolytic enzymes, (2) pancreatic amylase, and (3) pancreatic lipase.
- The three major proteolytic enzymes secreted by the pancreas are: (1) trypsinogen, (2) chymotrypsinogen, and (3) procarboxypeptidase.
- When trypsinogen is secreted into the duodenal lumen, it is activated to its active enzyme form, trypsin, by enterokinase.
- Chymotrypsinogen and procarboxypeptidase are converted to their active forms, chymotrypsin and carboxypeptidase within the duodenal lumen.
- Each of the proteolytic enzymes attacks different peptide linkages.
- Pancreatic amylase plays an important role in carbohydrate digestion by converting polysaccharides into disaccharides.
- Pancreatic lipase is extremely important because it is the only enzyme secreted throughout the entire digestive system that can accomplish digestion of fat.
- Pancreatic enzymes function best in a neutral or slightly alkaline environment.
- The alkaline fluid secreted by the pancreas into the duodenal lumen serves the important function of neutralizing the acidic chyme.

Pancreatic exocrine secretion is regulated by secretin and CCK.

- A small amount of parasympathetically induced pancreatic secretion occurs during the cephalic phase of digestion.
- The release of two major enterogastrones, secretin and cholecystokinin, in response to chyme in the duodenum plays the central role in the control of pancreatic secretion.
- The primary stimulus specifically for secretin release is acid in the duodenum.
- Secretin is carried by the blood to the pancreas, where it stimulates the duct cells to markedly increase their secretion of a

sodium bicarbonate—rich aqueous fluid into the duodenum.
- This mechanism provides a control system for maintaining neutrality of the chyme in the intestine.
- Cholecystokinin is important in the regulation of pancreatic enzyme secretion.
- The circulatory system transports cholecystokinin to the pancreas where it stimulates the pancreatic acinar cells to increase secretion.

The liver performs various important functions including bile production.

- The biliary system includes the liver, the gallbladder, and associated ducts.
- The liver's importance to the digestive system is its secretion of bile salts, but it performs a wide variety of other functions, including the following: (1) metabolic processing of the major categories of nutrients; (2) detoxification or degradation of body wastes and hormones as well as drugs and other foreign compounds; (3) synthesis of plasma proteins; (4) storage of glycogen, fats, copper, iron, and many vitamins; (5) activation of vitamin D; (6) removal of bacteria and worn-out red blood cells; (7) secreting the hormones thrombopoietin and insulin-like growth factor-I and (8) excretion of cholesterol and bilirubin.
- Venous blood enters the liver by means of a unique and complex vascular connection between the digestive tract and the liver that is known as the "hepatic portal system."
- The veins draining the digestive tract do not directly join the inferior vena cava.
- The veins from the stomach and intestine enter the hepatic portal vein, which carries the products absorbed from the digestive tract directly to the liver for processing, storage, or detoxification.

The liver lobules are delineated by vascular and bile channels.

- The liver is organized into functional units known as "lobules."
- At the outer edge of the lobule are three vessels: a branch of the hepatic artery, a branch of the portal vein, and a bile duct.
- Blood from the branches of both the hepatic artery and the portal vein flows from the periphery of the lobule in large sinusoids, which run between rows of liver cells to the central vein.
- The central vein of all liver lobules converge to form the hepatic vein.
- Hepatocytes continuously secrete bile into thin channels, the bile canaliculi, which carry the bile to a bile duct at the periphery of the lobule.

Bile is continuously secreted by the liver and is diverted to the gallbladder between meals.

- The liver continuously secretes bile, even between meals
- The opening of the bile duct into the duodenum is guarded by the sphincter of Oddi.
- When the sphincter is closed, the bile is diverted back up into the gallbladder.

Bile salts are recycled through the enterohepatic circulation.

- Bile consists of an aqueous alkaline fluid as well as several organic constituents, including bile salts, cholesterol, lecithin, and bilirubin.
- Bile salts are derivatives of cholesterol.
- Most bile salts are reabsorbed back into the blood by special active transport mechanisms located in the terminal ileum.
- This recycling of bile salts between the small intestine and the liver is referred to as the "enterohepatic circulation."

Bile salts aid fat digestion and absorption.

- Detergent actions refer to the bile salt's ability to convert large fat globules into a liquid emulsion that consists of many small fat droplets suspended in the aqueous chyme, thus increasing the surface area available for attack by pancreatic lipase.
- Bile salts, along with cholesterol and lecithin, play an important role in facilitating fat absorption through micellar formation.
- Micelles, being water soluble by virtue of their hydrophilic shells, can dissolve water-insoluble substances in their lipid-soluble cores.

Bile salts are the most potent stimulus for increased bile secretion; CCK promotes gallbladder emptying

- Any substance that increases bile secretion is called a choleretic.
- The most potent choleretic is bile salts themselves.
- Secretin stimulates an aqueous alkaline bile secretion by the liver.
- Vagal stimulation of the liver plays a minor role in bile secretion during the cephalic phase of digestion.

Bilirubin is a waste product excreted in the bile.

- Bilirubin does not play a role in digestion but is one of the few waste products excreted in the bile.
- Bilirubin is the end product resulting from the degradation of the heme portion of hemoglobin.
- Bilirubin is a yellow pigment that gives bile its yellow color.
- If bilirubin is formed more rapidly than it is excreted, it accumulates in the body and causes jaundice.
- Jaundice can be brought about in three different ways: (1) prehepatic or hemolytic jaundice, (2) hepatic jaundice, and (3) posthepatic or obstructive jaundice.

SMALL INTESTINE

- The small intestine is the site where most digestion and absorption take place.

- The small intestine is arbitrarily divided into three segments: (1) the duodenum, (2) the jejunum, and (3) the ileum.

Segmentation contractions mix and slowly propel the chyme.

- Segmentation consists of oscillating, ringlike contractions of the circular smooth muscle along the length of the small intestine.
- The contractile rings occur every few centimeters, dividing the small intestine into segments like a chain of sausages.
- After a brief period of time, the contracted segments relax, and ringlike contractions appear in the previously relaxed areas.
- Shortly thereafter, the areas of contraction and relaxation alternate again.
- In this way, the chyme is chopped, churned, and thoroughly mixed.
- Segmentation contractions are initiated by the smaller intestine's pacesetter cells, which produce a basic electrical rhythm.
- Parasympathetic stimulation enhances segmentation.
- Sympathetic stimulation depresses segmental activity.

The migrating motility complex sweeps the intestine clean between meals.

- When most of the meal has been absorbed, segmentation contractions cease and are replaced between meals by the migrating motility complex.
- The waves start at the stomach and migrate down the intestine, "sweeping" any remnants of the preceding meal plus mucosal debris and bacteria forward toward the colon.

The ileocecal juncture prevents contamination of the small intestine by colonic bacteria.

- The anatomical arrangement is such that valvelike folds of tissue protrude from the ileum into the lumen of the cecum.

- This ileocecal valve is easily pushed open, but the folds of tissue are forcibly closed when the cecal contents attempt to move backward.
- The smooth muscle within the last several centimeters of the ileal wall is thickened, forming a sphincter, the ileocecal sphincter, which is under neural and hormonal control.

Small intestine secretions do not contain any digestive enzymes.

- The exocrine glands located in the small intestine mucosa secrete into the lumen an aqueous salt and mucus solution.."
- The mucus in the secretion provides protection and lubrication.

The small intestine enzymes complete digestion within the brush-border membrane.

- Digestion within the small intestine lumen is accomplished by the pancreatic enzymes, with fat digestion being enhanced by bile secretion.
- Fat digestion is completed within the small intestine lumen.
- Hairlike projections on the luminal surface of the small intestine epithelial cells form the brush border, which contains three different categories of enzymes: (1) enterokinase, (2) the disaccharidases, and (3) the amino peptidases.

The small intestine is remarkably well adapted for its primary role in absorption.

- Most absorption occurs in the duodenum and jejunum.
- The small intestine has an abundant reserve absorptive capacity.
- The following special modifications of the small intestine mucosa greatly increase the surface area available for absorption: (1) The inner surface of the small intestine is thrown into circular folds. (2) Projecting from this folded surface are microscopic, fingerlike projections known as "villi." (3) Even

smaller hairlike projections known as "microvilli" or brush border arise from the luminal surface of the epithelial cells.

- Malabsorption (impairment of absorption) may be caused by damage to or reduction of the surface area of the small intestine. One of the most common causes is gluten enteropathy, also known as celiac disease.
- Each villus has the following major components: (1) epithelial cells that cover the surface of the villus—these cells possess carriers for absorption of specific nutrients and electrolytes from the lumen, as well as the intracellular digestive enzymes that complete carbohydrate and protein digestion; (2) a connective tissue core formed by the lamina propria; (3) a capillary network; and (4) a terminal lymphatic vessel known as a "central lacteal."
- Like renal transport, intestinal absorption may be an active or passive process, with active absorption involving energy expenditure during at least one of the steps in the transepithelial transport process.

The mucosal lining experiences rapid turnover.

- The epithelial cells lining the small intestine slough off and are replaced at a rapid rate as a result of high mitotic activity in the crypts of Lieberkühn.
- The old cells that are sloughed off into the lumen are digested, with the cell constituents being absorbed into the blood and reclaimed for synthesis of new cells, among other things.
- Energy-dependent Na+ absorption drives passive water absorption. Sodium may be absorbed both passively and actively.
- Most water absorption in the digestive tract depends on the active carrier that pumps sodium into the lateral spaces.

- Water entering the space reduces the osmotic pressure but raises the hydrostatic pressure.

Carbohydrate and protein are both absorbed by secondary active transport and enter the blood.

- Absorption of the digestion end products of both carbohydrates and proteins is accomplished by Na^+- energy-dependent secondary active transport
- Dietary carbohydrate is presented to the small intestine for absorption mainly in the forms of the disaccharides, maltose, sucrose, and lactose.
- Glucose and galactose are both absorbed by secondary active transport.
- The operation of these symporters carriers depends on the sodium concentration gradient established by the energy-consuming basolateral sodium-potassium pump.
- Glucose leaves the cell down its own concentration gradient to enter the blood.
- Not only are ingested proteins digested and absorbed, but endogenous proteins are digested and absorbed as well.
- Amino acids are absorbed across the intestinal cells by secondary active transport.
- Fat absorption is quite different from carbohydrate and protein absorption because the insolubility of fat in water presents a special problem.
- Biliary components facilitate absorption of these fatty end products through formation of micelles.
- Once the micelles reach the luminal membranes of the epithelial cells, the monoglycerides and free fatty acids passively diffuse from the micelles through the lipid component of the epithelial cell membranes to enter the interior of these cells.
- Once within the interior of epithelial cells, the mono glycerides and free fatty acids are resynthesized into triglycerides and move into chylomicrons

- The large, coated fat droplets, known as chylomicrons, enter the central lacteals rather than the capillaries because of the structural differences between these two vessels.
- Fat can be absorbed into the lymphatics but not directly into the blood.
- Fat absorption is a passive process.
- Water-soluble vitamins are primarily absorbed with water, whereas fat-soluble vitamins are carried in the micelles and absorbed passively with the end products of fat digestion.
- Dietary iron and calcium may not be absorbed completely because their absorption is subject to regulation, depending on the body's needs for these electrolytes

Most absorbed nutrients immediately pass through the liver for processing.
- Anything absorbed into the digestive capillaries first must pass through the hepatic biochemical factory before entering the general circulation.
- Fat is picked up by the central lacteal and enters the lymphatic system instead.

Extensive absorption by the small intestine keeps pace with secretion.
- The digestive juices are not lost from the body.
- After the constituents of the juices are secreted into the digestive tract lumen and perform their function, they are returned to the plasma.
- The only secretory product that escapes from the body is bilirubin, a waste product that must be eliminated.

Diarrhea results in loss of fluid and electrolytes.
- The other common digestive tract disturbance that can lead to loss of fluid and an acid-base imbalance is diarrhea.
- The most common cause of diarrhea is excessive intestinal motility, which arises either from local irritation of the gut wall by bacteria or viral infection of the intestine, or from emotional stress.

LARGE INTESTINE
The large intestine is primarily a drying and storage organ.
- The large intestine consists of the colon, cecum, appendix, and rectum.
- Since most digestion and absorption have been accomplished in the small intestine, the contents delivered to the colon consist of indigestible food residues, unabsorbed biliary components, and the remaining fluid.
- The colon extracts more water and salt from the contents.
- What remains to be eliminated is known as "feces."

Haustral contractions slowly shuffle the colonic contents back and forth.
- The colon's primary method of motility is haustral contractions initiated by the autonomous rhythmicity of colonic smooth muscle cells.

Mass movements propel colonic contents long distances.
- Haustral contractions are largely controlled by locally mediated reflexes involving the intrinsic plexuses.
- Slow colonic movement allows bacteria time to grow and accumulate in the large intestine.
- 3-4 times a day a marked increase in motility takes place during which large segments of the ascending and transverse colon contract simultaneously, driving the feces one third to three fourths of the length of the colon in a few seconds. These massive contractions, appropriately called mass movements
- When food enters the stomach, the gastrocolic reflex mediated by gastrin and by the extrinsic autonomic nerves initiates mass movements

Feces are eliminated by the defecation reflex.
- When mass movements of the colon move fecal material into the rectum, the resultant distention of the rectum

stimulates stretching receptors in the rectal wall, thus initiating the defecation reflex.
- The defecation reflex causes the internal anal sphincter to relax and the rectum and sigmoid colon to contract more vigorously.
- If the external anal sphincter is also relaxed, defecation occurs.

Constipation occurs when the feces become too dry.
- If defecation is delayed too long, constipation may result.
- When the colonic contents are retained for longer periods of time than normal, more than the usual amount of water is absorbed.

Large intestine secretion is entirely protective.
- Colonic secretion consists of an alkaline mucus solution whose function is to protect the large intestine mucosa from mechanical and chemical injury.
- No digestion takes place in the large intestine because there are no digestive enzymes.

The colon contains myriad beneficial bacteria.
- About 500-1000species of bacteria reside in the colon.
- They are not only harmless but many are beneficial.

The large intestine absorbs salt and water, converting the luminal contents into feces.
- The colon normally absorbs some salt and water.
- The main waste product excreted in the feces is bilirubin.

OVERVIEW OF GASTROINTESTINAL HORMONES
- Chyme in the stomach stimulates the release of gastrin, which acts on parietal and chief cells to increase secretion of HCl and pepsinogen.
- Gastrin enhances gastric motility, stimulates ileal motility, relaxes the ileocecal sphincter, and induces mass movements in the colon.
- Secretion performs four major interrelated functions: (1) It inhibits gastric emptying to prevent further acid from entering the duodenum until the acid that is present is neutralized. (2) It inhibits gastric secretion to reduce the amount of acid being produced. (3) It stimulates the pancreatic duct cells to produce a large volume of aqueous sodium bicarbonate secretion, which is emptied into the duodenum to neutralize the acid. (4) It stimulates secretion by the liver of a sodium bicarbonate-rich bile, which likewise is emptied into the duodenum to assist in the neutralization process.
- Cholecystokinin also performs several important interrelated functions: (1) It inhibits gastric motility and secretion, thereby allowing adequate time for the nutrients already in the duodenum to be digested and absorbed. (2) It stimulates the pancreatic acinar cells to increase secretion of pancreatic enzymes, which continue the digestion of these nutrients in the duodenum. (3) It causes contraction of the gallbladder and relaxation of the sphincter of Oddi so that bile is emptied into the duodenum to aid fat digestion and absorption.
- Cholecystokinin is an important regulator of food intake.
- Cholecystokinin plays a key role in satiety, the sensation of having had enough to eat.
- Gastric inhibitory peptide plays an important role in inhibiting gastric motility and secretion in response to fat, acid, hypertonicity, and distention in the duodenum.
- Gastric inhibitory peptide initiates the release of insulin in anticipation of the absorption of the meal.

Key Terms

Absorption
Accessory digestive organs
Basic electrical rhythm (BER)
Bile salts
Bilirubin
Bolus
Carboxypeptidase
Cholecystokinin
Choleretic
Chyme
Chymotrypsin
Chymotrypsinogen
Digestion
Digestive system
Digestive tract
Emetics
Enteric amylase
Enteric nervous system
Enterogastric reflex
Enterokinase
Gallbladder
Gallstones
Gastric filling
Gastric inhibitory peptide
Gastrin
Haustra
Heartburn
Hepatic portal system
Hepatocyte
Hernias
Hydrolysis
Interstitial cells of Cajal
Intrinsic factor
Islets of Langerhans
Long reflex
Malabsorption
Mass movements
Mesentery
Micelle
Microvilli
Migrating motility complex
Motility
Mucosa
Mucous membrane
Mucus
Muscularis externa
Myenteric plexus
Oropharyngeal phase

Pancreatic lipase
Parotid glands
Pepsin
Pepsinogen
Peptic ulcer
Peristalsis
Plasticity
Procarboxypeptidase
Saliva
Salivary amylase
Secretin
Secretion
Segmentation
Serosa
Short reflexes
Slow wave potential
Sublingual glands
Submucosa
Succus entericus
Swallowing
Taeniae coli
Villi
Vomiting center
Xerostomia

Review Exercises
Answers are in the appendix.

True/False-if an answer is false, underline the incorrect portion of the statement and rewrite it to make the statement true.

_____ 1. The intrinsic nerve plexus of the digestive tract controls all aspects of digestive activity.

_____ 2. Starch, glycogen, and cellulose are all disaccharides.

_____ 3. The end products of fat digestion are triglycerides and free fatty acids.

_____ 4. Digestion is accomplished by enzymatic hydrolysis.

_____ 5. The stomach can expand 2 times its empty volume to accommodate food intake. .

_____ 6. The submucosa is the outermost layer of the digestive tract.

_____ 7. The liver continues to secrete bile constantly, even between meals.

_____ 8. The palate separates the mouth from the nasal passages.

_____ 9. The tongue is composed of voluntarily controlled smooth muscle.

_____ 10. The stem cells of the stomach turnover at a rate so that the entire stomach mucosa is replaced every 3 days.

_____ 11. Large intestine bacteria are generally beneficial to our overall health.

_____ 12. Bicarbonate buffers are found in saliva.

_____ 13. Salivary amylase breaks down polysaccharides.

_____ 14. The hard outer covering or the crown of a tooth is enamel.

_____ 15. Swallowing is known as "deglutition."

_____ 16. Motilin release is inhibited by feeding.

_____ 17. The antrum is that portion of the stomach that lies above the esophageal opening.

_____ 18. The interior of the stomach is thrown into deep folds known as rugae.

_____ 19. Gastrin influences the rate of gastric emptying.

_____ 20. Cholecystokinin is responsible for the contraction of the gallbladder.

_____ 21. GIP increases insulin release.

_____ 22. Chief cells secrete mucus.

_____ 23. "Oxyntic cells" is another name for mucous neck cells.

_____ 24. Hydrochloric acid activates pepsinogen into the active enzyme pepsin.

_____ 25. The major digestive constituent of gastric secretion is trypsinogen.

_____ 26. Peptic ulcers frequently occur when the gastric mucosal barrier is severely eroded or broken.

_____ 27. The stomach absorbs aspirin.

_____ 28. Water is absorbed from the stomach contents.

_____ 29. The exocrine portion of the pancreas is known as the "islets of Langerhans."

_____ 30. Trypsinogen is activated to its active enzyme form, trypsin, by HCl.

_____ 31. Pancreatic juice is an aqueous acidic solution.

_____ 32. Cholecystokinin is important in the regulation of pancreatic enzyme secretion.

_____ 33. Decrease is saliva production is known as celiac disease.

_____ 34. Macrophages of the liver are called "Kupffer cells."

_____ 35. Bile is produced in the gallbladder.

_____ 36. Most absorption occurs in the duodenum and jejunum.

_____ 37. Small intestine secretions do not contain any digestive enzymes.

_____ 38. Villi, the fingerlike projections in the small intestine, are also known as the "brush border."

_____ 39. The acid-base balance of the body is not altered by the digestive processes.

_____ 40. The colon's primary method of motility is Haustral contractions.

_____ 41. Cholecystokinin inhibits gastric motility.

_____ 42. Secretin inhibits gastric secretion.

Fill in the Blank

43. _____ is produced by the stomach and is necessary for Vitamin B 12 absorption.

44. _____ activates the pancreatic enzyme trypsinogen.

45. _____ hydrolyze the small peptide fragments into their amino acid components, completing protein digestion.

46. Projecting from the small intestines' folded surface are microscopic fingerlike projections known as _____.

47. _____ may be caused by damage to or reduction of the surface area of the small intestine.

48. The neurotransmitter _____ stimulates production of large volumes of watery saliva.

49. Vitamin _____ is necessary for proper absorption of _____.

50. The first step in the digestive process is _____.

51. Saliva is composed of about _____ H$_2$O and _____ protein and electrolytes.

52. _____ is a condition characterized by salivary secretions.

53. With _____, salivation occurs without oral stimulation.

54. The _____ separates the mouth from the nasal passages.

55. Some therapeutic agents can be absorbed by the oral mucosa. An example is _____.

56. A(n) _____ is a ball of food.

57. The _____ consists of moving the bolus from the mouth through the pharynx and into the esophagus.

58. The lower esophageal sphincter is the _____.

59. _____ refers to ringlike contractions of the circular smooth muscle that move progressively forward with a stripping motion.

60. _____ are housekeeping contractions that occur in the small intestines between meals.

61. _____ are the pacemaker cells of the digestive tract.

62. The small fingerlike projection at the bottom of the cecum is the _____.

63. The colon's primary method of motility is _____.

64. The large intestine consists of the _____, _____, _____, and _____.

65. Malabsorption related to gluten enteropathy is known as _____ disease.

66. _____ plays an important role in inhibiting gastric motility and secretion in response to fat, acid, hypertonicity, and distention in the duodenum.

67. The _____ is the portion of the stomach that lies above the esophageal opening.

68. The lower portion of the stomach is called the _____.

69. The mucosal gland cells are found in _____.

70. Endocrine cells in the PGA secrete the hormone _____ into the blood.

71. Pepsinogen is activated by _____.

72. _____ phase of gastric secretion occurs when food actually reaches the stomach.

73. A(n) _____ occurs when the gastric wall is injured by its acidic and enzymatic contents.

74. _____ is a condition associated with elevated levels of bilirubin in the blood.

75. The _____ is the largest and most important metabolic organ in the body.

76. The opening of the bile duct into the duodenum is guarded by the _____.

77. _____ are derivatives of cholesterol.

78. Any substance that increases bile secretion by the liver is called a(n) _____.

79. The presence of food in the duodenal lumen triggers the release of _____.

80. The _____ phase of controlling stomach activity occurs before food is ingested.

81. Digestion is accomplished by enzymatic _____.

Matching
Match the following characteristics to the condition/situations listed below.

a. xerostomia
b. heartburn
c. gallstones
d. emesis
e. pernicious anemia
f. peptic ulcer
g. steatorrhea

h. jaundice
i. constipation
j. conditioned reflex
k. lactose intolerance
    malabsorption
m. gluten enteropathy
n. entero-gastric reflex

_____ 82. This situation occurs when we see and smell food and begin to secrete saliva.

_____ 83. This condition is characterized by diminished salivary secretion.

_____ 84. This is the forceful expulsion of gastric contents out through the mouth.

_____ 85. This clinical manifestation is excessive undigested fat in the feces.

_____ 86. This condition is impaired absorption in the intestines.

_____ 87. This situation occurs when signals from the duodenum inhibit stomach emptying.

_____ 88. This condition occurs when the gastric contents flow back into the esophagus.

_____ 89. In this condition the small intestine is abnormally sensitive to gluten.

_____ 90. This condition is characterized by malabsorption of Vitamin B12.
_____ 91. This condition occurs when bilirubin accumulates in the body.

_____ 92. This fairly common disorder involves a deficiency of lactose.

_____ 93. . This condition results from slower than normal motility in the large intestines.

_____ 94. This condition is characterized by an erosion of the stomach wall.

_____ 95. This condition is characterized by accumulation of bile in the biliary ducts.

Multiple Choice

96. Which category of foodstuffs is characterized by having peptide bonds?
    a. starches
    b. carbohydrates
    c. proteins
    d. fats
    e. triglycerides

97. Where is the salivary center located within the brain?
    a. pons
    b. medulla
    c. cerebellum
    d. cerebrum
    e. hypothalamus

98. Which of the following is absorbed in the mouth?
    a. starch
    b. amino acids
    c. lipids
    d. ethyl alcohol
    e. nitroglycerin

99. Which of the hormones trigger gastric motility?
    a. gastrin
    b. pepsin
    c. trypsin
    d. gastrokinin
    e. villikinin

100. Which of the following pancreatic proteolytic enzymes are converted to their active form by trypsin?
    a. trypsinogen
    b. pepsinogen
    c. enterokinase
    d. pancreatic lipase
    e. procarboxypeptidase

101. Which of the following has specialized cells known as "Kupffer cells"?
    a. islets of Langerhans
    b. gallbladder
    c. liver
    d. salivary glands
    e. small intestines

102. The opening of the bile duct into the duodenum is guarded by this sphincter.
    a. esophageal sphincter
    b. pyloric sphincter
    c. ileocecal sphincter
    d. sphincter of Acinius
    e. sphincter of Oddi

103. Which of the following play an important role in facilitating fat absorption through micellar formation?
    a. apoenzymes
    b. bile salts
    c. cholecystokinin
    d. disaccharidases
    e. enterokinase

104. Approximately how much fluid is absorbed by the small intestine per day?
    a. 500 ml
    b. 1,000 ml
    c. 1,200 ml
    d. 9,000 ml
    e. 9,500 ml

105. Which of the following is not associated with fat digestion and absorption?
    a. micelles
    b. Enterogastro antrumi chylomicrons
    c. Clestridium pepti bile
    d. Bacillus pylori peptidase
    e. Helicobacter pepti lipase

Modified Multiple Choice
Indicate which hormone(s) is (are) being described by writing the appropriate letter in the blank using the following answer code.

A = both secretin and CCK
B = secretin
C = CCK
D = gastrin

106. _____ Secreted by the pyloric gland area.
107. _____ Stimulated primarily by the presence of fat (fatty acids).
108. _____ Stimulated primarily by the presence of protein.
109. _____ Secretion is inhibited by a low pH.
110. _____ Inhibit(s) gastric secretion.
111. _____ Stimulate(s) gastric secretion.
112. _____ Inhibits gastric motility.
113. _____ Stimulate(s) pancreatic digestive enzyme secretion.
114. _____ Stimulate(s) pancreatic aqueous $NaHCO_3$ secretion.
115. _____ Stimulate(s) gallbladder contraction.
116. _____ Known as "enterogastrone."
117. _____ Secreted by the small intestine mucosa.
118. _____ Stimulate(s) the pancreatic acinar cells.
119. _____ Stimulate(s) the pancreatic duct cells.
120. _____ Stimulated primarily by the presence of acid.

Identify which part of the digestive system is associated with each item in question by writing the appropriate letter in the blank using the following answer code.

A = small intestine
B = large intestine
C = mouth
D = esophagus
E = stomach

121. _____ Absorbs weak acids such as aspirin.
122. _____ Primary motility is haustrations.
123. _____ Receptive relaxation occurs.
124. _____ Secondary peristaltic waves occur.
125. _____ Primary motility is segmentation.
126. _____ BER occurs at a rate of three contractions per minute.
127. _____ Absorbs nitroglycerin.
128. _____ Involved with secondary stage of swallowing.
129. _____ Involved with mastication.
130. _____ Food is converted into chyme.
131. _____ Chyme is converted into feces.
132. _____ Main site of digestion and absorption.
133. _____ Secretion is entirely mucus.
134. _____ Protein digestion begins here.
135. _____ Carbohydrate digestion begins here.
136. _____ Fat digestion begins here.

Points to Ponder

1. How selective is the digestive tract in its absorption process? How would you explain the absorption of indigestible compounds such as inulin? What about the absorption of known carcinogens?

2. What happens to the pH in your mouth when you sleep with your mouth open?

3. After studying this chapter how do you feel about the disorder discussed in television advertisements known as irregularity?

4. In terms of digestion what is happening when food is cooked?

5. Why do dental procedures usually stimulate salivation?

6. What are the different types of muscular action that occur during peristalsis?

7. What is the purpose of pepsin in the stomach?

8. What are "hunger pangs"?

9. What pacemakers occur in the digestive tract?

10. Why is the absorption of fats more complex than the absorption of proteins and carbohydrates?

Clinical Perspectives

1. Are there serious consequences when a person strains to have a bowl movement while constipated?

2. How do you suppose the bile flows through the ducts of the biliary systems?

3. What causes the dryness of "morning mouth"?

4. How would you explain heartburn to your roommate?

5. Your friend suffers from lactose intolerance. She wants to know why she has a problem. After reading this chapter on the digestive system, how would you explain lactose intolerance to her?

6. Your father has had his gall bladder removed (cholecystectomy). He knows that you are taking a physiology course and asks you if the gallbladder is essential for the digestive process. What would you tell him?

7. Your friend wants to know what gastroesophageal reflux disease is and how it can be treated. How would you explain this problem to him?

8. What are some factors that can cause cirrhosis of the liver? Jaundice?

9. What causes gallstones?

10. Why is the enteric nervous system sometimes called the "enteric brain"?

16

Energy Balance and Temperature Regulation

Chapter Overview

Food intake is essential to power cell activities. Energy balance and thus body weight are maintained by controlling food intake. The thousands of chemical reactions within the body are frequently referred to as the "metabolic processes" of the organism. Metabolism requires energy and has an optimum temperature range. The survival of cells depends on certain intracellular chemical reactions. In maintaining homeostasis the body systems utilize chemical processes for the benefit of all cells. In this chapter, the balance of energy and the regulation of temperature necessary for homeostasis are considered.

The first law of thermodynamics states that energy cannot by created or destroyed but can be converted to another form. The second law of thermodynamics states that energy will be lost as energy as it is converted from one form to another. Thus, energy balance is interrelated to temperature production. Inasmuch as energy is not supplied in a form that can be utilized directly, heat is produced as the body converts the energy from one form to another. To meet the energy needs of the organism, and at the same time maintain an optimum temperature, various body systems interact to provide an energy balance and adequate methods of thermoregulation. The controls are the same; i.e., nervous and endocrine. The phenomenon is homeostasis.

Chapter Outline

ENERGY BALANCE
Most food energy is ultimately converted into heat in the body.
*   According to the first law of thermodynamics, energy can neither be created nor destroyed, but it can be converted from one form to another.
*   The energy in nutrient molecules that is not used to energize work is transformed into thermal energy, or heat.
*   The energy in ingested foodstuffs constitutes energy input to the body.
*   Energy output or expenditure falls into two categories: (1) external work, and (2) internal work.

The metabolic rate is the rate of energy use.
*   The rate at which energy is expended by the body during both external and internal work is known as the "metabolic rate."
*   The so-called basal metabolic rate is a reflection of the body's "idling speed," or the minimal waking rate of internal energy expenditure.

Energy input must equal output to maintain a neutral energy balance.
*   Since energy cannot be created or destroyed, energy input must equal energy output.

- There are three possible states of energy balance: (1) neutral energy balance, (2) positive energy balance, and (3) negative energy balance.
- To maintain a constant body weight, energy acquired through food intake must equal energy expenditure by the body.

Food intake is controlled primarily by the hypothalamus.

- Control of food intake is primarily a function of the hypothalamus.
- The hypothalamus is considered to house a pair of appetite centers located in the lateral regions, one on each side, and another pair of satiety centers located in the ventromedial area.
- The appetite centers tell us to eat, whereas the satiety centers tell us when we have had enough.
- Neuropeptide Y and melanocortin are two chemical messengers that influence food intake and energy homeostasis.
- Adipose cells secrete hormones called adipokines that play a role in energy balance. One of the most important is leptin.
- In addition, insulin also plays a role in long-term control of body weight.
- Secondary hypothalamic regions receive signals from the arcuate nucleus and release hormones to affect appetite including orexins that increase food intake and CRH that decreases food intake.
- In addition to the key role the hypothalamus plays in maintaining energy balance, a satiety center in the brain stem processes signals important in the feeling of being full and thus contributes to short-term control of meals
- The following factors are among those that have been hypothesized as contributing to the control of food intake: (1) the size of fat stores; (2) the extent of gastrointestinal distention; (3) the extent of glucose utilization; (4) the intensity of cell power production; (5) the level of cholecystokinin; (6) the influences of neurotransmitters; and (7) psychosocial influences.

Obesity occurs when more kilocalories are consumed than are burned up.

- Obesity is defined as excessive fat content in the adipose tissue stores.
- The number of obese adults in the United States is 75% higher now than it was 15 years ago.
- The causes of obesity are many and obscure. Some factors that may be involved with obesity include: (1) emotional disturbances, where overeating replaces other gratifications; (2) development of an excessive number of fat cells as a result of overfeeding; (3) certain endocrine disorders such as hypothyroidism; (4) disturbances of the satiety-appetite regulatory centers in the hypothalamus; (5) hereditary tendencies; (6) the palatability of available food; (7) lack of physical exercise (8) too little sleep; and (8) possible viral link.
- It is known that obesity, especially of the android type, can predispose an individual to illness and premature death from a multitude of diseases.

Persons suffering from anorexia nervosa have a pathological fear of gaining weight.

- Chronic diseases such as renal failure, cancer, and tuberculosis are commonly accompanied by a lack of appetite.
- Researchers have identified a protein called cachetin, which is released by the immune system and is believed to contribute to the development of emaciation.
- Patients with anorexia nervosa, most commonly adolescent girls and young women, have a morbid fear of becoming fat.

- Because they have an aversion to food, they eat very little and consequently lose considerable weight, perhaps even starving themselves to death.
- Other characteristics of the condition include altered secretion of many hormones, absence of menstrual periods, and low body temperature.

TEMPERATURE REGULATION
Internal core temperature is homeostatically maintained at 100° F.
- Humans are usually in environments cooler than their bodies, so they must constantly generate heat internally to maintain their body temperatures.
- Most people suffer convulsion when the internal body temperature reaches about 106° F.
- The upper limit is considered to be 110° F.
- Most of the body's tissues can transiently withstand substantial cooling.
- The cooled tissues need less nourishment than they do at normal body temperature because of their pronounced reduction in metabolic activity.
- Normal body temperature has traditionally been considered 98.6° F (37° C).
- A recent study indicates that normal body temperature varies among individuals and varies throughout the day, ranging from 96.0° F in the morning to 99.9° F in the evening.
- From a thermoregulatory viewpoint, the body may be conveniently viewed as a central core surrounded by an outer shell.

- Skin temperatures may fluctuate between 68° and 104° F without damage.
- We are accustomed to thinking in terms of oral, rectal, or axillary temperature because these are easy sites for monitoring body temperature.
- Recently available is a temperature-monitoring instrument that scans the heat generated by the eardrum and converts this temperature into an oral equivalent.
- Most people's core temperature normally varies during the day, with the lowest level occurring early in the morning before rising and the highest point occurring in late afternoon.
- The core temperature may vary slightly with exposure to extremes of temperatures.

Heat input must balance heat output to maintain a stable core temperature.
- Balance between heat input and output is frequently disturbed by: (1) changes in internal heat production by exercise, and (2) changes in the external environmental temperature.
- If the core temperature starts to fall, heat production is increased and heat loss is minimized so that normal temperature can be restored.
- If the temperature starts to rise above normal, it can be corrected by increasing heat loss and reducing heat production.

Heat exchange takes place by radiation, conduction, convection, and evaporation.
- Heat always moves down its own concentration gradient; that is, down a thermal gradient from a warmer to a cooler region.
- Radiation is the emission of heat energy from the surface of a warm body in the form of electromagnetic waves.
- Conduction is the transfer of heat between objects of differing temperatures that are in direct contact with each other.
- Convection refers to the transfer of heat energy by air currents.

- When water evaporates from the skin surface, the heat required to transform water from a liquid to a gaseous state is absorbed from the skin, thereby cooling the body.
- Sweating is an active evaporative heat-loss process under sympathetic nervous control.
- Sweat is a dilute salt solution that is actively extruded to the surface of the skin by sweat glands dispersed all over the body.
- The most important factor determining the extent of evaporation of sweat is the relative humidity of the surrounding air.

The hypothalamus integrates a multitude of thermosensory inputs.

- The hypothalamus serves as the body's thermostat.
- The hypothalamus must be continuously apprised of both the skin temperature and the core temperature by means of specialized temperature-sensitive receptors called "thermoreceptors."
- Peripheral thermoreceptors monitor skin temperature.
- The core temperature is monitored by central thermoreceptors.

Shivering is the primary involuntary means of increasing heat production.

- The hypothalamus takes advantage of the fact that increased skeletal muscle activity generates more heat.
- Shivering consists of rhythmic, oscillating skeletal muscle contractions that occur at a rapid rate of 10–20 per second.

- Shivering is a very effective mechanism in increasing heat production.
- Nonshivering thermogenesis also plays a role in thermoregulation.
- Nonshivering thermogenesis is mediated by the hormones epinephrine and thyroid hormone.

The magnitude of heat loss can be adjusted by varying the flow of blood through the skin.

- Most of the flow of blood through the skin is for purposes of temperature regulation.
- Skin vasomotor responses are coordinated by the hypothalamus.
- Increased sympathetic activity to the skin vessels produces heat-conserving vasoconstriction in response to cold exposure, whereas decreased sympathetic activity produces heat-losing vasodilation of the skin vessels in response to heat exposure.

The hypothalamus simultaneously coordinates heat-production and heat-loss mechanisms.

- In response to cold exposure, the posterior region of the hypothalamus directs increased heat production such as by shivering, while simultaneously decreasing heat loss by skin vasoconstriction and other measures. Behavioral changes such as hunching over reduces surface area from which heat can escape and putting on blankets and warm clothing to trap insulating air.
- During heat exposure the anterior part of the hypothalamus reduces heat production by decreasing skeletal muscle activity and promotes increased heat loss by inducing skin vasodilation.
- When even the maximal skin vasodilation is inadequate to rid the body of excess heat, sweating is brought into play to accomplish further heat loss through evaporation.

During a fever, the hypothalamic thermostat is "reset" at an elevated temperature.

- Fever refers to an elevation in body temperature as a result of infection or inflammation.

- In response to microbial invasion, white blood cells release endogenous pyrogen, which acts on the hypothalamic thermoregulatory center to raise the setting of the thermostat.
- The hypothalamus now maintains the temperature at the new set level instead of maintaining normal body temperature.
- Endogenous pyrogen raises the set point of the hypothalamic thermostat during fever production by triggering the local release of prostaglandins.
- Aspirin reduces a fever by inhibiting the synthesis of prostaglandins.

Key Terms

Adipocytes
Anorexia nervosa
Arcuate nucleus
Basal metabolic rate (BMR)
Brown fat
Calorie
Cholecystokinin (CCK)
Conduction
Convection
Core temperature
Electromagnetic waves (heat waves)
Endogenous pyrogen
Energy balance
Energy equivalent of O_2
Evaporation
External work
Feeding (appetite) centers
Ghrelin
Indirect calorimetry
Internal work

Kilocalorie (Calorie)
Leptin
Melanocortins
Metabolic rate
Neuropeptide Y (NPY)
Non exercise activity thermogenesis (NEAT)
Nonshivering (chemical) thermogenesis
Obesity
Orexins
Radiation
Satiety
Satiety centers
Shivering
Sweat
Sweating
Temperature regulation
Thermal energy (heat)
Thermoreceptors

Review Exercises

Answers are in the appendix.

True/False-if an answer is false, underline the incorrect portion of the statement and rewrite it to make the statement true.

_____ 1. Internal work refers to the energy expended when skeletal muscles are contracted to move external objects.

_____ 2. Energy cannot be created or destroyed or converted from one form to another form.

_____ 3. The rate at which energy is expended by the body is known as the "metabolic rate."

_____ 4. The appetite centers are located in the lateral regions of the thalamus.

_____ 5. Overall about 75% of the energy from our food is lost to heat during digestion and cellular respiration.

_____ 6. Cachetin is released by the immune system.

_____ 7. Orexins increase food intake. .

_____ 8. Skin temperature may fluctuate between 68° and 104° F without damage.

_____ 9. Rectal temperatures are the same as oral temperatures.

_____ 10. Heat moves down a thermal gradient from a warmer to a cooler region.

_____ 11. Conduction is the transfer of heat using air currents.

_____ 12. The hypothalamus serves as the body's thermostat.

_____ 13. Sweating is an active evaporative heat-loss process.

_____ 14. Most of the blood flow through the skin is for purposes of temperature regulation.

Fill in the Blank

15. _____ sympathetic activity to the skin vessels produces heat-conserving vasoconstriction in response to cold exposure.

16. _____ is made by the stomach and stimulates hunger.

17. Satiety centers are located in the _____.

18. In response to heat exposure the _____ part of the hypothalamus reduces heat production by _____ skeletal muscle activity and promotes increased heat loss by inducing skin _____.

19. Aspirin reduces a fever by inhibiting the synthesis of _____.

20. The most important adipokine is _____.

21. Sweating is under _____ nervous control.

22. _____ refers to the transfer of heat energy by air currents.

23. The prominent feature in anorexia nervosa is a(n) _____.

24. Obesity levels in the US have increased _____ percent over the past 15 years..

25. Energy output or expenditure falls into two categories: _____ and
_____.

26. _____ energy balance exists if the amount of energy in food intake is greater than the amount of energy expended.

27. _____ is the transfer of heat between objects of differing temperatures that are in direct contact with each other.

28. _____ is the emission of heat energy from the surface of a warm body.

Matching

Match the situation or condition to the approximate temperature in degrees Fahrenheit.

_____ 29. Upper limit compatible with life   a. $100°$ F

_____ 30. Central core temperature   b. $68°$ to $104°$ F

_____ 31. Oral temperature   c. $99.7°$ F

_____ 32. Skin temperature   d. $110°$ F

_____ 33. Rectal temperature   e. $106°$ F

_____ 34. Convulsions at this temperature and higher   f. $98.6°$ F

_____ 35. Core temperature during hard exercise   g. $104°$ F

Multiple Choice

36. Which of the following are released in response to endogenous pyrogens?
    a. cachetin
    b. cholecystokinin
    c. prostaglandins
    d. brown fat
    e. electromagnetic waves

37. Which of the following nervous activities is utilized to control the skin vasomotor responses when the body is subjected to heat exposure?
    a. decreased parasympathetic activity
    b. increased sympathetic activity
    c. increased parasympathetic activity
    d. increased thermoreceptor activity
    e. decreased sympathetic activity

38. Which of the following nervous activities is utilized to control the skin vasomotor responses when the body is subjected to cold exposure?
    a. decreased parasympathetic activity
    b. increased sympathetic activity
    c. increased parasympathetic activity
    d. increased thermoreceptor activity
    e. decreased sympathetic activity

39. Which part of the brain houses the body's thermostat?
    a. cerebral cortex
    b. cerebellum
    c. pons
    d. hypothalamus
    e. medulla

40. Which part of the brain houses the satiety centers?
    a. pons
    b. thalamus
    c. hypothalamus
    d. medulla
    e. pars nervosa

41. Which of the following activates shivering in response to cold exposure?
    a. pons
    b. thalamus
    c. hypothalamus
    d. medulla
    e. cerebellum

42. Which of the following is mediated by epinephrine?
    a. shivering
    b. nonshivering thermogenesis
    c. increased glucose utilization
    d. decreased thyroid hormone production
    e. decreased in physical exercise

43. Which of the following increase carbohydrate consumption?
    a. cachetin
    b. serotonin
    c. dopamine
    d. neuropeptide
    e. insulin

44. Which of the following decrease carbohydrate consumption?
    a. glucagon
    b. neuropeptide
    c. dopamine
    d. epinephrine
    e. cachetin

45. Which of the following is characterized by a low body temperature?
    a. hard exercise
    b. microbial invasion
    c. obesity
    d. anorexia nervosa
    e. cancer

46. Changes in BMR are tied directly to changes in what hormone?
    a. growth hormone
    b. thyroid hormone
    c. parathyroid hormone
    d. insulin
    e. none of the above

47. Which type of obesity is characterized by abdominal fat distribution around the waist?
    a. glucostatic obesity
    b. android obesity
    c. ischymetric obesity
    d. psychosocial obesity
    e. gynoid obesity

48. Which of the following represents the normal variation in most people's core temperature during the day?
    a. 0° C
    b. 5° C
    c. 3° C
    d. 1° C
    e. 7° C

Modified Multiple Choice

Indicate which form of heat loss is being described by writing the appropriate letter in the blank using the following answer code.

> A = convection
> B = evaporation
> C = radiation
> D = conduction

49. _____ Emission of heat energy from the surface of a warm body in the form of electromagnetic waves.
50. _____ Heat loss by means of conversion of water from a liquid to a gaseous state.
51. _____ Transfer of heat between objects of differing temperatures that are in direct contact with each other.
52. _____ Transfer of heat by air or water currents.

Indicate which form of thermal descriptor is being described by writing the appropriate letter in the blank using the appropriate code.

A = energy equivalent
B = external work
C = calorie
D = basal metabolic rate
E = internal work

53. _____ Energy expended by contraction of skeletal muscles to move external objects to move the body.
54. _____ All forms of biological energy expenditure that do not accomplish work outside the body.
55. _____ Affected by thyroid hormone levels..
56. _____ Amount of heat required to raise the temperature of one gram of water 1°C.
57. _____ Quantity of heat produced per liter of oxygen consumed.
58. _____ Minimal waking rate of internal energy expenditure.

Points to Ponder

1. Using common sense and simple logic how do you know that the hypothalamus raises the thermostat setting during fever?
2. Should you take aspirin at the onset of fever? Why do you have fever in response to microbial invasions?

3. What would you do if your roommate or member of your family developed anorexia nervosa? Could you help their situation?

4. Should we begin to treat obesity as a disorder or continue with the present social attitude?

5. In the body, what do anabolic reactions accomplish? Catabolic reactions?

6. Under what major condition is gluconeogenesis employed?

7. In what form do lipids enter the lymphatic system?

8. How is the brain's glucose supply kept constant?

9. Why are some amino acids called "essential"?

10. Contrast the metabolic rate with the basal metabolic rate.

Clinical Perspectives

1. What is the relationship between diet and cancer?

2. How does dieting affect one's metabolic rate?

3. Why do you wake up cold when you fall asleep uncovered?

4. Your friend asks you why she feels warmer on hot, humid days than on hot, dry days. How would you answer her question?

5. What are some of the ways in which the normal body temperature can vary?

6. Your friend has stopped smoking and has gained weight. How would you explain this physiological weight gain to your friend?

7. Why do surgeons put a patient into a state of hypothermia during open heart or brain surgery?

8. How would you explain the fact that ketoacidosis and dehydration are produced in a person with type I diabetes?

9. How can a high fat diet in childhood lead to increased numbers of adipocytes? Explain how this process is related to the ability of adipocytes to regulate the insulin sensitivity of skeletal muscles in adults.

10. Explain the relationship between insulin resistance, obesity, exercise, and non-insulin diabetes mellitus.

Experiment of the Day

1. Using your automobile as an "experimental animal," list as many analogies as you can concerning energy balance and thermoregulation.

17

The Endocrine System

Chapter Overview

The two major control systems within the body are the nervous system and the endocrine system. The nervous system usually deals with rapid responses while the endocrine system generally provides prolonged sustained control. Endocrine glands release hormones, blood-borne chemical messengers that act on target cells located long distances from the endocrine glands. The nervous system controls the endocrine system primarily through the hypothalamus. The hypothalamus secretes tropic hormones, which act on endocrine glands, and nontropic hormones, which act on nonendocrine target cells.

The hypothalamus and endocrine glands secrete hormones or chemical messengers, which evoke responses in target cells elsewhere in the body. Synthesis, storage, secretion, transportation, and mechanism of action all depend on the chemical properties of the hormones. There are chemically three types of hormones. Hormone secretion is controlled by negative feedback. Hormones, regardless of their source, utilize the blood to reach their targets. Hypothalamic neurons secrete two nontropic hormones into capillaries in the posterior pituitary.

The hypothalamus also secretes seven tropic hormones, which are transported to their target, the anterior pituitary, via the hypothalamic-hypophyseal portal system. Growth, which is under endocrine control, is a typical example of how the nervous system ultimately controls prolonged events. Growth also demonstrates the interaction of hormones, which brings about attainment of genetic potential. Homeostasis is a very complex phenomenon.

The glands in this part of the endocrine system include the thyroid, adrenal, endocrine, pancreas, and parathyroid. These glands respond indirectly to the central nervous system and through their various hormones control the metabolic rate, fuel metabolism, and calcium and phosphate balances. In most instances there are dual controls for increased and decreased activity.

Maintaining adequate nutrient levels to meet the cellular requirements of the organism involves a complex set of hormonal controls. The peripheral endocrine glands not only meet the needs but respond to excesses and shortages by stimulating the various body systems involved with the storage, release, conversion, and excretion of nutrients and minerals. Homeostasis is maintained and the cells survive.

Chapter Outline

GENERAL PRINCIPLES OF ENDOCRINOLOGY
Hormones exert a variety of regulatory effects throughout the body.
- Endocrinology is the study of the homeostatic chemical adjustments and other activities accomplished by hormones.

- The endocrine system is one of the body's two major control systems. Endocrine controls require duration rather than speed.
- The endocrine system primarily controls activities that require duration rather than speed, including the following: (1) regulatory organic metabolism and water and electrolyte balance; (2) inducing adaptive changes to help the body cope with stressful situations; (3) promoting smooth, sequential growth and development; (4) controlling reproduction; (5) regulating red cell production; and (6) along with the autonomic nervous system, controlling and integrating both circulation and the digestion and absorption of food.
- A hormone that has as its primary function the regulation of hormone secretion by another endocrine gland is classified functionally as a tropic hormone.
- A nontropic hormone primarily exerts its effects on nonendocrine target tissues.
- Some organs are exclusively endocrine in function whereas other organs called mixed glands perform nonendocrine functions in addition to secreting hormones.

The effective plasma concentration of a hormone is normally regulated by changes in its rate of secretion.

- The plasma concentration of free, biologically active hormone depends on several factors: (1) the hormone's rate of secretion into the blood by the endocrine gland; (2) its rate of removal from the blood by metabolic inactivation and excretion in the urine; (3) for lipophilic hormones, its extent of binding to plasma proteins; and (4) for a few hormones, its rate of metabolic activation.
- Negative feedback is a prominent feature of hormonal control systems.
- Many endocrine control systems involve neuroendocrine reflexes, which include neural as well as hormonal components.
- The purpose of such reflexes is to produce a sudden increase in hormone secretion in response to a specific stimulus, frequently a stimulus external to the body.
- The secretion rates of all hormones rhythmically fluctuates up and down as a function of time.
- The most common endocrine rhythm is the diurnal or circadian rhythm, which is characterized by repetitive oscillations in hormone levels that are very regular and have a frequency of one cycle every 24 hours.
- Some endocrine cycles operate on time scales other than a circadian rhythm—some much shorter than a day and some much longer. Even though the plasma concentration of a hormone is normally regulated by adjusting its rate of secretion, alterations in its rate of inactivation and excretion in the urine can also influence the hormone's plasma concentration.

Endocrine disorders result from hormone excess or deficiency or decreased target-cell responsiveness.

- Endocrine disorders most commonly result from abnormal plasma concentrations of a hormone due to inappropriate secretion rates; that is, hyposecretion or hypersecretion.
- If an endocrine gland is secreting too little of its hormone because of an abnormality within that gland, the condition is referred to as "primary hyposecretion."
- If the endocrine gland is normal but is secreting too little hormone because of a deficiency of its tropic hormone, the condition is known as "secondary hyposecretion."

The responsiveness of a target cell can be varied by regulating the number of its hormone-specific receptors.

- A target cell's response to a hormone is correlated with the number of the cell's receptors occupied by molecules of that hormone.
- A hormone can influence the activity of another hormone at a given target cell in one of three ways: permissiveness, synergism, and antagonism.

HYPOTHALAMUS AND PITUITARY

The pituitary gland consists of anterior and posterior lobes.

- The pituitary gland, or hypophysis, is a small endocrine gland located in a bony cavity at the base of the brain just below the hypothalamus.
- The pituitary has two anatomically and functionally distinct lobes, the posterior pituitary and the anterior pituitary.
- The posterior pituitary is also termed the "neurohypophysis."
- The anterior pituitary is also known as the "adenohypophysis."
- They arise from different tissues embryonically, serve different functions, and are subject to different control mechanisms

The hypothalamus and posterior pituitary act as a unit to secrete vasopressin and oxytocin.

- The hypothalamus and posterior pituitary form a neuroendocrine system that consists of a population of neurosecretory neurons whose cell bodies lie in two well-defined clusters in the hypothalamus and whose axons pass down through the connecting stalk to terminate on capillaries in the posterior pituitary.
- The posterior pituitary does not actually produce any hormones.
- It simply stores and releases into the blood two small peptide hormones, vasopressin and oxytocin, which are synthesized by the neuronal cell bodies in the hypothalamus.
- Vasopressin has two major effects that correspond to its two names: (1) it enhances the retention of water by the kidneys, and (2) it causes contraction of arteriolar smooth muscle.
- Oxytocin stimulates contraction of uterine smooth muscle to aid in expulsion of the baby during childbirth, and it promotes ejection of milk from t mammary glands during breast-feeding.

Most anterior pituitary hormones are tropic.

- Unlike the posterior pituitary, the anterior pituitary itself synthesizes the hormones that it releases into the blood.
- Different cell populations within the anterior pituitary produce and secrete six established peptide hormones: (1) somatotropes secrete growth hormone, (2) thyrotropes thyroid-stimulating hormone, (3) corticotropes secrete adrenocorticotropic hormone, (4) gonadotropes secrete both follicle-stimulating hormone and (5) luteinizing hormone, and (6) lactotropes secrete prolactin.

Hypothalamic releasing and inhibiting hormones help regulate anterior pituitary hormone secretion.

- The two most important factors that regulate anterior pituitary hormone secretion are: (1) hypothalamic hormones, and (2) feedback from target-gland hormones.
- The secretion of each of the anterior pituitary hormones is stimulated or inhibited by one or more of the seven generally accepted hypophysiotropic hormones.
- Depending on their actions, these small peptides are called "releasing hormones" or "inhibiting hormones."

- The anatomical and functional link between the hypothalamus and the anterior pituitary is an unusual capillary-to-capillary connection, the hypothalamic-hypophyseal portal system.

Target-gland hormones inhibit hypothalamic and anterior pituitary hormone secretion via negative feedback.

- In most cases, hypophysiotropic hormones initiate a three-hormone sequence: (1) hypothalamic releasing hormone, (2) anterior pituitary tropic hormone, and (3) peripheral target hormone.
- The target-gland hormone also acts to suppress secretion of the tropic hormone that is driving it.

ENDOCRINE CONTROL OF GROWTH
Growth depends on growth hormone but is influenced by other factors as well.

- Growth requires net synthesis of proteins and includes lengthening of long bones as well as increases in the size and number of cells in the soft tissues throughout the body.
- An individual's maximum growth capacity is genetically determined.
- Attainment of this full growth potential further depends on: (1) an adequate diet, (2) freedom from chronic disease and stressful environmental conditions, and (3) a normal levels of growth-influencing hormones.

Growth hormone is essential for growth, but it also directly exerts metabolic effects not related to growth.

- The overall metabolic effect of growth hormone is to mobilize fat stores as a major energy source while conserving glucose for glucose-dependent tissues such as the brain.
- To exert its metabolic effects, GH binds directly with its target organs
- Growth hormone increases fatty acid levels; increases blood glucose levels; stimulates amino acid uptake and protein synthesis and spares glucose for the brain by promoting other cells to use fatty acids as an energy source.
- Growth hormone exerts its growth-promoting effects indirectly by stimulating insulin-like growth factor

Growth hormone/IGF promote growth of soft tissues by stimulating hyperplasia and hypertrophy.

- GH (acting through IGF) stimulates growth of both soft tissues and the skeleton. GH promotes growth of soft tissues by (1) increasing the number of cells (hyperplasia) and (2) increasing the size of cells (hypertrophy).
- Growth of the long bones resulting in increased height is the most dramatic effect of growth hormone.
- Bones grow in length as a result of proliferation of the cartilage cells in the epiphyseal plates.
- Growth in thickness of bone is achieved by the addition of new bone on top of the already existing bone on the outer surface.

Bone grows in thickness and in length by different mechanisms, both stimulated by growth hormone.

- Bone structures affected by GH include osteocytes, osteoblasts, osteoclasts, epiphyseal plate and cartilage.

Abnormal growth hormone secretion results in aberrant growth patterns.

- Diseases related to both deficiencies and excesses of growth hormone can occur. Growth hormone deficiency may be caused by a pituitary defect or occur secondarily to hypothalamic dysfunctions.

- Hyposecretion of growth hormone in a child results in dwarfism.
- Growth may be thwarted because the tissues fail to respond normally to growth hormone (Laron dwarfism).
- The symptoms of this condition resemble those of severe growth hormone deficiency even though blood levels of growth hormone are actually high.
- If overproduction of growth hormone begins in childhood the condition is known as "gigantism."
- If growth hormone hypersecretion occurs after adolescence a disproportionate growth pattern produces a disfiguring condition known as "acromegaly."

PINEAL GLAND AND CIRCADIAN RHYTHMS

- The pineal gland, located in the center of the brain, secretes the hormone melatonin.
- One of melatonin's most widely accepted roles is helping to keep the body's inherent circadian rhythms in synchrony with the light/dark cycle.

The suprachiasmatic nucleus is the master biological clock.

- The eyes cue the pineal gland about the presence or absence of light by means of a neural pathway that passes through the suprachiasmatic nucleus.
- Specific self-starting genes within the nuclei of the SCN neurons set in motion a series of events that bring about the synthesis of clock proteins.

- The biological clock is also synchronized with environmental cues.

Melatonin helps keep the body's circadian rhythms in time with the light-dark cycle.

- Based on recent evidence, scientists suspect that melanopsin is the likely receptor for light that keeps the body in tune with external time.
- This is the major way the internal clock is coordinated to a 24-hour day.
- Melatonin induces a natural sleep without the side effects that accompany hypnotic sedatives.
- Melatonin inhibits the hormones that stimulate reproductive activity.
- In some species, melatonin plays a role in the breeding season, migration, and hibernation.
- Melatonin is a very beneficial antioxidant.
- Melatonin can enhance immunity.

THYROID GLAND

The major cells that secrete thyroid hormones are organized into colloid-filled follicles.

- The chief constituent of the colloid is a large, complex molecule known as "thyroglobulin," within which are incorporated the thyroid hormones in their various stages of synthesis.
- The follicular cells produce two iodine-containing hormones derived from the amino acid tyrosine: tetraiodothyronine (thyroxine) and triiodothyronine.
- These two hormones, collectively referred to as thyroid hormone, are important regulators of overall basal metabolic rate.
- Interspersed in the interstitial spaces between the follicles is another secretory type cell, the C cells, so called because they secrete the peptide hormone calcitonin, which plays a role in calcium metabolism.

Thyroid- hormone synthesis and storage occur on the thyroglobulin molecule.

- The synthesis, storage, and secretion of thyroid hormone involve the following steps: (1) Tyrosine becomes incorporated in the much larger thyroglobulin molecules, which are exported from the follicular cells into the colloid by exocytosis. (2) The thyroid captures iodine from the blood and transfers it into the colloid by means of a very active "iodine pump." (3) Within the colloid, iodine is quickly attached to a tyrosine within the thyroglobulin molecule. Attachment of one iodine to tyrosine yields monoiodotyrosine (MIT). Attachment of two iodines to tyrosine yields diiodotyrosine (DIT). (4) Coupling of two DITs yields tetraiodothyronine (thyroxine). Coupling of one MIT and one DIT yields triiodothyronine.

To secrete thyroid hormone, the follicular cells phagocytize thyroglobulin-laden colloid.

- Upon appropriate stimulus for thyroid hormone secretion, the follicular cells internalize a portion of the thyroglobulin-hormone complex by phagocytizing a piece of colloid.
- The biologically active thyroid hormones are split off.
- The MIT and DIT are of no endocrine value.

Most of the secreted T_4 is converted into T_3 outside the thyroid.

- About 90% of the secretory product released from the thyroid gland is in the form of T_4, yet T_3 is about 10 times more potent in its biological activity

Thyroid hormone is the main determinant of the basal metabolic rate and exerts other effects as well.

- Virtually every tissue in the body is affected either directly or indirectly by thyroid hormone.
- Thyroid hormone increases the body's overall basal metabolic rate.
- Closely related to thyroid hormone's overall metabolic effect is its calorigenic effect.
- Thyroid hormone increases target cell responsiveness to catecholamines, the chemical messengers used by the sympathetic nervous system and its hormonal reinforcements from the adrenal medulla.
- Thyroid hormone increases heart rate and force of contraction, thus increasing cardiac output.
- Thyroid hormone not only stimulates growth hormone secretion but also promotes the effects of growth hormone on the synthesis of new structural proteins and on skeletal growth.
- Thyroid hormone plays a crucial role in the normal development of the nervous system.

Thyroid hormone is regulated by the hypothalamus-pituitary-thyroid axis.

- Thyroid-stimulating hormone, the thyroid tropic hormone from the anterior pituitary, is the most important physiological regulator of thyroid hormone secretion.
- Thyroid hormone, in negative-feedback fashion, "turns off" TSH secretion, whereas the hypothalamic thyrotropin-releasing hormone (TRH), in tropic fashion, "turns on" TSH secretion by the anterior pituitary.

Abnormalities of thyroid function include both hypothyroidism and hyperthyroidism.

- Hypothyroidism can result: (1) from primary failure of the thyroid gland itself; (2) secondary to a deficiency of TRH, TSH, or both; or (3) from an inadequate dietary supply of iodine.
- The most common cause of hyperthyroidism is Grave's disease, an autoimmune disease in which the body erroneously produces thyroid-stimulating immunoglobulin (TSI), an antibody whose target is the TSH receptors on the thyroid cells.

- Less frequently, hyperthyroidism occurs secondary to excess TRH or TSH or in association with a hypersecreting thyroid tumor.

A goiter develops when the thyroid gland is over stimulated.

- A goiter refers to an enlarged thyroid gland.
- Hypothyroidism secondary to hypothalamic or anterior pituitary failure will not be accompanied by a goiter because the thyroid is not being adequately stimulated.
- With hypothyroidism caused by thyroid gland failure or lack of iodine, a goiter does develop because the circulating level of thyroid hormone is so low that there is little negative-feedback inhibition on the anterior pituitary, and TSH secretion is therefore elevated.
- Excessive TSH secretion resulting from hypothalamic or anterior pituitary defect would obviously be accompanied by a goiter because of over stimulation of thyroid growth.
- In Grave's disease, a hypersecreting goiter occurs because TSI promotes growth of the thyroid as well as enhances secretion of thyroid hormone.
- Hyperthyroidism resulting from overactivity of the thyroid in the absence of over stimulation, such as that caused by an uncontrolled thyroid tumor, is not accompanied by a goiter.

ADRENAL GLANDS
Each adrenal gland consists of a steroid-secreting cortex and a catecholamine-secreting medulla.

- Each adrenal is actually composed of two endocrine organs, the adrenal cortex surrounding the adrenal medulla.

The adrenal cortex secretes mineralocorticoids, glucocorticoids, and sex hormones.

- The adrenal cortex produces a number of different adrenocortical hormones, all of which are steroids derived from the common precursor molecule, cholesterol.
- Mineralocorticoids primarily aldosterone,. influence mineral balance,
- Glucocorticoids, primarily cortisol, play a major role in glucose metabolism as well as in protein and lipid metabolism.
- The sex hormones are identical or similar to those produced by the gonads.

The major effect of mineralocorticoids are on Na+ and K+ balance and blood pressure homeostasis.

- Mineralocorticoids are essential for life.
- Without aldosterone, a person rapidly dies from circulatory shock because of a marked fall in plasma volume.
- Aldosterone secretion is increased by: (1) activation of the renin-angiotensin-aldosterone system by factors related to a reduction in sodium and a fall in blood pressure, and (2) direct stimulation of the adrenal cortex by a rise in plasma potassium concentration.

Glucocorticoids exert metabolic effects and play a key role in adaptation to stress.

- Cortisol stimulates hepatic gluconeogenesis.
- Cortisol inhibits glucose uptake and use by many tissues, but not the brain, thus sparing glucose for use by the brain, which absolutely requires it as metabolic fuel.
- Cortisol stimulates protein degradation in many tissues, especially muscle.
- Cortisol facilitates lipolysis, the breakdown of lipid stores in adipose tissue, thus releasing free fatty acids into the blood.
- Cortisol must be present in adequate amounts to permit catecholamines to induce vasoconstriction.
- Cortisol plays a key role in adaptation to stress.

- Cortisol is also known to alter mood and behavior.
- Administration of large amounts of glucocorticoids inhibits almost every step of the inflammatory response.

Cortisol secretion is regulated by the hypothalamus-pituitary cortex axis.

- Cortisol secretion by the adrenal cortex is regulated by a long-loop negative-feedback system involving the hypothalamus and anterior pituitary.
- Adrenocorticotropic hormone from the anterior pituitary stimulates the adrenal cortex to secrete cortisol.
- ACTH enhances many steps in the synthesis of cortisol.
- The ACTH-producing cells in turn secrete only at the command of corticotropin-releasing hormone from the hypothalamus.
- The other major factor that is independent of, and in fact can override, the stabilizing negative-feedback control is stress.

The adrenal cortex secretes both male and female sex hormones in both sexes.

- No hormones are unique to either males or females (except those from the placenta during pregnancy) because small amounts of the sex hormone of the opposite sex are produced by the adrenal cortex in both sexes.
- The only adrenal sex hormone that has any biological importance is the androgen dehydroepiandrosterone (DHEA).

- This androgen is responsible for androgen-dependent processes in the female, such as growth of pubic and axillary hair, enhancement of the pubertal growth spurt, and development and maintenance of the female sex drive.
- ACTH controls adrenal androgen secretion.
- DHEA inhibits gonadotropin-releasing hormone, just as testicular androgens do.
- Some scientists suspect that the age-related decline of DHEA and other hormones, such as human growth hormone and melatonin, plays a role in some of the problems associated with getting older.
- Early studies with DHEA replacement therapy demonstrated some physical improvement, but the most pronounced effect was a marked increase in psychological well-being and an improved ability to cope with stress.

The adrenal gland may secrete too much or too little of any of its hormones.

- Excess mineralocorticoid secretion may be caused by: (1) a hypersecreting adrenal tumor made up of aldosterone-secreting cells (Conn's syndrome), or (2) inappropriately high activity of the renin-angiotensin system.
- Excessive cortisol secretion can be caused by: (1) overstimulation of the adrenal cortex by excessive amounts of CRH and/or ACTH (Cushing's syndrome); (2) adrenal tumors that uncontrollably secrete cortisol independent of ACTH; or (3) ACTH-secreting tumors located in places other than the pituitary, most commonly in the lung.
- Excess adrenal androgen secretion, a masculinizing condition, is more common than the extremely rare feminizing condition of excess adrenal estrogen secretion.
- The adrenogenital syndrome is most commonly caused by an inherited enzymatic defect in the cortisol steroidogenic pathway.
- In primary adrenocortical insufficiency, also known as Addison's disease, all layers of the adrenal cortex are undersecreting.

- Secondary adrenocortical insufficiency may occur because of a pituitary or hypothalamic abnormality, resulting in insufficient ACTH secretion.

The adrenal medulla consists of modified sympathetic postganglionic neuron.

- The adrenal medulla is actually a modified part of the sympathetic nervous system.
- The adrenal medulla is composed of modified postganglionic sympathetic neurons.
- Postganglionic sympathetic neurons in the adrenal medulla do not possess axonal fibers that terminate on effector organs.
- The ganglionic cell bodies within the adrenal medulla release their chemical transmitter directly into the circulation upon stimulation by the preganglionic fiber.
- Epinephrine accounts for 80% of total release and norepinephrine 20%.

Epinephrine and norepinephrine vary their affinities for the different adrenergic receptor types.

- Like sympathetic fibers, the adrenal medulla does release norepinephrine, but its most abundant secretory output is epinephrine.
- Epinephrine and norepinephrine both accelerate the heart rate.
- Both induce generalized arteriolar vasoconstriction of the skin, digestive tract, and kidneys.
- Epinephrine brings about vasodilation of the blood vessels that supply skeletal muscles and the heart.
- Epinephrine is able to exert some unique effects, such as its metabolic effects, because it reaches places not supplied by sympathetic fibers.
- Epinephrine secretion always accompanies a generalized sympathetic discharge, so sympathetic activity indirectly exerts control over actions performed by epinephrine.

Epinephrine reinforces the sympathetic nervous system and exerts additional metabolic effects.

- The sympathetic system and epinephrine both exert widespread effects on organ systems that are ideally suited for fight-or-flight responses.

The stress response is a generalized pattern of reactions to any situation that threatens homeostasis.

- Stress is the generalized, nonspecific response of the body to any factor that overwhelms, or threatens to overwhelm, the body's compensatory abilities to maintain homeostasis.
- Besides epinephrine, a number of other hormones are involved in the overall stress response.
- The predominant hormonal response is activation of the CRH-ACTH-cortisol system.
- The sympathetic nervous system and the epinephrine secreted at its bidding both inhibit insulin and stimulate glucagon.

The multifaceted stress response is coordinated by the hypothalamus.

- The hypothalamus receives input concerning physical and emotional stressors from virtually all areas of the body and from many receptors throughout the body.
- In response, the hypothalamus directly activates the sympathetic nervous system, secretes CRH to stimulate ACTH and cortisol release, and triggers the release of vasopressin.
- Sympathetic stimulation brings about the secretion of epinephrine.

ENDOCRINE PANCREAS AND
CONTROL OF FUEL METABOLISM
*Fuel metabolism includes
anabolism, catabolism, and
interconversions among energy-
rich organic molecules.*

- Metabolism refers to all the
  chemical reactions that occur
  within the cells of the body.
- Anabolism refers to the
  buildup or synthesis of larger
  organic macromolecules from
  the small organic molecular
  subunits.
- Catabolism refers to the
  breakdown or degradation of
  large, energy-rich organic
  molecules within cells.
- Mobilized glucose, fatty acid,
  and amino acid molecules can
  be used as needed for energy
  production or cellular synthesis
  elsewhere in the body.
- Essential nutrients such as the
  essential amino acids and
  vitamins cannot be formed in
  the body by conversion from
  another type of organic
  molecule.

*Because food intake is intermittent,
nutrients must be stored between
meals.*

- Dietary intake is intermittent,
  not continuous.
- Stored energy fills in the gaps
  between meals
- Excess circulating glucose is
  stored as glycogen.
- Additional glucose is
  transformed into fatty acids
  and glycerol, which are used to
  synthesize triglycerides.
- Excess circulating fatty acids
  derived from dietary intake
  also become incorporated into
  triglycerides.
- Excess circulating amino acids
  not needed for protein

synthesis are converted to glucose and fatty
acids.

- The major site of energy storage for excess
  nutrients of all three classes is adipose
  tissue.
- Glycogen and triglyceride reservoirs serve
  solely as energy depots.
- A substantial amount of energy is stored as
  structural protein, primarily in muscle, the
  most abundant protein mass in the body.

*The brain must be continuously supplied with
glucose.*

- The brain normally depends on the delivery
  of adequate blood glucose as its sole source
  of energy.
- When new dietary glucose is not entering
  the blood, tissues not obligated to use
  glucose shift their metabolic gears to burn
  fatty acids instead, thus sparing glucose for
  the brain.

*Metabolic fuels are stored during the absorptive
state and are mobilized during the postabsorptive state.*

- Ingested nutrients are being absorbed and
  entering the blood during the absorptive
  state, or fed state.
- During this time, glucose is plentiful and
  serves as the major energy source.
- The average meal is completely absorbed in
  about four hours.
- On a typical three-meals-a-day diet, no
  nutrients are being absorbed from the
  digestive tract during late morning, late
  afternoon, and throughout the night.
- These times constitute the postabsorptive
  state, or fasting state.
- During the absorptive state, the glut of
  absorbed nutrients is swiftly removed from
  the blood and placed into storage; during the
  postabsorptive state, these stores are
  catabolized to maintain the blood
  concentrations at levels necessary to sustain
  tissue energy demands.

*The pancreatic hormones insulin and glucagon
are the most important in regulating fuel
metabolism.*

- Scattered throughout the pancreas are clusters of endocrine cells known as the "islets of Langerhans."
- The most abundant pancreatic endocrine cell type is the beta cell, the site of insulin synthesis and secretion.
- Next most important are the alpha cells, which produce glucagon.
- The delta cells are the pancreatic site of somatostatin synthesis, whereas the least common islet cells, the PP cells, secrete pancreatic polypeptide.
- The overall effect of pancreatic somatostatin is to inhibit digestion of nutrients and to decrease nutrient absorption.

Insulin lowers blood glucose, amino acid, and fatty acid levels and promotes their storage.
- Insulin lowers the blood levels of glucose, fatty acids, and amino acids and promotes their storage.
- Insulin facilitates glucose transport into most cells.
- Insulin stimulates glycogenesis and inhibits glycogenolysis.
- Insulin further decreases hepatic glucose output by inhibiting gluconeogenesis, the conversion of amino acids into glucose in the liver.
- Insulin increases the transport of glucose into adipose tissue cells by means of GLUT-4 recruitment.
- Glucose serves as a precursor for the formation of fatty acids and glycerol.
- Insulin activates enzymes that catalyze the production of fatty acids from glucose derivatives.

- Insulin promotes the entry of fatty acids from the blood into adipose tissue cells.
- Insulin inhibits lipolysis, thus reducing the release of fatty acids from adipose tissue into the blood.
- Insulin promotes the active transport of amino acids from the blood into muscles and other tissues.
- Insulin increases the rate of amino acid incorporation into protein by stimulating the cells' protein synthesizing machinery.
- Insulin inhibits protein degradation.

The primary stimulus for increased insulin secretion is an increase in blood glucose concentration.
- The primary control of insulin secretion is a direct negative-feedback system between the pancreatic beta cells and the concentration of glucose in the blood flowing to them.
- An elevated plasma amino acid level directly stimulates the beta cells to increase insulin secretion.
- The major gastrointestinal hormones, especially gastric inhibitory peptide, stimulate pancreatic insulin secretion.
- The increase in parasympathetic activity that occurs in response to food in the digestive tract stimulates insulin release.
- Sympathetic stimulation and the concurrent increase in epinephrine both inhibit insulin secretion.

The symptoms of diabetes mellitus are characteristic of an exaggerated post-absorptive state.
- Diabetes mellitus is by far the most common of all endocrine disorders.
- The most prominent feature of diabetes mellitus is hyperglycemia.
- A large urine volume occurs both in diabetes mellitus due to insulin insufficiency and in diabetes insipidus due to vasopressin deficiency.
- Type I diabetes is characterized by a lack of insulin secretion.

- Type II diabetes is characterized by normal or even increased insulin secretion but reduced sensitivity of insulin's target cells to its presence. The acute consequences of diabetes mellitus can be grouped according to the effects of inadequate insulin action on carbohydrate, fat, and protein metabolism
- Numerous long-range complications occur despite treatment to prevent the short-term effects. These chronic complications, which account for the shorter life expectancy of diabetics, primarily involve degenerative disorders of the blood vessels and nervous system
- The treatment for Type 1 diabetes is a controlled balance of regular insulin injections timed around meals, management of the amounts and types of food consumed, and exercise.
- Dietary control and weight reduction may be all that is necessary to completely reverse the symptoms in Type 2 diabetics. Various blood-glucose lowering oral medications are available if needed.

Insulin excess causes brain-starving hypoglycemia.

- Insulin excess can occur in a diabetic patient when too much insulin has been injected for the person's caloric intake and exercise level, resulting in so-called insulin shock.
- An abnormally high blood insulin level may occur in a nondiabetic individual with a beta cell tumor or in whom the beta cells are over-responsive to glucose.
- The consequences of insulin excess are primarily manifestations of the effects of hypoglycemia on the brain.

Glucagon in general opposes the actions of insulin.

- Many physiologists view the insulin-secreting beta cells and the glucagon-secreting alpha cells as a coupled endocrine system whose combined secretory output is a major factor in the regulation of fuel metabolism.
- Glucagon affects many of the same metabolic processes that are influenced by insulin, but in most cases glucagon's actions are opposite to those of insulin.

Glucagon secretion is increased during the postabsorptive state.

- Glucagon secretion is increased during the postabsorptive state and decreased during the absorptive state.
- The major factor regulating glucagon secretion is a direct effect of the blood glucose concentration on the endocrine pancreas.
- There is a direct negative-feedback relationship between blood glucose concentration and the alpha cells' rate of secretion.
- An elevated blood glucose level inhibits glucagon secretion.

Insulin and glucagon work as a team to maintain blood glucose and fatty acid levels.

- A negative-feedback relationship exists between blood glucose concentration and both the beta and alpha cells' rates of secretion.

Epinephrine, cortisol, growth hormone, and thyroid hormone also exert direct metabolic effects.

- The stress hormones, epinephrine and cortisol, both increase blood levels of glucose and fatty acids.
- Cortisol appears to contribute to the maintenance of blood glucose concentration during long-term starvation.

- Growth hormone, like cortisol, appears to contribute to the maintenance of blood glucose concentrations during starvation.

CONTROL OF CALCIUM METABOLISM

Plasma calcium must be closely regulated to prevent changes in neuromuscular excitability.

- Most ECF calcium is bound to some other molecule and only a small percentage is considered "free" to perform essential activities.
- Three hormones—parathyroid hormone, calcitonin, and vitamin D—control calcium and phosphate metabolism.Extracellular fluid calcium plays a vital role in a number of essential activities including the following: (1) neuromuscular excitability, (2) excitation-contraction coupling in cardiac and smooth muscle, (3) stimulus-secretion coupling, (4) maintenance of tight junctions between cells, and (5) clotting of blood.

Parathyroid hormone raises free plasma calcium levels by its effects on bone, kidneys, and intestine.

- Parathyroid hormone (PTH) is a peptide hormone secreted by the parathyroid glands.
- The overall effect of PTH is to increase the calcium concentration of plasma.
- Parathyroid hormone also acts to lower plasma phosphate concentration.
- Parathyroid hormone stimulates calcium conservation and promotes phosphate elimination by the kidneys during the formation of urine.

- The primary regulator of PTH secretion is the plasma concentration of free calcium.
- Parathyroid hormone secretion is increased in response to a fall in plasma calcium concentration and decreased by a rise in plasma calcium levels.
- Calcitonin lowers the plasma calcium concentration but is not important in the normal control of calcium metabolism.
- Calcitonin, the hormone produced by the C cells of the thyroid gland, also exerts an influence on plasma calcium levels.
- Since calcitonin reduces plasma calcium levels, this system constitutes a second simple negative-feedback control over plasma calcium concentration, one that is opposed to the PTH system.
- The rates of bone formation and reabsorption are about equal throughout most of adult life.
- A reduction in bone mass is known as "osteoporosis."

PTH raises plasma calcium by withdrawing calcium from the bone bank.

- Parathyroid hormone uses bone as a "bank" from which it withdraws calcium as needed to maintain the plasma calcium level.
- As an immediate effect, PTH promotes rapid movement of Ca^{2+} into the plasma from Ca^{2+}-rich bone fluid
- Under conditions of chronic hypocalcemia, such as may occur with dietary Ca^{2+} deficiency, PTH stimulates actual localized dissolution of bone

PTH acts on the kidneys to conserve calcium and eliminate phosphate.

- PTH stimulates calcium conservation and promotes phosphate elimination by the kidneys during the formation of urine.

Calcitonin lowers plasma calcium concentration but is not important in the normal control of calcium metabolism.

- Calcitonin exerts its effect on plasma calcium levels.
- Although calcitonin protects against hypercalcemia, this condition rarely occurs under normal circumstances

Vitamin D is actually a hormone that increases calcium absorption in the intestine.

- The final factor involved in the regulation of calcium metabolism is cholecalciferal or vitamin D.
- The skin is actually an endocrine gland and vitamin D a hormone.
- The most dramatic and biologically important effect of activated vitamin D is to increase calcium absorption in the intestine

- The major consequence associated with vitamin D deficiency is impaired intestinal absorption of calcium.
- Recent research Vitamin D also influence muscle strength, energy metabolism and immune health.

Disorders in calcium metabolism may arise from abnormal levels of PTH or vitamin D.

- Excess PTH secretion is called "hyperparathyroidism" and is caused by a tumor.
- This causes hypercalcemia and hypophosphatemia.

Key Terms

Absorptive (fed) state
Acromegaly
Addison's disease
Adenohypophysis
Adrenal cortex
Adrenal genital syndrome
Adrenal glands
Adrenal medulla
Adrenocorticotropic hormone (ACTH)
Aldosterone
Alpha cells
Anabolism
Antagonism
Anterior pituitary gland
Antioxidant
Beta cells
Bone
Bone deposition
Bone fluid
Bone reabsorption
C cells
Calcitonin
Calorigenic effect
Catabolism
Cholecalciferol (vitamin D)
Clock proteins
Colloid
Conn's syndrome
Cretinism
Cushing's syndrome

Dehydroepiandrosterone (DHEA)
Delta cells
Diabetes mellitus
Diurnal (circadian) rhythm
Down regulation
Dwarfism
Endocrine pancreas
Endocrine system
Epinephrine
Essential nutrients
Exophthalmos
Follicle
Follicle-stimulating hormone (FSH)
Gigantism
Glucagon
Glucocorticoids
Gluconeogenesis
Glucose transporters (GLUT)
Glycogenesis
Glycogenolysis
Goiter
Grave's disease (hyperthyroidism)
Growth hormone (GH)
Hirsutism
Hormones
Hyperparathyroidism
Hyperplasia
Hypersecretion
Hypertrophy
Hypoparathyroidism

Hypophysiotropic hormone
Hyposecretion
Hypothalamic-hypophyseal portal system
Hypothalamus
Insulin
Insulin-like growth factor (IGF)
Insulin shock
Intermediary (fuel) metabolism
Islets of Langerhans
Ketogenesis
Luteinizing hormone (LH)
Melanin
Melanopsin
Melanocyte-stimulating hormone (MSH)
Melatonin
Metabolism
Mineralocorticoids
Myxedema
Neuroendocrine reflex
Ossification
Osteoblasts
Osteoclasts
Osteocytes
Osteomalacia

Pancreas
Parathyroid glands
Permissiveness
Pineal gland (hypophysis)
Postabsorptive (fasting) state
Posterior pituitary gland
Precocious pseudopuberty
Prolactin (PRL)
Reactive hypoglycemia
Rickets
Sex hormones
Stress
Suprachiasmatic nucleus (SCN)
Sympathomimetic effect
Synergism
Tetraiodothyronine (TA)
Thyroglobulin
Thyroid gland
Thyroid stimulating hormone (TSH)
Thyroxine
Transporter recruitment
Triiodothyronine (T_3)
Tropic hormone

Review Exercises

Answers are in the appendix.

True/False-if an answer is false, underline the incorrect portion of the statement and rewrite it to make the statement true.

_____ 1. Mixed glands secrete hormones and also have non-endocrine functions.

_____ 2. Bones grow in length as a result of proliferation of the cartilage cells in the epiphyseal plates.

_____ 3. Growth hormone deficiency may be caused by a pituitary defect.

_____ 4. Androgens are made only by the reproductive glands.

_____ 5. The posterior pituitary is also known as the "adenohypophysis."

_____ 6. Prolactin enhances breast development and milk production in females.

_____ 7. Aldosterone release is affected by the rennin-angiotensin system and plasma levels of potassium.

_____ 8. A single hormone may be secreted by only one endocrine gland.

_____ 9. The so called endocrine axis includes the hypothalamus, pituitary gland and one other endocrine gland.

_____ 10. Negative feedback exists when the output of a system opposes a change in input.

_____ 11. The MIT and DIT are of no endocrine value.

_____ 12. POMC can be cleaved into ACTH, MSH and endorphin.

_____ 13. Thyroid hormone decreases target-cell responsiveness to catecholamines.

_____ 14. A hypersecreting goiter occurs because TSI promotes growth of the thyroid as well as enhancing secretion of thyroid hormone.

_____ 15. About 20% of the hormone released by the adrenal medulla is epinephrine.

_____ 16. Cortisol stimulates protein degradation in many tissues, except muscle.

_____ 17. Myxedema is characteristic of hypothyroidism.

_____ 18. Epinephrine dilates the respiratory airways.

_____ 19. Excess circulating glucose is stored as glycogen.

_____ 20. Growth hormone exerts it growth effects through IGF and its metabolic effects directly.

_____ 21. The beta cell is the site of insulin synthesis and secretion.

_____ 22. Diabetes mellitus is by far the most common of all endocrine disorders.

_____ 23. Type II diabetes is characterized by a lack of insulin secretion.

_____ 24. An elevated blood glucose level inhibits glucagon secretion.

_____ 25. Parathyroid hormone uses bone as a "bank."

_____ 26. Calcitonin increases plasma calcium levels.

_____ 27. Phosphate is regulated directly by vitamin D.

_____ 28. Hypercalcemia reduces the excitability of muscle and nervous tissue.

Fill in the Blank

29. Conn's syndrome is characterized by _____ levels of the hormone _____.

30. The _____ is a small endocrine gland located in a bony cavity at the base of the brain just below the hypothalamus.

31. _____ refers to a pituitary hormone's action at the hypothalamus to inhibit the release of its stimulatory neurohormone.

32. The condition in which the eyes bulge out of the face is known as _____ and is associated with excess levels of _____.

33. A long bone basically consists of a fairly uniform cylindrical shaft, the _____ with a flared articulating knob at either end, a(n) _____.

34. Hyposecretion of growth hormone in a child results in _____.

35. _____ are long-range chemical messengers that are specifically secreted into the blood by endocrine glands in response to an appropriate signal.

36. _____ regulates skin coloration by controlling the dispersion of granules containing the pigment _____.

37. The _____, which secretes blood-borne chemical messengers known as hormones, specializes in intracellular communications.

38. _____ are local chemical messengers whose effect is exerted only on neighboring cells in the immediate environment of their site of secretion.

39. _____ which include the hormones secreted by the adrenal cortex, gonads, and most placental hormones, are neutral lipids derived from cholesterol.

40. Glut _____ transporters are not present in the membrane of cells in the absence of insulin.

41. With _____, one hormone must be present in adequate amounts for the full exertion of another hormone's effect.

42. _____ occurs when one hormone causes the loss of another hormone's receptors, reducing the effectiveness of the second hormone.

43. The anatomical and functional link between the _____ and _____ is an unusual capillary-to-capillary connection called the "hypothalamic-hypophyseal portal system."

44. Metabolic syndrome or syndrome X is a precursor to _____.

45. _____ refers to the build-up or synthesis of larger organic macromolecules from the small organic molecular subunits.

46. Scattered throughout the pancreas are clusters of endocrine cells known as the

    _____.

47. _____ is the production of glycogen.

48. _____ are layers of osteocytes entombed within the bone that they have deposited around.

49. _____, the hormone produced by the C-cells of the thyroid gland, exerts an influence on plasma calcium levels.

50. _____ energy stores the gap between meals.

51. The chief constituent of the colloid is a large, complex molecule known as

    _____.

52. _____ is an autoimmune disease in which the body erroneously produces thyroid-stimulating immunoglobulin (TSI).

Matching

Match the endocrine gland to its hormone.

_____ 53. Cortisol   a. adrenal medulla

_____ 54. Oxytocin   b. posterior pituitary

_____ 55. Vitamin D   c. islets of Langerhans

_____ 56. Estrogen   d. pineal gland

_____ 57. Melatonin   e. hypothalamus

_____ 58. Glucagon   f. thyroid follicular cells

_____ 59. Inhibin   g. kidneys

_____ 60. Testosterone   h. duodenum

_____ 61. Thymosin   i. heart

_____ 62. Prolactin   j. placenta

_____ 63. TRH   k. skin

_____ 64. Growth hormone   l. thymus

_____ 65. Calcitonin   m. liver

_____ 66. Chorionic gonadotropin   n. anterior pituitary

_____ 67. Gastrin   o. stomach

_____ 68. Atrial natriuretic peptide   p. parathyroid

_____ 69. FSH   q. thyroid C cells

_____ 70. CRH   r. gonads

_____ 71. Thyroxine   s. adrenal cortex

_____ 72. Aldosterone

_____ 73. Progesterone

_____ 74. PTH

_____ 75. Insulin

_____ 76. Somatomedins

_____ 77. LH

_____ 78. GHIH

_____ 79. Epinephrine

_____ 80. TSH

_____ 81. Renin

_____ 82. Somatostatin

_____ 83. Androgens

_____ 84. ACTH

_____ 85. Vasopressin

_____ 86. PIH

_____ 87. Norepinephrine

_____ 88. Cholecystokinin

_____ 89. Erythropoietin

_____ 90. PRH

_____ 91. GHRH

_____ 92. Secretin

Multiple Choice

93. Which of the following is the hormone of the pineal gland?
    a. melanocyte
    b. melanin
    c. melatonin
    d. MSH
    e. PGH

94. Which of the following hormones has as its target an endocrine gland?
    a. paracrines
    b. tropic hormones
    c. nontropic hormones
    d. neurotransmitters
    e. neuromodulators

95. Which of the following promotes breast development and stimulates milk secretion?
    a. luteinizing hormone
    b. somatomedins
    c. renin
    d. oxytocin
    e. prolactin

96. Which of the following decreases plasma calcium concentration?
    a. parathyroid hormone
    b. glucagon
    c. cortisol
    d. calcitonin
    e. oxytocin

97. Which of the following hormones shows limited physiological activity?
    a. insulin
    b. glucagon
    c. PTH
    d. Vitamin D
    e. calcitonin

98. A decrease in secretion of cortisol but not aldosterone is characteristic of:
    a. Laron dwarfism
    b. Conn's syndrome
    c. Cushing's syndrome
    d. secondary adrenal insufficiency
    e. exophthalmic goiter

99. Which of the following hormones increase uterine contractility?
    a. oxytocin
    b. vasopressin
    c. PRH
    d. GHRH
    e. none of the above

100. Which of the following hormones produces vasoconstriction in arterioles?
    a. oxytocin
    b. aldosterone
    c. cortisol
    d. ACTH
    e. vasopressin

101. Comparing T4 and T3:
    a. T4 is more potent
    b. T3 is released in a higher concentration
    c. T4 is converted to T3
    d. T 4 is made by the thyroid gland, T3 by the parathyroid gland
    e. none of the above is correct

102. Which of the following conditions is associated with decreased adrenal cortical function across all three layers?
     a. dwarfism
     b. gigantism
     c. Addison's syndrome
     d. Laron gigantism
     e. Conn's syndrome

Modified Multiple Choice
Indicate the effect the item on the left has on the secretion rate of the hormone on the right by circling the appropriate letter using the answer code below.

> A = increases
> B = decreases
> C = has little or no effect on

103. Increased TSH _____ thyroid hormone.
104. Increased TRH _____ TSH.
105. Decreased thyroid hormone _____ TSH.
106. Decreased iodine in the diet _____ thyroid hormone.
107. Increased ACTH _____ aldosterone.
108. Increased ACTH _____ cortisol.
109. Stress _____ cortisol.
110. Stress _____ epinephrine.
111. Increased renin-angiotensin _____ aldosterone.
112. Increased plasma potassium ion _____ aldosterone.
113. Increased blood glucose _____ insulin.
114. Increased blood glucose _____ glucagon.
115. Increased plasma calcium _____ parathyroid hormone.
116. Increased plasma calcium _____ calcitonin.
117. Increased TSH _____ calcitonin.
118. Increased plasma phosphate ion _____ parathyroid hormone.

Indicate the metabolic effects on each of the following hormones by writing the appropriate letter using the answer code below.

A = increases
B = decreases
C = has no effect on

119. Insulin _____ blood glucose.
120. Insulin _____ fatty acids.
121. Insulin _____ blood amino acids.
122. Insulin _____ muscle protein.
123. Glucagon _____ blood glucose.
124. Glucagon _____ blood fatty acids.
125. Glucagon _____ blood amino acids.
126. Glucagon _____ muscle protein.
127. Epinephrine _____ blood glucose.
128. Epinephrine _____ blood fatty acids.
129. Epinephrine _____ blood amino acids.
130. Epinephrine _____ muscle protein.
131. Cortisol _____ blood glucose.
132. Cortisol _____ blood fatty acids.
133. Cortisol _____ blood amino acids.
134. Cortisol _____ muscle protein.
135. Growth hormone _____ blood glucose.
136. Growth hormone _____ blood fatty acids.
137. Growth hormone _____ blood amino acids.
138. Growth hormone _____ muscle protein.

Points to Ponder

1. Why do hypothalamic neurons extend into the posterior pituitary but not into the anterior pituitary?

2. Why must there be releasing and inhibiting hormones secreted by the hypothalamus? Why not evolve a system whereby the endocrine gland responds in the negative feedback loop?

3. How can LH and FSH affect both male and female organs?

4. How can the adenohypophysis be specialized to produce so many hormones?

5. How do glucagon and insulin work together to regulate the concentration of blood glucose?

6. Explain why the pituitary gland is sometimes referred to as the "master gland" in the body and why this reference is misleading.

7. What relationship does the adrenal medulla have to the autonomic nervous system?

8. Recent research indicates that diabetes mellitus is genetic. If this is true then the use of insulin only increases the frequency of the disease in the human population. How do we reduce the frequency of the disease and still help those who have the disorder?

9. What do you think causes osteoporosis? Is it an estrogen deficiency? Do men have estrogen levels?

10. Explain the statement that glucocorticoids have a catabolic effect on protein metabolism.

Clinical Perspectives

1. How would you determine that the negative feedback system has failed for a particular hormone? What would you test for?

2. Does milk production cease once the menstrual cycle resumes after pregnancy? Explain your answer.

3. Why do diabetics have to inject insulin instead of taking it orally?

4. Why would oxytocin be given to a pregnant woman? How does it function?

5. Why do individuals take melatonin pills? Do you think they work? Explain your answers.

6. What effects do abnormally high levels of glucocorticoids have on glucose metabolism?

7. What are the factors that contribute to hypoglycemia in Addison's disease?

8. What is the purpose of measuring fasting blood glucose levels?

9. Why is it important to identify individuals who are at risk for developing diabetes mellitus?

10. Why is ketoacidosis harmful to a diabetic?

18

The Reproductive System

Chapter Overview

The reproductive system contributes very little to the maintenance of homeostasis of the body. An individual can survive quite normally without reproducing. However, the survival of the species depends on the reproductive system. This is the major function of the system.

While the reproductive system is not a primary factor in overall homeostasis of the body there are unique situations in the reproductive process that involve homeostasis. These situations include Sertoli cells, endometrial cells, and the hormonal controls within the reproductive system.

The testes and ovaries produce gametes with reduced chromosome numbers necessary for zygote formation through fusion of the egg and sperm. The reproductive system also produces hormones, which play a role in initiation and completion of this reproductive function of the species. This body system is controlled by the central nervous system and the endocrine system. The various body systems maintain homeostasis in the body and cells of the body to survive. But, if it were not for a pair of reproductive systems there would be no body.

Chapter Outline

INTRODUCTION
The reproductive system includes the gonad, reproductive tract, and accessory sex glands, all of which are different in males and females.
- The primary reproductive organs, or gonads, perform the dual function of: (1) gametogenesis, and (2) secreting sex hormones.
- Testosterone in the male and estrogen in the female are responsible for the secondary sex characteristics.
- The essential reproductive functions of the male are: (1) spermatogenesis, and (2) delivery of sperm to the female.
- The essential female reproductive functions include: (1) oogenesis, (2) reception of sperm, (3) fertilization, (4) gestation, (5) parturition, and (6) lactation.
- In addition to the goands, the reproductive system includes the reproductive tract, accessory sex glands and external genitalia.

Reproductive cells each contain a half set of chromosomes.
- Somatic cells contain 46 chromosomes (the diploid number).
- Gametes contain only one of each homologous pair for a total of 23 chromosomes (the haploid number).

Gametogenesis is accomplished by meiosis, resulting in genetically unique sperm and ova
- During meiosis, a specialized diploid germ cell undergoes one chromosome replication followed by two nuclear divisions.
- Thus, sperm and ova each have a unique haploid number of chromosomes.

- When fertilization takes place, a sperm and ovum fuse to form the start of a new individual with 46 chromosomes.

The sex of an individual is determined by the combination of sex chromosomes.

- 22 of the chromosome pairs are autosomal chromosomes.
- The remaining pair of chromosomes are the sex chromosomes.
- Genetic males have both an X and a Y sex chromosome; genetic females have two X sex chromosomes

Sex differentiation along male or female lines depends on the presence or absence of masculinizing determinants.

- Differences between males and females exist at three levels: genetic, gonadal, and phenotypic (anatomic) sex
- Gonadal specificity appears during the seventh week of intrauterine life.
- The sex-determining region of the Y chromosome "masculinizes" the gonads by
- by coding for production of testis-determining factor (TDF) within primitive gonadal cells. TDF directs a series of events that leads to differentiation of the gonads.Because genetic females lack this gene they never receive a signal for testes formation.
- Phenotypic sex, the apparent anatomic sex of an individual, is hormonally mediated and depends on the genetically determined gonadal sex

- By 10–12 weeks of gestation, the sexes can be easily distinguished by the anatomical appearance of the external genitalia.
- Development of the reproductive tract along Occasionally, discrepancies occur between genetic and anatomic sexes because of errors in sexual differentiation

MALE REPRODUCTIVE PHYSIOLOGY
The scrotal location of the testes provides a cooler environment essential for spermatogenesis.

- In the last months of fetal life, the testes begin a slow descent, passing out of the abdominal cavity through the inguinal canal into the scrotum.
- Descent of the testes into this cooler environment is essential because spermatogenesis is temperature sensitive and cannot occur at normal body temperature.

The testicular Leydig cells secrete masculinizing testosterone.

- About 80% of the testicular mass consists of highly coiled seminiferous tubules, within which spermatogenesis takes place.
- The endocrine cells that produce testosterone—the Leydig cells or interstitial cells—are located in the connective tissue between the seminiferous tubules.
- Before birth, testosterone secretion by the fetal testes is responsible for masculinizing the reproductive tract and external genitalia and for promoting descent of the testes into the scrotum.
- At puberty, the Leydig cells start secreting testosterone, and spermatogenesis is initiated in the seminiferous tubules.
- Testosterone is responsible for growth and maturation of the entire male reproductive system.
- In both sexes, the reproductive changes that take place during puberty are (1) enlargement and maturation of gonads, (2) development of secondary sexual characteristics, (3) achievement of fertility (gamete production), (4) growth and maturation of the reproductive tract, and (5) attainment of libido (sex drive).

- Ongoing testosterone secretion is essential for spermatogenesis and for maintaining a mature male reproductive tract throughout adulthood.
- Testosterone is responsible for development of sexual libido at puberty and helps to maintain the sex drive in the adult male.
- All male secondary sexual characteristics depend on testosterone for their development and maintenance.
- Testosterone has a general protein anabolic effect and promotes bone growth.
- Testosterone secretion and spermatogenesis occur continuously throughout the male's life.
- In males, testosterone can be converted to estrogen by the enzyme aromatase.

Spermatogenesis yields an abundance of highly specialized, mobile sperm.

- Two functionally important cell types are present in the seminiferous tubules: germ cells and Sertoli cells.
- During meiosis, spermatogoiun divide mitotically to produce primary spermatocytes. Each primary spermatocyte forms two secondary spermatocytes during the first meiotic division, finally yielding four spermatids as a result of the second meiotic division.
- Sperm are essentially "stripped-down" cells in which most of the cytosol and the organelles not needed for the task of delivering the sperm's

genetic information to an ovum have been extruded.
- A spermatozoon has four parts: a head, an acrosome, a midpiece, and a tail.

Throughout their development, sperm remain intimately associated with Sertoli cells.

- The Sertoli cells form a ring that extends from the outer basement membrane to the lumen of the tubule.
- The supportive Sertoli cells perform the following functions essential for spermatogenesis: (1) The tight junctions between adjacent Sertoli cells form a blood-testes barrier. (2) Sertoli cells provide nourishment for sperm cells. (3) Sertoli cells engulf cytoplasm extruded from spermatids during their remodeling and destroy defective germ cells that fail to successfully complete all stages of spermatogenesis.(4) The Sertoli cells secrete into the lumen seminiferous tubule fluid, which "flushes" sperm into the epididymis. (5)Sertoli cells secrete androgen-binding protein. The Sertoli cells are the site of action for control of spermatogenesis by both testosterone and FSH.
- Androgen-binding protein retains testosterone within the tubules because the lipid soluble steroid would diffuse easily out of the lumen.
- The Sertoli cells release inhibin, which acts in negative-feedback fashion to regulate FSH secretion.

LH and FSH from the anterior pituitary control testosterone secretion and spermatogenesis.

- The testes are controlled by two gonadotropic hormones secreted by the anterior pituitary, luteinizing hormone (LH), and follicle-stimulating hormone (FSH).
- Luteinizing hormone acts on the Leydig cells to regulate testosterone secretion.
- Follicle-stimulating hormone acts on the seminiferous tubules, especially the Sertoli cells, to enhance spermatogenesis.
- Secretion of both LH and FSH from the anterior pituitary is stimulated by a single hypo-thalamic hormone, gonadotropin-releasing hormone (GnRH).
- Follicle-stimulating hormone is required for spermatid remodeling.

Gonadotropin-releasing hormone activity increases at puberty.

- The pubertal process is initiated by an increase in GnRH activity sometime between eight and twelve years of age.
- At puberty, the hypothalamus becomes less sensitive to feedback inhibition by testosterone.

The reproductive tract stores and concentrates sperm and increases their fertility.

- The remainder of the male reproductive system consists of: (1) a tortuous pathway of tubes that transport sperm from the testes to the outside of the body; (2) several glands, which contribute secretions that are important to the viability and motility of the sperm; and (3) the penis, which is designed to penetrate and deposit the sperm within the vagina of the female.
- As they leave the testes, the sperm are incapable of either movement or fertilization.
- Sperm gain both capabilities during their passage through the epididymis.
- The epididymis concentrates the sperm by absorbing most of the fluid that enters from the seminiferous tubules.
- The ductus deferens serves as an important site for sperm storage.

The accessory sex glands contribute the bulk of the semen.

- The seminal vesicles: (1) supply fructose, (2) secrete prostaglandins, (3) provide more than half of the semen, and (4) secrete fibrinogen.
- The prostate gland: (1) secretes an alkaline fluid that neutralizes the acidic vaginal secretions, and (2) provides clotting enzymes and fibrinolysin.
- The bulbourethral glands secrete a mucuslike substance that provides lubrication for sexual intercourse.

Prostaglandins are ubiquitous, locally acting chemical messengers.

- Prostaglandins are produced in virtually all tissues from arachidonic acid, a fatty acid constituent of the phospholipids within the plasma membrane.
- Besides enhancing sperm transport in semen, prostaglandins are known, or suspected, to exert other actions in the female reproductive system and in the respiratory, urinary, digestive, nervous, and endocrine systems, in addition to having effects on platelet aggregation, fat metabolism, and inflammation.
- Specific prostaglandins have been medically administered in such diverse situations as inducing labor, treating asthma, and treating gastric ulcers.

SEXUAL INTERCOURSE BETWEEN MALES AND FEMALES
The male sex act is characterized by erection and ejaculation.

- The male sex act involves two components: (1) erection, and (2) ejaculation.
- The sexual response cycle encompasses broader physiological responses that can be divided into four phases: (1) the excitement phase, (2) the plateau phase, (3) the orgasmic phase, and (4) the resolution phase.

Erection is accomplished by penis vasocongestion.

- During sexual arousal, the arterioles reflexly dilate and the erectile tissue fills with blood, causing the penis to enlarge both in length and width and to become more rigid.
- A larger buildup of blood and further enhancement of erection is achieved by a reduction in venous outflow.

- Tactile stimulation of the glans reflexly triggers increased parasympathetic vasodilator activity and decreased sympathetic vasoconstrictor activity to the penile arterioles.
- This response is the major instance of direct parasympathetic control over blood vessel caliber in the body.
- Parasympathetic impulses promote secretion of lubricating mucus from the bulbourethral glands and the urethral glands in preparation for coitus.
- The erection reflex is a spinal reflex that can be either facilitated or inhibited by higher brain centers through descending pathways that also terminate on the autonomic nerves supplying the penile arterioles.
- Psychic stimuli can induce an erection.

Ejaculation includes emission and expulsion.
- Ejaculation is accomplished by a spinal reflex.
- Sympathetic impulses cause sequential contraction of smooth muscles in the prostate, reproductive ducts, and seminal vesicles.
- This contractile activity delivers prostatic fluid, then sperm, and finally seminal vesicle fluid into the urethra. These events are emission.
- The filling of the urethra with semen triggers nerve impulses that activate a series of skeletal muscles at the base of the penis.
- Rhythmic contractions of these muscles forcibly expel

the semen through the urethra to the exterior. These events are expulsion.
- The average volume of semen is 2.75 ml, ranging from 2 to 6 ml, the higher volumes following periods of abstinence. An average human ejaculate contains about 180 million sperm (66 million/ml), but some ejaculates contain as many as 400 million sperm.During the resolution phase following orgasm, sympathetic vasoconstrictor impulses slow the inflow of blood into the penis so that the erection subsides.

The female sexual cycle is very similar to the male cycle.
- Both sexes experience the same four phases of sexual cycle.
- The excitement phase in females can be initiated by either physical or psychological stimuli.
- Tactile stimulation of the clitoris and surrounding perineal area is an especially powerful stimulus.
- Spinal reflexes bring about parasympathetically induced vasodilation of arterioles throughout the vagina and external genitalia.
- Vasocongestion of the vaginal capillaries forces fluid out of the vessels into the vaginal lumen.
- This fluid serves as the primary lubricant for intercourse.
- Further vasocongestion of the lower third of the vagina reduces its inner capacity so that it tightens around the thrusting penis.
- Simultaneously, the uterus raises upward, lifting the cervix and enlarging the upper two-thirds of the vagina to create a space for ejaculate deposition.
- The sexual response culminates in orgasm as sympathetic impulses trigger rhythmic contractions of the pelvic musculature.

FEMALE REPRODUCTIVE PHYSIOLOGY
Complex cycling characterizes female reproductive physiology.
- The release of ova is intermittent, and secretion of female sex hormones displays wide cyclical swings.

- During each menstrual cycle, the female reproductive tract is prepared for the fertilization and implantation of an ovum released from the ovary at ovulation.
- The ovaries perform the dual function of oogenesis and secreting the female sex hormones, estrogen and progesterone.
- Estrogen in the female is responsible for maturation and maintenance of the entire female reproduction system and establishment of female secondary sexual characteristics.
- Estrogen is essential for ova maturation and release, development of physical characteristics that are sexually attractive to males, and transport of sperm from the vagina to the site of fertilization in the oviduct.
- Estrogen contributes to breast development in anticipation of lactation.
- Progesterone is important in preparing a suitable environment for nourishing a developing embryo/fetus and for contributing to the breasts' ability to produce milk.

The steps of gametogenesis are the same in both sexes but the timing and outcome differ sharply.

- The undifferentiated primordial germ cells in the fetal ovaries divide mitotically to give rise to millions of oogonia by the fifth month of gestation, at which time mitotic proliferation ceases.

- Known now as "primary oocytes," they contain 46 replicated chromosomes.
- Before birth, each primary oocyte is surrounded by a single layer of granulosa cells to form the primary follicle.
- No new oocytes or follicles appear after birth.
- A follicle is destined for one of two fates once it starts to develop: it will reach maturity and ovulate, or it will degenerate to form scar tissue.
- By menopause, few, if any, primary follicles remain.
- The primary oocyte within a primary follicle is still a diploid cell.
- Just before ovulation, the primary oocyte, whose nucleus has been in meiotic arrest for years, completes its first meiotic division.
- This division yields two daughter cells.
- Almost all of the cytoplasm remains with one of the daughter cells, now called the "secondary oocyte," which is destined to become the ovum.
- The chromosomes of the other daughter cell together with a small share of cytoplasm form the first polar body.
- The nutrient-poor polar body soon degenerates.
- Sperm entry into the secondary oocyte is needed to trigger the second meiotic division.
- 23 unpaired chromosomes remain in what is now the mature ovum.
- These 23 maternal chromosomes unite with the 23 paternal chromosomes of the penetrating sperm to complete fertilization.

The ovarian cycle consists of alternating follicular and luteal phases.

- After the onset of puberty, the ovary constantly alternates between two phases: (1) the follicular phase, which is dominated by the presence of maturing follicles, and (2) the luteal phase, which is characterized by the presence of the corpus luteum.

The follicular phase is characterized by the development of maturing follicles.

- The follicle operates in the first half of the cycle to produce a mature egg ready for ovulation at midcycle.

- The corpus luteum takes over during the second half of the cycle to prepare the female reproductive tract for pregnancy if fertilization of the released egg occurs.
- The single layer of granulosa cells in a primary follicle proliferates to form several layers that surround the oocyte.
- Follicular cells function as a unit to secrete estrogen.
- The hormonal environment that exists during the follicular phase promotes enlargement and development of the follicular cells' secretory capacity, converting the primary follicle into a secondary, or antral, follicle capable of estrogen secretion.
- This stage is characterized by the formation of a fluid-filled antrum.
- The antrum occupies most of the space in a mature follicle.
- The oocyte, surrounded by the zona pellucida and a single layer of granulosa cells, is displaced asymmetrically at one side of the growing follicle in a little mound that protrudes into the antrum.
- The greatly expanded mature follicle bulges on the ovarian surface, creating a thin area that ruptures to release the oocyte at ovulation.
- Just before ovulation, the oocyte completes its first meiotic division.
- The ovum, still surrounded by its tightly adhering zona pellucida and granulosa cells, is swept out of the ruptured follicle into the abdominal cavity by the leaking antral fluid.
- The released ovum is quickly drawn into the oviduct.
- Rupture of the follicle at ovulation signals the end of the follicular phase.

The luteal phase is characterized by the presence of a corpus luteum.
- Old follicular cells undergo dramatic structural transformations to form the corpus luteum, in a process called "luteinization."
- The follicular-turned luteal cells hypertrophy and are converted into very active steroid-producing tissue known as the "corpus luteum."
- Estrogen secretion in the follicular phase followed by progesterone secretion in the luteal phase is essential for preparing the uterus to be a suitable site for implantation of a fertilized ovum.
- If the released ovum is not fertilized, the corpus luteum degenerates within 14 days after its formation.
- If fertilization and implantation do take place, the corpus luteum continues to grow and produce increasing quantities of progesterone and estrogen.

The ovarian cycle is regulated by complex hormonal interactions.
- Gonadal function in the female is directly controlled by the anterior pituitary gonadotropic hormones, FSH and LH.
- These hormones are regulated by pulsatile episodes of hypothalamic gonadotropin-releasing hormone and feedback actions of gonadal hormones.
- Estrogen exerts negative-feedback effects during part of the cycle and positive-feedback effects during another part of the cycle, depending on the concentration of estrogen.
- Antrum formation is induced by FSH.
- Both FSH and estrogen stimulate proliferation of granulosa cells.
- Both LH and FSH are required for synthesis and secretion of estrogen by the follicle.
- LH stimulates the thecal cells to synthesize androgen, which diffuses into the adjacent granulosa cells. FSH stimulates the granulosa cells to convert this androgen into

estrogen Part of the estrogen produced by the growing follicle is secreted into the blood and is responsible for steadily increasing plasma estrogen levels during the follicular phase.

- The secreted estrogen inhibits the hypothalamus and anterior pituitary in negative-feedback fashion.
- The plasma FSH level declines during the follicular phase as the estrogen level rises.
- The LH secretion continues to rise slowly during the follicular phase despite inhibition of GnRH secretion.
- Ovulation and subsequent luteinization of the ruptured follicle are triggered by an abrupt, massive increase in LH secretion.
- This LH surge brings about four major changes in the follicle: (1) It halts estrogen synthesis by the follicular cells. (2) It reinitiates meiosis in the oocyte (3) It triggers production of locally acting prostaglandins. (4) It causes differentiation of follicular cells into the luteal cells.
- The LH surge is triggered by a positive-feedback effect.
- Thus, LH enhances estrogen production by the follicle, and the resultant peak estrogen concentration stimulates LH secretion.
- Luteinizing hormone "maintains" the corpus luteum.
- No progesterone is secreted during the follicle phase.
- The follicular phase is dominated by estrogen and the luteal phase by progesterone.
- Even though a high level estrogen stimulates LH secretion, progesterone, which dominates the luteal phase, powerfully inhibits LH secretion as well as FSH secretion.
- Inhibition of FSH and LH by progesterone prevents new follicular maturation and ovulation during the luteal phase.
- Under progesterone's influence, the reproductive system is gearing up to support the just-released ovum, should it be fertilized.

Cyclic uterine changes are caused by hormonal changes during the ovarian cycle.

- The menstrual cycle consists of three phases: (1) the menstrual phase, (2) the proliferative phase, and (3) the secretory or progestational phase.
- The menstrual phase is characterized by discharge of blood and endometrial debris from the vagina.
- The first day of menstruation is considered to be the start of a new cycle and coincides with termination of the ovarian luteal phase and onset of the follicular phase.
- As the corpus luteum degenerates, circulating levels of estrogen and progesterone drop precipitously.
- The fall in ovarian hormone levels also stimulates release of a uterine prostaglandin that causes vasoconstriction of the endometrial vessels.
- The entire uterine lining sloughs during each menstrual period except for a deep, thin layer of epithelial cells and glands from which the endometrium will regenerate.
- Local uterine prostaglandin stimulates mild rhythmic contractions of the uterine myometrium to help expel the blood and endometrial debris from the uterine cavity.
- Excessive uterine contractions caused by overproduction of prostaglandin are responsible for the menstrual cramps.
- Withdrawal of estrogen and progesterone upon degeneration of the corpus luteum leads simultaneously to sloughing of the endometrium and development of new follicles in the ovary under the influence of rising gonadotropic hormone levels.

- The proliferative phase of the uterine cycle begins concurrent with the last portion of the ovarian follicular phase as the endometrium starts to repair.
- The estrogen-dominant proliferative phase lasts from the end of menstruation to ovulation.
- Following ovulation, a new corpus luteum is formed.
- Progesterone acts on the thickened, estrogen-primed endometrium to convert it to a richly vascularized, glycogen-filled tissue.
- This period is called the "secretory phase," or the "progestational phase."
- If fertilization and implantation do not occur, the corpus luteum degenerates and a new follicular phase and menstruation phase begin once again.
- Oral contraceptives prevent ovulation and thus unwanted pregnancy primarily by suppressing gonadotropin secretion.

Pubertal changes in females are similar to those in males.
- Unlike the fetal testes, the fetal ovaries do not need to be functional, because feminization of the female reproductive system automatically takes place in the absence of fetal testosterone secretion without the presence of female sex hormones.
- The resultant secretion of estrogen by the activated ovaries induces growth and maturation of the female reproductive tract as well as development of the female secondary sexual characteristics.
- Three other pubertal changes in females: (1) growth of axillary and pubic hair, (2) the pubertal growth spurt, and (3) development of libido—are attributed to a spurt in adrenal androgen secretion at puberty, not to estrogen.
- The pubertal rise in estrogen closes the epiphyseal plates.

Menopause is unique to females.
- The cessation of a woman's menstrual cycles at menopause is attributable to the limited supply of ovarian follicles present at birth.
- The termination of reproductive potential in a middle-aged woman is "preprogrammed" at her own birth.
- Menopause is preceded by a period of progressive ovarian failure characterized by increasingly irregular cycles and dwindling estrogen levels.
- Postmenopausal women are not completely devoid of estrogen, however, because adipose tissue, the liver, and the adrenal cortex continue to produce estrogen.
- The absence of ovarian estrogen is responsible for the physical postmenopausal changes that occur in the reproductive tract, including vaginal dryness and gradual atrophy of the genital organs.
- Postmenopausal women still have a sex drive because of their adrenal androgens

The oviduct is the site of fertilization.
- Fertilization, the union of male and female gametes, normally occurs in the ampulla, the upper third of the oviduct.
- The dilated end of the oviduct cups around the ovary and contains fimbriae, fingerlike projections that contract in a sweeping motion to guide the released ovum into the oviduct.
- Fertilization must occur within 24 hours after ovulation.
- Sperm can survive up to five days in the female reproductive tract.
- Once sperm are deposited in the vagina at ejaculation, they must travel through the cervical canal, through the uterus, and then

- up to the egg in the upper third of the oviduct.
- Once the sperm have entered the uterus, contractions of the myometrium churn them around in "washing-machine" fashion.
- Sperm transport in the oviduct is facilitated by antiperistaltic contractions of the oviduct smooth muscle.
- These myometrial and oviduct contractions that facilitate sperm transport are induced by the high estrogen level that exists prior to ovulation, perhaps aided by seminal prostaglandins.
- To fertilize an ovum, a sperm must first pass through the corona radiata and the zona pellucida surrounding it.
- The acrosomal enzymes, which are exposed as the acrosomal membrane disrupts on contact with the corona radiata, enable the sperm to tunnel a path through these protective barriers.
- After passing through the corona, sperm bind to ZP3 receptors on the zona pellucida. These receptors are species specific. Once binding occurs, the acrosome reactions take place and the sperm can pass though the zona.
- The first sperm to reach the ovum itself fuses with the plasma membrane of the ovum, triggering a chemical change in the ovum's surrounding membrane that makes the outer layer impenetrable to the entry of

- any more sperm, thus preventing polyspermy
- The head of the fused sperm is gradually pulled into the ovum's cytoplasm by a growing cone that engulfs it.
- Penetration of the sperm into the cytoplasm triggers the final meiotic division of the secondary oocyte.
- Fertilized egg is known as a zygote.

The blastocyst implants in the endometrium through the action of its trophoblastic enzymes.

- The zygote is rapidly undergoing a number of mitotic cell divisions to form a solid ball of cells called the "morula."
- About three or four days after ovulation, progesterone is being produced in sufficient quantities to relax oviduct constriction, thus permitting the morula to be rapidly propelled into the uterus by oviductal peristaltic contractions and ciliary activity.
- During the first six to eight days after ovulation, while the developing embryo is in transit in the oviduct and floating in the uterine lumen, the uterine lining is simultaneously being prepared for implantation under the influence of luteal-phase progesterone.
- By the time the endometrium is suitable for implantation, the morula has descended to the uterus and continued to proliferate and differentiate into a blastocyst capable of implantation.
- Implantation begins when the trophoblastic cells overlying the inner cell mass release protein-digesting enzymes upon contact with the endometrium.
- The trophoblast performs the dual functions of: (1) accomplishing implantation as it carves out a hole in the endometrium for the blastocyst, and (2) making metabolic fuel and raw materials available for the developing embryo as the advancing trophoblastic projections break down the nutrient-rich endometrial tissue.
- After the blastocyst burrows into the decidua by means of trophoblastic activity, a layer of endometrial cells covers over the surface of the hole, completely burying the blastocyst within the uterine lining.

The placenta is the organ of exchange between maternal and fetal blood.

- To sustain the growing embryo/fetus for the duration of its intrauterine life, the placenta is rapidly developed.
- By day 12, the trophoblastic layer is two cell-layers thick and is called the "chorion."
- Fingerlike projections of chorionic tissue extend into the pools of maternal blood.
- The developing embryo sends out capillaries into these chorionic projections to form placental villi.
- Each placental villus contains embryonic capillaries surrounded by a thin layer of chorionic tissue, which separates the embryonic/fetal blood from the maternal blood in the intervillus spaces.
- The placenta is well established and operational by five weeks after implantation.
- During the time of implantation and early placental development, the inner cell mass forms a fluid-filled amniotic cavity between the trophoblast/chorion and the portion of the inner cell mass destined to become the fetus.
- The fluid in the amniotic cavity, the amniotic fluid, which is similar in composition to normal ECF, surrounds and cushions the fetus throughout gestation By this time, the heart of the developing embryo is pumping blood into the placental villi as well as to the embryonic tissues.

- During intrauterine life, the placenta performs the functions of the digestive system, the respiratory system, and the kidneys for the "parasitic" fetus.

Hormones secreted by the placenta play a critical role in the maintenance of pregnancy.

- The placenta has the remarkable capacity to secrete a number of peptide and steroid hormones essential for the maintenance of pregnancy.
- The most important are human chorionic gonadotropin, estrogen, and progesterone.
- One of the first events following implantation is secretion by the developing chorion of human chorionic gonadotropin (hCG), a peptide hormone that prolongs the life span of the corpus luteum.
- The maintenance of a normal pregnancy depends on high concentrations of progesterone and estrogen.
- By the tenth week of pregnancy, hCG output declines to a low rate of secretion that is maintained for the duration of gestation.
- The fall in hCG occurs at a time when the corpus luteum is no longer needed for its steroidal hormone output because the placenta has begun to secrete substantial quantities of estrogen and progesterone.
- The placenta is unable to produce sufficient quantities of either estrogen or progesterone in the first trimester of pregnancy.
- The placenta is able to convert the androgen hormone produced by the fetal adrenal cortex, DHEA, into estrogen.
- The placenta is unable to produce estrogen until the fetus has developed to the point that its adrenal cortex is secreting DHEA into the blood.
- The placenta extracts DHEA from the fetal blood and converts it into estrogen, which it then secretes into the maternal blood.
- The placenta is simply too small in the first 10 weeks of pregnancy to produce sufficient quantities of progesterone to maintain the endometrial tissue.

Maternal body systems respond to the increased demands of gestation.

- During gestation, a number of physical changes take place within the mother to accommodate the demands of pregnancy.
- The most obvious change is uterine enlargement.
- The breasts enlarge and develop the capability of producing milk.
- The volume of blood increases and the cardiovascular system responds to the increasing demands of the growing placental mass.
- Respiratory activity is increased.
- Urinary output increases, and the kidneys excrete the additional wastes from the fetus.

Changes during late gestation prepare for parturition.

- Parturition requires: (1) dilation of the cervical canal to accommodate passage of the fetus from the uterus through the vagina and to the outside, and (2) contractions of the uterine myometrium that are sufficiently strong to expel the fetus.
- Cervical softening is believed to be caused by relaxin, a peptide hormone produced by the corpus luteum of pregnancy and by the placenta.
- Relaxin also "relaxes" the pelvic bones.
- Rhythmic, coordinated contractions begin at the onset of true labor.
- The contractions occur with increasing frequency and intensity.

- After having dilated the cervix sufficiently for passage of the fetus, these contractions force the fetus out through the birth canal.
- A dramatic, progressive increase in the concentration of myometrial receptors for oxytocin, probably induced by the increasing levels of estrogen during pregnancy, is ultimately responsible for initiating labor.

Scientists are closing in on the factors that trigger the onset of parturition.

- Oxytocin is a powerful uterine muscle stimulant and is known to play a role in the progression of labor.
- The uterine responsiveness to oxytocin is 100 times greater at term than in nonpregnant women.
- This gives rise to speculation that labor is initiated when the oxytocin receptor concentration reaches a critical threshold level.
- Evidence suggests that corticotropin-releasing hormone (CRH) secreted by the fetal portion of the placenta into both the maternal and fetal circulations not only drives the manufacture of placental estrogen, thus ultimately dictating the timing of the onset of labor, but also promotes changes in the fetal lungs needed for breathing.
- Recent research suggests that inflammation plays a central role in the labor process. Key to this inflammatory response is activation of nuclear factor κB (NF-κB) in the uterus. NF-κB boosts production of inflammatory cytokines and prostaglandins that increase the sensitivity of the uterus to contraction-inducing chemical messengers and help soften the cervix.

Parturition is accomplished by a positive-feedback cycle.

- Once uterine responsiveness to oxytocin reaches a critical level and regular uterine contractions begin, myometrial contractions progressively increase in frequency, strength, and duration throughout labor until they expel the uterine contents.
- Myometrial contractions incessantly increase as labor progresses because of a positive feedback cycle involving oxytocin and prostaglandins.

- Labor is divided into three stages: (1) cervical dilation, (2) delivery of the baby, and (3) delivery of the placenta.
- After delivery, the uterus shrinks to its pregestational size, a process known as "involution."
- Involution occurs largely because of the precipitous fall in circulating estrogen and progesterone when the placental source of these steroids is lost at delivery.

Lactation requires multiple hormonal inputs.

- During gestation, the mammary glands, or breasts, are prepared for lactation.
- The high concentration of estrogen promotes extensive duct development, whereas the high level of progesterone stimulates abundant alveolar-lobular formation.
- Elevated concentrations of prolactin and a placental hormone similar to prolactin contribute to mammary gland development by inducing the synthesis of enzymes needed for milk production.
- The high estrogen and progesterone concentrations during the last half of pregnancy prevent lactation by blocking prolactin's stimulatory action on milk secretion.
- Prolactin is the primary stimulant of milk secretion.
- Two hormones are critical for maintaining lactation: (1) prolactin, and (2) oxytocin.
- Release of both these hormones is stimulated by a neuroendocrine reflex triggered by suckling.
- Suckling of the breast by the infant stimulates sensory nerve endings in the nipple, resulting in initiation of action potentials that travel up the spinal cord to the hypothalamus.
- The hypothalamus triggers a burst of oxytocin release from the posterior pituitary.
- Oxytocin stimulates contractions of the myoepithelial cells in the breasts to bring about milk ejection.

Breast-feeding is advantageous to both the infant and the mother.

- Milk is composed of water, triglyceride fat, the carbohydrate lactose, a number of proteins, vitamins, and the minerals calcium and phosphate.
- Colostrum, the milk produced for the first five days postpartum, contains lower concentrations of fat and lactose but higher concentrations of proteins, most notably lactoferrin and immunoglobulins.
- All human babies acquire some passive immunity during gestation through the passage of antibodies across the placenta.
- Breast-feeding is also advantageous to the mother.
- Oxytocin release triggered by nursing hastens uterine involution.
- Suckling suppresses the menstrual cycle by inhibiting LH and FSH secretion.
- Lactation prevents ovulation and serves as a means of preventing pregnancy, thus permitting all the mother's resources to be directed toward the newborn instead of being shared with a new embryo.

The end is a new beginning.

- The single cell resulting from the union of male and female gametes divides mitotically and differentiates into a multicellular individual made up of a number of different organ systems that interact cooperatively to maintain homeostasis.
- All of the life-supporting homeostatic processes introduced throughout this book begin all over again at the start of a new life.

Key Terms

Acrosome	Luteal phase
Adolescence	Luteinization
Ampulla	Meiosis
Androgens	Menarche
Aromatase	Menstrual (uterine) cycle
Atresia	Mitosis
Autosomal chromosomes	Morning sickness
Blastocyst	Morula
Blood-testes barrier	Oogonia
Cervix	Orgasm
Chorion	Ova
Climacteric	Oviducts
Colostrum	Ovulation
Corpus albicans	Parturition
Corpus luteum	Penis
Corticotropin-releasing hormone	Placenta
Cryptorchidism	Polyspermy
Decidua	Primary follicle
Dihydrotestosterone (DHT)	Primary oocyte
Diploid	Primary spermatocytes
Embryo	Progestational phase
Emission	Proliferative phase
Endocrine disrupters	Prostaglandins
Erectile tissue	Puberty
Estrogen receptor modulators (ERM)	Reproduction
Expulsion	Reproductive system
Fertilization	Secondary oocyte
Fetoplacental unit	Secondary sexual characteristics
Follicular phase	Seminiferous tubules
Gametes (reproductive, germ cells)	Sex chromosomes
Gametogenesis	Sperm
Gestation	Spermatids
Gonads	Spermatogenesis
Graafian follicle	Spermeogenesis
Granulosa cells	Theca cells
Haploid	Trophoblast
Homologous chromosomes	Tubal pregnancy
Hymen	Uterus
Inhibin	Vasectomy
Involution	Vulva
Leydig cells	ZP3 receptors
LH surge	Zygote

Review Exercises

Answers are in the appendix.

True/False-if an answer is false, underline the incorrect portion of the statement and rewrite it to make the statement true.

_____ 1. Formation of the gonads in the embryo is known as "gametogenesis."

_____ 2. All fetuses regardless of their genetic sex are exposed to female androgens.

_____ 3. Fallopian tubes are the same as oviducts.

_____ 4. Somatic cells are haploid.

_____ 5. Gametes are haploid.

_____ 6. During meiosis, a diploid germ cell undergoes one chromosome replication followed by two nuclear divisions.

_____ 7. Gonadal specificity appears during the seventh week of intrauterine life.

_____ 8. Androgen binding protein concentrates testosterone in the blood.

_____ 9. Estrogen exerts both positive and negative feedback during the ovarian cycle.

_____ 10. Sertoli cells secrete testosterone.

_____ 11. Each primary spermatocyte has 23 doubled chromosomes.

_____ 12. The acrosome is an enzyme-filled vesicle at the tip of the head of the sperm.

_____ 13. The tight junctions between adjacent Leydig cells form a blood-testes barrier.

_____ 14. Sertoli cells phagocytize cytoplasm extruded from spermatids.

_____ 15. Seminiferous tubule fluid is secreted by Sertoli cells.

_____ 16. FSH acts on the Leydig cells to regulate testosterone secretion.

_____ 17. FSH is required for spermatid remodeling.

_____ 18. The bulbourethral glands secrete an alkaline fluid that neutralizes the acidic vaginal secretions.

_____ 19. Prostaglandins are an important chemical messenger during the process of labor.

_____ 20. Increased parasympathetic activity to the penile arterioles produces an erection.

_____ 21. Progesterone in the female is responsible for maturation and maintenance of the entire female reproductive system.

_____ 22. Amniotic fluid is similar in composition to intracellular fluid.

_____ 23. Sperm entry into the secondary oocyte is needed to trigger the second meiotic division.

_____ 24. The corpus luteum operates in the first half of the cycle to produce a mature egg ready for ovulation at midcycle.

_____ 25. LH stimulates granulosa cells to convert androgens into estrogen.

_____ 26. Estrogen secretion in the follicular phase and progesterone secretion in the luteal phase are essential for preparing the uterus for implantation.

_____ 27. Antrum formation is induced by LH.

_____ 28. The LH surge causes differentiation of follicular cells into luteal cells.

_____ 29. No progesterone is secreted during the follicular phase.

_____ 30. Under the influence of estrogen, the mucus secreted by the cervix is abundant, clear, and thin.

Fill in the Blank

31. In females, reproductive potential ceases during middle age at a time known as _____.

32. Before birth, each primary oocyte is surrounded by a single layer of _____ to form a(n) _____.

33. Oral contraceptives function by preventing _____.

34. The thecal and granulosa cells, collectively known as _____, function as a unit to secrete estrogen.

35. _____ is the principle ovarian estrogen.

36. The uterus consists of two layers: the _____, which is the outer smooth-muscle layer; and the _____, which is the inner lining that contains numerous blood vessels and glands.

37. The enzyme _____ converts estrogen to testosterone.

38. Sperm interact with the zona pellucida by way of _____ receptors.

39. Fertilization normally occurs in the _____.

40. The _____: (1) supply fructose, which serves as the primary energy source for ejaculated sperm, and (2) secrete prostaglandins.

41. The _____ phase returns the genitalia and body systems to their prearousal state.

42. The penis consists almost entirely of _____ made up of three columns of spongelike vascular spaces.

43. After fertilization a barrier is created around the newly fertilized egg to prevent _____.

44. The orgasmic contractions occur most intensely in the engorged lower third of the vaginal canal called the _____.

45. Reproduction depends on the union of male and female _____.

46. _____ contain 46 chromosomes.

47. 22 of the chromosome pairs are
    _____.

48. The placenta makes estrogen from _____ which it
    receives from fetal tissue.

49. When a testis remains undescended into adulthood,
    this condition is known as _____.

50. The endocrine cells that produce testosterone in
    males are _____ or
    _____.

51. Production of extremely specialized, mobile
    spermatozoa from spermatids requires extensive
    remodeling, or packaging, of cellular elements, a
    process known as _____.

52. A spermatozoon has four parts: _____,
    _____, _____, and
    _____.

53. The tight junctions between adjacent Sertoli cells
    form a(n) _____.

54. The testes are controlled by the two gonadotropic
    hormones secreted by the anterior pituitary,
    _____ and
    _____.

55. A(n) _____ is a single-layered sphere
    of cells encircling a fluid-filled cavity.

56. The endometrial tissue so modified at the
    implantation site is called the _____.

57. When the trophoblastic layer is two cell-layers thick
    it is then called _____.

58.
    _____
    _____ is a peptide hormone that prolongs the
    life span of the corpus luteum.

59. The period of gestation is about _____ weeks from
    conception.

60. _____ is a peptide hormone that is
    produced by the hypothalamus and is a key role in
    the progression of labor.

61. After delivery, the uterus shrinks to its pregestational size, a process known as _____.

62. _____ is the milk produced for the first five days postpartum.

Matching
Match the function to the placental hormone using the code on the right. Some functions may require multiple hormones

a. estrogen
b. progesterone
c. hCG
d. relaxin
e. CRH

_____ 63. Helps prepare mammary glands for lactation

_____ 64. Loosens connective tissue between pelvic bones in preparation for parturition

_____ 65. Promotes formation of cervical mucus plug

_____ 66. Maintains corpus luteum

_____ 67. Softens cervix in preparation for cervical dilation at parturition

_____ 68. Suppresses uterine contractions to provide quiet environment for fetus

_____ 69. Stimulates secretion of testosterone by developing testes in XY embryos

_____ 70. This hypothalamic factor is thought to play a role in labor onset.

Multiple Choice

71. Which of the following hormones is the primary stimulant of milk secretion?
    a. human chorionic somatomammotropin
    b. prolactin
    c. estrogen
    d. progesterone
    e. prostaglandins

72. Which of the following hormones promotes fat deposition?
    a. prolactin
    b. human chorionic somatomammotropin
    c. estrogen
    d. progesterone
    e. none of the above

73. Which of the following hormones secreted by the corpus luteum softens the cervix?
    a. estrogen
    b. progesterone
    c. oxytocin
    d. relaxin
    e. none of the above

74. Which of the following is not associated with puberty?
    a. enlargement and maturation of gonads
    b. development of secondary sexual characteristics
    c. achievement of fertility
    d. growth and maturation of the reproductive tract
    e. all of the above are associated with puberty

75. Which of the following hormones promotes the conversion of progesterone into estrogen?
    a. ACTH
    b. fetal cortisol
    c. FSH
    d. CRH
    e. LH

76. This fluid is similar in composition to ECF:
    a. semen
    b. amniotic fluid
    c. fluid in the antrum
    d. cervical secretions
    e. none of the above

77. Which of the following hormones is directly
   responsible for uterine contractions?
   a. PIH
   b. estrogen
   c. CRH
   d. progesterone
   e. oxytocin

78. Which of the following hormones maintains the
   corpus luteum?
   a. CRH
   b. ACTH
   c. LH
   d. FSH
   e. hCG

79. Which of the following hormones enhances uterine
   contractility?
   a. progesterone
   b. estrogen
   c. CRH
   d. fetal cortisol
   e. BST

80. Which of the following hormones shields
   premenopausal women from osteoporosis?
   a. estrogen
   b. progesterone
   c. CRH
   d. ACTH
   e. relaxin

81. Which of the following hormone secretions is
   suppressed by suckling?
   a. ACTH
   b. CRH
   c. FSH
   d. hCG
   e. IgA

82. Which of the following hormones directly
   stimulates testosterone production?
   a. progesterone
   b. cortisol
   c. LH
   d. FSH
   e. estrogen

83.   Which of the following is not thought to play a role in labor initiation and onset?
   a. prostalgandins
   b. CRH
   c. fetal surfactant
   d. inflammation
   e. all of the above are involved

Modified Multiple Choice

Write the appropriate letter in the blank by using the following answer code to indicate the relative magnitude of the two items in question.

A = A is greater than B
B = B is greater than A
C = A and B are approximately equal

84. _____ A. The length of the luteal phase of the average menstrual cycle.
       B. The length of the follicular phase of the average menstrual cycle.

85. _____ A. The length of the luteal phase of the ovarian cycle.
       B. The length of the secretory phase of the menstrual cycle.

86. _____ A. The number of chromosomes in am immature sperm.
       B. The number of chromosomes in a spermatogonium.

87. _____ A. The number of primary follicles present in the ovary at any time during the menstrual cycle.
       B. The number of mature follicles just before ovulation.

88. _____ A. Plasma concentration of testosterone.
       B. Concentration of testosterone in tubular fluid. .

89. _____ A. Plasma concentration of progesterone during the follicular phase of the ovarian cycle.
       B. Plasma concentration of progesterone during the luteal phase of the ovarian cycle.

Indicate which item in the following answer code best matches the description in question by writing the appropriate letter in the blank. There is only one correct answer per question. Each answer may be used more than once. Some answers may not be used at all.

A = erection
L = prostate gland
B = acrosome
M = trophoblast
C = Leydig cells
N = corpus luteum
D = seminal vesicles
O = epididymis & ductus deferens
E = oviduct
P = ovarian follicle
F = uterus

Q = ejaculation
G = inner cell mass
R = tail of sperm
H = seminiferous tubules
S = Sertoli cells
I = midpiece of sperm
T = endometrium
J = head of sperm
U = Placenta
K = vagina

90. _____ Formation is stimulated by LH after ovulation.
91. _____ Site of spermatogenesis.
92. _____ Store sperm prior to ejaculation.
93. _____ Occurs as a result of vascular engorgement.
94. _____ Site of fertilization.
95. _____ Destined to become the fetus.
96. _____ Fills with glycogen and becomes richly vascularized under the influence of progesterone.
97. _____ Tissue that secretes testosterone.
98. _____ Contains enzymes utilized for penetration of the ovum.
99. _____ Secretes HCG.
100. _____ Secretes prostaglandins into semen.
101. _____ Responsible for final maneuvering for penetration of ova.
102. _____ Supplies fructose utilized by sperm for energy.
103. _____ Destined to become the fetal portion of the placenta.
104. _____ Semen expelled from penis.
105. _____ Contains primarily the nucleus that contains the sperm's genetic information.
106. _____ Serves as a route by which nutrients reach developing sperm cells.
107. _____ Organ of exchange between mother and fetus.
108. _____ Secretes alkaline fluid to neutralize acidic vaginal secretions.
109. _____ Portion of blastocyst responsible for accomplishing implantation.
110. _____ Structure within the ovum develops prior to ovulation.

111. _____ Site of initial maturation of sperm for motility and fertility in the male.
112. _____ Contains mitochondria for energy production within sperm.
113. _____ Secretes estrogen but not progesterone.
114. _____ Secretes estrogen and progesterone during the first 10 weeks of gestation.
115. _____ Secretes estrogen and progesterone during the last seven months of gestation.

Points to Ponder

1. It is very commonplace to discuss abnormalities in hormone secretion of all endocrine glands and tissues except the gonads. Is it because such abnormalities concerning the gonads do not exist or is it due to a social attitude?

2. After carefully studying this chapter, do you think gay people could have abnormal hormone levels?

3. Why do you think most oral contraceptives are for women? Is there a physiological reason?

4. Is it possible for a woman to become pregnant if the male does not ejaculate during sexual intercourse?

5. Which contraceptive methods are the most effective? Explain why. The least effective? Explain why.

6. Explain the vascular mechanism by which penile erection occurs. How does Viagra help this process?

7. What are some of the first signs of pregnancy?

8. When can the sex of a fetus first be detected?

9. How is hCG related to the various pregnancy tests?

10. Why are some babies born prematurely?

Clinical Perspectives

1.   Explain why birth control pills are prescribed to women who have complications at the onset of menopause.

2.   Explain how an infant girl's breasts can begin to enlarge.

3.   During a routine physical examination of a male, why does a physician assistant insert a finger in the inguinal rings of the scrotum and ask the male to cough?

4.   Why do nursing babies sometimes refuse milk after the mother has exercised?

5.   Do women with small breasts produce less milk than large-breasted women? Explain your answer.

6.   After missing a menstrual period, your roommate purchases a home pregnancy test. The test is positive and she makes an appointment with the school health physician. The physician tells her to bring a urine sample in with the appointment. After giving her a pelvic exam, the physician told her she is pregnant and should begin taking vitamins and minerals and to return in one month. Why is a missed menstrual cycle one of the first signs of pregnancy? What is in the urine that indicates a positive test for pregnancy? Why was your roommate told to take vitamins and minerals?

7.   Based on what you know about female physiology, what is the principle behind the surgical removal of the ovaries as a treatment for breast cancer?

8.   What serum enzyme is likely to be elevated in advanced prostate cancer?

9.   Your sister asks you to explain the proposed mechanisms whereby the act of a mother nursing her baby results in lactation. How would you answer this question? By what mechanisms might the sound of a baby crying elicit the milk-ejection reflex?

10. According to your college roommate, there is a female birth control pill and not a male birth control pill because the medical establishment is run by men. Do you agree with this theory? Explain your answer.

Appendix: Answers

Chapter 1: Homeostasis: The Foundation of Physiology

True/False

1. True
2. False—It's reversed.
3. False—Some organs are in several systems.
4. True
5. False—Extrinsic control.
6. True
7. True
8. False—Physiologist focus on why something happens rather than emphasizing the purpose.
9. True
10. True
11. False—Cells make life-sustaining changes directly with the internal environment,
12. True
13. True
14. False—Exocrine glands.
15. True
16. True
17. True
18. True
19. True
20. False—Four.
21. True
22. True
23. False—Three.
24. True
25. True

Fill in the Blank

26. Exocrine
27. Lumen
28. Local
29. Homeostasis
30. Interstitial
31. Digestive
32. Negative feedback
33. Cell
34. Internal environment
35. Interstitial fluid
36. Negative feedback
37. Pathophysiology
38. Cell
39. Endocrine
40. Glands
41. Body systems
42. Circulatory, digestive, respiratory, urinary, skeletal, muscular, integumentary, immune, nervous, endocrine, and reproductive
43. Physiology
44. Extracellular
45. Positive
46. Stress
47. Connective
48. Lymph
49. Cytosol
50. Lymphatic

Matching

51. e
52. g
53. k
54. h
55. j
56. a
57. d
58. f
59. b
60. i
61. c
62. a
63. e
64. b

65. d
66. c

Multiple Choice

67. c
68. d
69. c
70. e
71. c
72. b
73. a
74. b
75. b
76. a
77. c
78. a
79. e
80. a
81. d
82. b
83. a
84. b
85. b
86. e
87. a
88. c
89. b
90. e
91. e
92. e
93. c
94. a
95. a
96. a
97. c
98. e
99. a
100. e

Chapter 2: Cellular Physiology

True/False-if an answer is false, underline the incorrect portion of the statement and rewrite it to make the statement true.

1. False— is associated with abnormal lysosome function.
2. True
3. True

4. False— Aerobic reactions are .
5. False—Another name for electron transport chain.
6. True
7. True
8. True
9. False—Plant cells, not human or other animal cells, are surrounded by a cell wall.
10. True
11. True
12. True
13. False—Single cell.
14. False—DNA.
15. True
16. False—carried out by ribosomes.
17. False-contain different number
18. False—Rough ER contains ribosomes.
19. True
20. True
21. True
22. True
23. True
24. False—Double membrane.
25. False—ATP.
26. False—Two molecules of ATP.
27. True
28. True
29. True
30. False—Coated vesicles capture specific cargo, not a representative mixture of proteins, before budding off from the Golgi sac.
31. True
32. False-reactions are only aerobic.
33. True
34. False—The protective, waterproof outer layer of skin is formed by the persistence of tough keratin. Keratin is a protein component of special intermediate filaments found in skin cells.
35. True
36. False—Many ribosomes can attach simultaneously to a single mRNA molecule, with each ribosome, one behind the other, synthesizing the same type of protein as specified by the mRNA code.

Fill in the Blank

37. Transport vesicles
38. Microtrabecular lattice
39. Cytosol
40. Kinesin
41. Hydrogens
42. Peroxisomes
43. Plasma membrane, nucleus, cytoplasm
44. Cytoskeleton
45. Proteins, lipids
46. pinocytosis, receptor-mediated endocytosis, and phagocytosis
47. Smooth
48. Exocytosis
49. Hydrolytic
50. Endocytosis
51. Catalase
52. DNA
53. ATP synthase
54. Microtubules

Matching

55. g
56. k
57. j
58. a
59. i
60. h
61. b
62. c
63. f
64. e
65. d

Multiple Choice

66. b
67. e
68. a
69. b
70. a
71. d
72. b
73. a
74. e
75. a
76. b
77. b
78. d
79. b
80. b
81. a
82. b
83. b
84. c
85. b
86. b
87. d
88. a
89. d
90. d
91. c
92. b
93. a
94. b
95. a
96. a
97. b
98. b

Chapter 3: The Plasma Membrane and Membrane Potential

True/False

1. True
2. False-Most cells do not have active transport mechanisms for Cl-.
3. True
4. False—Phagocytosis is the process.
5. True
6. True
7. False—There is no net movement; at equilibrium an equal number of molecules are moving in opposite directions across the membrane.
8. True
9. True
10. True
11. True
12. True
13. True
14. True
15. False—Is possible.
16. False—salt is the solute.
17. True
18. False—Fibroblast instead of fibronectin.
19. False—Tight junction.
20. True

21. True
22. False—Heart muscle.
23. False—Both charged and uncharged molecules are soluble in water. The size of the particle is the second property that influences permeability.
24. False—8.
25. False—The faster the rate of net diffusion.
26. True
27. True
28. True.
29. True
30. True
31. True
32. True
33. True
34. False—Opposite charges not like charges.
35. True
36. False—the greater the ions equilibrium potential.

Fill in the Blank

37. Cholesterol
38. Cell adhesion molecules in the cells' plasma membranes, the extracellular matrix, and specialized cell junctions
39. Spot desmosomes
40. Collagen, elastin, fibronectin
41. Fibroblasts
42. Desmosomes, tight junctions, gap junctions
43. Luminal border, basolateral border
44. Tight junctions
45. Collagen
46. Plasma membrane
47. Membrane potential, negative
48. Phospholipids
49. Hydrophilic, hydrophobic
50. Proteins
51. Lipid bilayer
52. Carrier molecules, receptor sites
53. Elastin
54. Gap junctions, connexons
55. Fibronectin
56. Potassium
57. Net diffusion
58. State of equilibrium, steady state
59. Impermeable
60. Carrier-mediated transport

61. Transport maximum
62. Primary active transport
63. Secondary active transport
64. Endocytosis
65. Phagocytosis
66. Membrane potential
67. Cations, negatively, anions, positively
68. Electrical gradient
69. Electrochemical gradient
70. Concentration
71. Osmosis
72. Osmotic pressure

Matching

73. d
74. e
75. a
76. d
77. d
78. a
79. b
80. d
81. f
82. a

Multiple Choice

83. c
84. b
85. b
86. c
87. d
88. a
89. d
90. b
91. e
92. c
93. c
94. b
95. c
96. c
97. a
98. b

Chapter 4: Principles of Neural and Hormonal Communication

True/False

1. False—Some neurotransmitters function through intracellular second messengers.
2. False—Synapses operate in one direction only.
3. False—A given synapse can be only excitatory or inhibitory.
4. True
5. True
6. False—Larger the graded potential.
7. True
8. False- voltage-gated sodium channels
9. True
10. False-Hydrophilic hormones
11. True
12. True
13. True
14. False—50 times.
15. False—Urgent information is carried by myelinated fibers, less urgent by unmyelinated.
16. False—Never regenerates.
17. True
18. True
19. True
20. False—The lowest
21. True
22. True
23. True

Fill in the Blank

24. Resting membrane potential
25. Dendrites, axon, nerve fibers
26. Ranvier
27. Gastrin, glucagon, gonadotropin-releasing hormone (GnRH)
28. Cytosol, endoplasmic reticulum/Golgi complex
29. Polarization
30. Saltatory conduction
31. Synaptic cleft
32. Adenylyl cyclase
33. Temporal summation, spatial summation

34. Voltage-gated, chemical messenger-gated, mechanically-gated
35. Axon hillock
36. Oligodendrocytes, Schwann cells
37. Hormone response element
38. Glycine, glutamate, gamma-aminobutyric acid (GABA)
39. preprohormones
40. Cell body, dendrites
41. Lipids
42. All-or-none law
43. Multiple Sclerosis
44. Acetylcholine
45. Divergence

Matching

46. b
47. a
48. a
49. b
50. b
51. b
52. a
53. b
54. b
55. a
56. a
57. a

Multiple Choice

58. b
59. a
60. d
61. c
62. c
63. b
64. d
65. c
66. b
67. e
68. a
69. d
70. e
71. e
72. a
73. d
74. c
75. d

76. e
77. b
78. e
79. e
80. e
81. b
82. c
83. c
84. a
85. b
86. a
87. b
88. e
89. d

Modified Multiple Choice

90. a
91. a
92. b
93. b
94. b
95. b
96. a
97. a
98. b
99. a
100. a

Chapter 5: The Central Nervous System

True/False

1. True
2. False--Oligodendrocytes are
3. True
4. True
5. True
6. False—left
7. True
8. True
9. False—Afferent.
10. True
11. False—Afferent neuron.
12. False—Efferent neuron.
13. True
14. True
15. True
16. True
17. False—Cerebrocerebellum.
18. False—Cerebellum.
19. False—Glial cells or neuroglia.

20. True
21. False—Oligodendrocytes.
22. True
23. True
24. False— brain and spinal cord
25. True
26. True
27. True
28. False—31 pairs
29. True
30. False—Descending tracts.
31. False—Afferent.
32. True
33. True
34. True
35. True
36. True
37. False—Decreasing.
38. False—Activity is not reduced.
39. False—20%.
40. False—Paradoxical.
41. True
42. False—Basal nuclei.
43. True
44. True
45. False—Long-term memory.
46. True
47. True
48. True
49. True
50. False—Rapid.
51. True
52. False—Nervous system.
53. False—Long term
54. False—Synaptic cleft.

Fill in the Blank

55. Spinocerebellum
56. Vestibulocerebellum
57. Cerebrocerebellum
58. Simple (basic reflexes) and acquired (conditioned reflexes)
59. Cervical, thoracic, lumbar, sacral, coccygeal
60. Cauda equina
61. Afferent, efferent
62. Stretch
63. Spinal nerve
64. Nerve
65. Hypothalamus, thalamus
66. Cerebellum
67. Cerebral cortex68. Grey, white
69. Parietal lobes, somesthetic sensations

70. Broca's, frontal lobe, Wernicke's, cortex
71. Glial cells, neuroglia
72. Astrocytes
73. Oligodendrocytes
74. Microglia
75. Pia mater
76. Choroid plexuses
77. Medulla, pons, midbrain
78. Reticular formation
79. Consciousness
80. Coma
81. Paradoxical
82. Slow-wave
83. REM (paradoxical)
84. Basal nuclei, thalamus, hypothalamus
85. Thalamus
86. Hypothalamus
87. Motivation
88. Learning
89. Consolidation
90. Nerve
91. Brain, spinal cord, peripheral nervous system
92. Afferent division
93. Afferent neurons, efferent neurons, interneurons
94. CNS

Matching

95. b
96. d
97. a
98. a
99. a
100. a
101. c
102. a
103. b
104. a
105. a

Multiple Choice

106. c
107. c
108. b
109. d
110. b
111. c
112. a
113. e
114. d
115. c

Modified Multiple Choice

116. b
117. e
118. d
119. d
120. b
121. c
122. d
123. b
124. a
125. b
126. a
127. e
128. c

Chapter 6: The Peripheral Nervous System: Afferent Division; Special Senses

True/False

1. True
2. True
3. True
4. True
5. False—carry subconscious
6. False—Afferent fibers.
7. True
8. True
9. True
10. True
11. False—caused by deficient levels
12. True
13. True
14. True
15. False—Open.
16. True
17. False—Phasic receptors.
18. True
19. True
20. False Cones have
21. True
22. True
23. True

24. True
25. True
26. False—Opens
27. True
28. True
29. False—Pinna.
30. True
31. False—Three.
32. True
33. False—The basilar membrane.
34. True
35. True
36. False—Cones are

Fill in the Blank

37. Sclera
38. Cornea
39. Accommodations
40. Cataract
41. Optic disc
42. Opsin
43. Left
44. Taste pore
45. Olfactory bulbs
46. Olfactory nerve
47. Taste buds
48. Subcortical route, thalamic-cortical route
49. Mitral cells
50. Sensory input
51. Phasic
52. Cornea
53. Visceral afferent
54. Sensory information
55. First-order sensory neuron
56. Phantom pain
57. Vestibular apparatus
58. Melanopsin
59. Endolymph
60. Vestibular
61. Auditory nerve
62. Basilar
63. Otolith organs
64. 10
65. Sensory Transduction
66. Substance P
67. Prostaglandins
68. Somatosensory cortex, thalamus, reticular formation
69. Opiate receptors
70. Ganglion

Matching

71. e
72. b
73. c
74. b
75. e
76. b
77. a
78. d
79. b
80. c

Multiple Choice

81. c
82. e
83. a
84. d
85. e
86. a
87. e
88. b
89. b
90. d
91. b
92. e
93. d

Modified Multiple Choice

94. d
95. a
96. g
97. b
98. f
99. c
100. h
101. e

Chapter 7: The Peripheral Nervous System: Efferent Division

True/False

1. True
2. False—Cells do not touch neuromuscular joint.
3. True
4. False—with one muscle cell
5. True
6. False—Just reversed.
7. False—Turned off.
8. True
9. False—Ventral horn.
10. True
11. False—Somatic nervous system.
12. True
13. False—releases less
14. False—Parasympathetic.
15. True
16. True
17. True
18. False—Sympathetic.

Fill in the Blank

19. Motor neurons
20. Common pathway
21. Neuromuscular junction
22. Muscle fiber
23. Terminal
24. Motor end-plate
25. ACh
26. Botulinum toxin
27. Organophosphates
28. Myasthenia gravis
29. Postganglionic fiber, preganglionic fiber
30. Adrenergic fibers
31. Varicosities
32. Epinephrine (adrenaline)
33. Nicotinic receptors
34. Muscarinic receptors

Matching

35. a
36. b
37. b
38. a
39. a
40. b
41. b
42. b
43. a
44. b

Multiple Choice

45. a
46. c
47. d
48. a
49. a
50. b
51. d
52. e
53. e
54. a
55. d
56. b
57. c
58. c
59. d
60. d
61. e
62. b
63. d
64. c
65. e
66. d
67. b
68. e
69. b
70. a

]Modified Multiple Choice

71. c
72. d
73. b
74. a
75. a
76. b
77. b
78. a
79. b
80. a
81. c
82. d

Multiple Choice

83. c
84. b
85. d
86. e
87. a
88. e
89. b
90. e
91. a

Chapter 8: Muscle Physiology

True/False

1. True
2. False—Intrafusal.
3. False--Alpha motor neurons.
4. False—Negative feedback.
5. False—Externally applied tension.
6. True
7. True
8. False—Single nucleus.
9. False--binds to calmodulin
10. True
11. True
12. True
13. False—Calcium concentration.
14. False—Myosin ATPase.
15. True
16. False—Myoglobin.
17. True
18. True
19. False—Oxidative fibers.
20. False—Muscle fiber.
21. True
22. True
23. True
24. True
25. True
26. False—Calcium binds.
27. True
28. True
29. False— contracts slower
30. False—Produces sustained contractions.
31. True
32. True
33. True
34. False—Origin.
35. False—Lengthens.
36. True
37. False—Sarcoplasmic reticulum.
38. False—Lateral sacs.
39. True
40. False—Latent period.
41. True
42. False—Calcium.

Fill in the Blank

43. Excitation-contraction coupling
44. Transverse tubule
45. Sarcoplasmic reticulum
46. Lateral sacs
47. Actin binding, ATPase
48. Rigor mortis
49. Flaccid
50. Calmodulin
51. Multiunit smooth muscle
52. Neurogenic
53. Functional syncytium
54. Slow-wave potentials
55. Latch-phenomenon
56. Muscle fiber
57. Myofibrils
58. A band
59. Sarcomere
60. Myosin, actin
61. Tropomyosin
62. Tropomyosin, troponin
63. Hemiplegia
64. Quadriplegia
65. Paraplegia
66. Gamma motor neuron, alpha motor neurons
67. Golgi tendon organs
68. Motor unit
69. Tension
70. Series-elastic
71. Origin, insertion
72. Isotonic
73. Concentric, eccentric
74. Sphincter
75. Spastic
76. Lever, fulcrum

77. Creatine phosphate
78. Oxidative phosphorylation
79. Myoglobin
80. Atrophy
81. Muscular dystrophy

Matching

82. b
83. a
84. c
85. a
86. c
87. a
88. c
89. a
90. b
91. b
92. a
93. c

Multiple Choice

94. c
95. a
96. a
97. e
98. d
99. b
100. b
101. e
102. b
103. c
104. c
105. a
106. e

Modified Multiple Choice

107. d
108. f
109. a
110. f
111. a
112. f
113. a
114. f
115. d
116. c
117. b

118. d
119. b
120. f
121. a
122. d
123. g
124. d
125. b
126. d
127. b
128. a
129. b
130. c
131. a
132. b, c
133. a

Chapter 9: Cardiac Physiology

True/False

1. False—Contractile cells.
2. True
3. False—originating in the SA node
4. True
5. True
6. False—inward calcium current
7. True
8. False—P wave.
9. True
10. False—Decrease.
11. False—Sympathetic.
12. True
13. True
14. False— have more mitochondria
15. True
16. True
17. False—parasympathetic controls
18. True
19. False—Late.
20. True
21. True
22. True
23. False—Produces no sound.
24. True
25. True
26. True
27. True
28. True
29. True
30. False—Systemic circulation.

31. False—Function as two separate pumps.
32. False—Right side.
33. False—under different pressure
34. False—Pulmonary circulation.
35. True
36. True

Fill in the Blank

37. AV delay
38. Chordae tendineae
39. Funny
40. Collateral circulation
41. Refractory period
42. Fibrillation
43. Interatrial
44. T
45. Tachycardia
46. Bradycardia
47. Atrial
48. Apex
49. Cardiopulmonary resuscitation
50. Ventricles
51. Veins
52. Aorta
53. Tricuspid valve
54. Epicardium
55. 40
56. Cardiac output
57. Cardiac reserve
58. Intrinsic control
59. Frank-Starling Law of the Heart
60. Heart failure
61. End-diastolic volume
62. Isovolumetric ventricular contraction
63. End-systolic volume (ESV)
64. Murmurs
65. Systolic murmur

Matching

66. c
67. a
68. b
69. b
70. a
71. a
72. a
73. a
74. b

75. a

Multiple Choice

76. e
77. d
78. b
79. c
80. b
81. c
82. b
83. c
84. d
85. a
86. c

Modified Multiple Choice

87. 1
88. 8
89. 11
90. 3
91. 12
92. 7
93. 2
94. 4
95. 9
96. 10
97. 6
98. 5
99. b
100. a
101. b
102. a
103. a
104. b
105. c
106. a
107. c
108. a
109. Increased
110. Increased
111. Greater than before
112. More than
113. Increases
114. a
115. d
116. b
117. b
118. a
119. c
120. a
121. b

Chapter 10: The Blood Vessels and Blood Pressure

True/False

1. False—is constant
2. True
3. False—Vary from organ to organ.
4. True
5. False—Exchanges are not made directly.
6. True
7. True
8. True
9. True
10. False—The greater the rate of flow.
11. False—systolic pressure and diastolic pressure
12. True
13. True
14. True
15. False—Transient hypotension.
16. True
17. True
18. False—Short term.
19. True
20. True
21. True
22. True
23. True
24. False—Can either be chemical or physical.
25. False- in the kidney or digestive system have fenestrations
26. True
27. False— Leads to an increase
28. True
29. True
30. True
31. True
32. False—Can only be driven forward.
33. True
34. True

Fill in the Blank

35. Hypertension
36. Hypotension
37. Ultrafiltration
38. Orthostatic hypertension
39. Sympathetic system
40. Viscosity
41. Irreversible shock
42. Capacitance vessels
43. Venous capacity
44. Venous return
45. Varicose veins
46. Baroreceptors
47. Carotid-sinus, aortic-arch baroreceptors
48. Arteries
49. Microcirculation
50. Flow rate
51. 120 mm Hg
52. Pulse pressure
53. Mean arterial pressure
54. Arterioles
55. Diffusion
56. Brain
57. Bulk flow
58. Capillary blood pressure
59. Lymphatic system
60. Edema
61. Vasoconstriction
62. Vascular tone
63. Paracrines
64. Endothelial cells
65. Cardiovascular control center

Matching

66. c
67. a
68. a
69. b
70. a
71. a
72. a
73. a
74. b
75. a
76. c
77. a
78. a
79. c

Multiple Choice

80. c
81. b
82. d
83. e
84. a

Modified Multiple Choice

85. c
86. a
87. d
88. b
89. c
90. d
91. a
92. b
93. e
94. a
95. d
96. b
97. e
98. a
99. a
100. a
101. a

Chapter 11: The Blood and Body Defenses

True/False

1. True
2. True
3. False—Testosterone.
4. False—Primary.
5. False—Thrombin.
6. False—by intrinsic pathway
7. True
8. True
9. True
10. True
11. True
12. False—they are partially responsible for changes.
13. True
14. False— Basophils release
15. True
16. False—Coumadin (warfarin) does
17. False—Lymphocytes.
18. False—Two: antibody-mediated immunity and cell-mediated immunity.
19. True
20. False—Five.
21. True
22. True
23. True
24. False—Autoimmune disease is

25. False—IgE antibodies.
26. True
27. False—Helper T cells.
28. True
29. True
30. True
31. False—Does have control.
32. True
33. False—Interferon.
34. True
35. True

Fill in the Blank

36. Hemostasis
37. Tissue thromboplastin
38. Plasmin
39. Hemophilia
40. Spleen
41. Aplastic
42. Hemolytic
43. Nutrients, waste products, dissolved gases, hormones
44. II
45. Globulins
46. Leukocytes
47. Lymphocytes
48. Neutrophils
49. Cell-mediated immune response
50. E
51. Complement system, phagocytes, killer cells
52. Active immunity
53. Cell-mediated immunity
54. Perforin
55. Helper t cells
56. AIDS
57. Anaphylactic shock
58. Lymphoid tissue
59. Fibrinogen, fibrin
60. Suppressor
61. Antibody-mediated, cell-mediated
62. Haptens
63. Skin

64. Sweat glands, the sweat pores
65. Keratinocytes
66. Granstein cells
67. Alveolar macrophages

Matching

68. h
69. a
70. j
71. b
72. c
73. i
74. d
75. f
76. g
77. e
78. b
79. c
80. b
81. a
82. d
83. a
84. b
85. c
86. d
87. d
88. b

Multiple Choice

89. b
90. b
91. e
92. e
93. d
94. d
95. a
96. e
97. c
98. b
99. c
100. e
101. d
102. e
103. c
104. e
105. d
106. e
107. a
108. e

Modified Multiple Choice

109. a
110. b
111. a
112. b
113. a
114. a
115. c
116. b
117. b
118. d
119. b
120. a
121. b
122. b
123. b
124. c
125. c
126. a
127. b

Chapter 12: The Respiratory System

True/False

1. True
2. False—When compliance is decreased.
3. True
4. False—5.7 liters.
5. True
6. True
7. True
8. False—when drops below 60 mmHg.
9. True
10. True
11. True
12. False—Between the alveoli and the blood.
13. True
14. False—DRG.
15. True
16. True
17. True
18. False—Increase.
19. True
20. True
21. True
22. True
23. False—Decreased.
24. False—97.5% saturated.

25. True
26. True
27. False—Circulatory hypoxia.
28. True
29. False—Higher pressure to lower pressure.
30. False—Lower.
31. True
32. False—Decreases.
33. False—Increases.
34. True
35. True

Fill in the Blank

36. Histotoxic hypoxia
37. Hypercapnia
38. Oxygen-hemoglobin dissociation curve
39. Glottis
40. Carboxyhemoglobin
41. Diaphragm
42. Hypoxic hypoxia
43. Pneumonia
44. Partial pressure
45. Partial pressure gradient
46. 60 mm Hg, 6 mm Hg
47. Functional residual capacity
48. Obstructive, restrictive
49. Anatomic dead space
50. Atmospheric (barometric) pressure
51. Bronchodilation
52. Emphysema
53. Compliance
54. Interdependence
55. Surfactant
56. Tidal volume
57. Medullary respiratory center
58. Apneustic center, pneumotaxic center
59. Dorsal respiratory group
60. Carbonic anhydrase
61. Carotid bodies, aortic bodies
62. Apnea
63. Dyspnea
64. Respiration
65. Trachea, Esophagus
66. Vocal folds
67. Type I alveolar cells
68. Type II alveolar cells
69. Central

Matching

70. a
71. f
72. e
73. a
74. g
75. d
76. b
77. h
78. c
79. d
80. g
81. c

Multiple Choice

82. c
83. e
84. a
85. b
86. e
87. b
88. c
89. e
90. b
91. a
92. d
93. b
94. a
95. b
96. b
97. b
98. a
99. b
100. b

Modified Multiple Choice

101. c
102. a
103. c
104. a
105. b
106. a

Chapter 13: The Urinary System

True/False

1. True
2. False—Arterial blood.
3. True
4. True
5. False—ANP decreases sodium
6. False—And the filtration coefficient.
7. True
8. True
9. True
10. False—Renal threshold.
11. True
12. False—Proximal tubule.
13. True
14. True
15. False—Hydrogen and potassium.
16. True
17. True
18. False—1.5 liters.
19. True
20. False—Greater.
21. True
22. False—also known as ADH

Fill in the Blank

23. Urea
24. Sodium
25. Phosphate and calcium
26. Urea, uric acid, creatine and bilirubin
27. Renin-angiotensin-aldosterone
28. Angiotensin I
29. Angiotensin II
30. Aldosterone
31. Transepithelial
32. Water
33. Acute, chronic
34. Proximal
35. Osmotic
36. Impermeable

37. a. Infectious organisms
    b. Toxic agents
    c. Inappropriate immune response
    d. Obstruction of urine flow
    e. Insufficient renal blood supply
38. Peristaltic
39. Internal urethral sphincter, external urethral sphincter
40. Countercurrent exchange
41. 300
42. Sodium-potassium pump
43. Potassium secretion

Matching

44. a
45. b, c
46. a, b, c
47. a
48. b
49. a
50. a
51. a
52. b
53. a
54. a

Multiple Choice

55. e
56. c
57. a
58. d
59. a
60. e
61. d
62. a
63. a
64. e
65. c
66. d
67. c
68. e
69. c
70. b
71. c
72. d

73. a
74. d
75. b
76. a
77. e
78. c
79. d
80. b
81. a
82. a
83. d
84. c
85. d
86. a
87. c
88. e
89. e
90. b
91. c
92. b
93. c
94. c
95. d
96. e
97. e
98. c
99. c
100. e

Modified Multiple Choice

101. a
102. c
103. a
104. a
105. b
106. a
107. a
108. c
109. b
110. e
111. a
112. c
113. d

Chapter 14: Fluid and Acid-Base Balance

True/False

1. False—Positive balance.
2. False—Two-thirds.
3. False—classified as sensible.
4. True
5. False—not important in the ECF
6. True
7. True
8. False—Acids.
9. True
10. True
11. False—Chemical buffer system.
12. False—Reversed; ammonia combines with hydrogen to form ammonium ions.
13. False—Diabetes mellitus.
14. True

Fill in the Blank

15. Stable balance
16. Fatty acids, lactic acid
17. Phosphate
18. Obligatory loss of salt in sweat, feces, urine
19. Renin-angiotensin-aldosterone
20. Insufficient water intake, excessive water loss, diabetes insipidus
21. Hypothalamic osmoreceptors
22. pH
23. Phosphate buffer system
24. Hypoventilation
25. Respiratory alkalosis
26. Metabolic alkalosis
27. Hydrogen
28. Hydrogen excretion, bicarbonate excretion, ammonia secretion

Matching

29. c
30. d
31. a
32. a
33. c
34. c
35. b
36. c

Multiple Choice

37. e
38. b
39. c
40. e
41. c
42. e
43. a
44. c
45. b
46. b

Modified Multiple Choice

47. b
48. a
49. a
50. b
51. a
52. d
53. b
54. b
55. a
56. c
57. b
58. b
59. a
60. a
61. a
62. b
63. a
64. b
65. d
66. b
67. b
68. c
69. c
70. a
71. c
72. a
73. d
74. c
75. a
76. c
77. c
78. c
79. c
80. c
81. b
82. b

83. b
84. b
85. c
86. b
87. b
88. b
89. c
90. c
91. b
92. b

Chapter 15: *The Digestive System*

True/False
1. True.
2. False—Polysaccharides not disaccharides.
3. False—Monoglycerides not triglycerides.
4. True
5. False—expands 20 times
6. False—the serosa is.
7. True
8. True
9. False—Skeletal, not smooth.
10. True
11. True
12. True
13. True
14. True
15. True
16. True.
17. False—Fundus, not antrum.
18. True
19. True
20. True
21. True
22. False—Pepsinogen, not mucus.
23. False—Parietal cells, not mucous neck cells.
24. True
25. False—Pepsinogen, not trypsinogen.
26. True
27. True
28. False—Water is not absorbed from the stomach contents.
29. False—Acini, not islets of Langerhans.
30. False—Enterokinase, not HCl.

31. False—Alkaline, not acidic.
32. True
33. False—known as xerostomia
34. True
35. False—Liver, not gallbladder.
36. True
37. True
38. False—Microvilli are known as brush border.
39. True
40. True
41. True
42. True

Fill in the Blank

43. Intrinsic factor
44. Enterokinase
45. Aminopeptidases
46. Villi
47. Malabsorption
48. Acetylcholine
49. D, calcium
50. Mastication
51. 99.5%, 0.5%
52. Xerostomia
53. Acquired (or conditioned) salivary reflex
54. Palate
55. Nitroglycerin
56. Bolus
57. Oropharyngeal stage
58. Gastroesophageal sphincter
59. Peristalsis
60. Migrating motility contractions
61. Intestinal cells of Cajal
62. Appendix
63. Haustral contractions
64. Colon, cecum, appendix, rectum
65. Celiac
66. GIP (gastric inhibitory peptide)
67. Fundus
68. Antrum
69. Gastric pits
70. Gastrin
71. HCl
72. Gastric
73. Peptide ulcer
74. Jaundice
75. Liver
76. Sphincter of Oddi
77. Bile salts

78. Choleretic
79. CCK
80. Cephalic
81. Hydrolysis

Matching

82. j
83. a
84. d
85. g
86. l
87. n
88. b
89. m
90. e
91. h
92. k
93. i
94. f
95. c

Multiple Choice

96. c
97. b
98. e
99. a
100. e
101. c
102. e
103. b
104. d
105. d

Modified Multiple Choice

106. d
107. c
108. d
109. d
110. a
111. d
112. a
113. c
114. b
115. c
116. a
117. a
118. c
119. b

120. b
121. a
122. b
123. e
124. d
125. a
126. e
127. c
128. d
129. c
130. e
131. b
132. a
133. d
134. e
135. c
136. a

Chapter 16: Energy Balance and Temperature Regulation

True/False

1. False—External work.
2. False—Energy can be converted to another form.
3. True
4. True
5. True
6. True
7. True
8. True
9. False—Normally rectal temperature is a degree higher.
10. True
11. False—Conduction uses direct contact.
12. True
13. True
14. True

Fill in the Blank

15. Increased
16. Ghrelin
17. Brainstem
18. Anterior, decreasing, vasodilation
19. Prostaglandins
20. Leptin
21. Sympathetic
22. Convection
23. Lack of appetite

24. 75
25. External work, internal work
26. Positive
27. Conduction
28. Radiation

Matching

29. d
30. a
31. f
32. b
33. c
34. e
35. g

Multiple Choice

36. c
37. e
38. b
39. d
40. c
41. c
42. b
43. d
44. c
45. d
46. b
47. b
48. d

Modified Multiple Choice

49. c
50. b
51. d
52. a
53. b
54. e
55. d
56. c
57. a
58. d

Chapter 17: The Endocrine System

True/False

1. True
2. True
3. True
4. False—also made by adrenal cortex
5. False—Neurohypophysis.
6. True
7. True
8. False—Several, not just one.
9. True
10. True
11. True
12. True
13. False—Increase.
14. True
15. False—about 80%.
16. False—Especially muscle.
17. True
18. True
19. True
20. True
21. True
22. True
23. False—Type I.
24. True
25. True
26. False—Decreases.
27. True
28. True

Fill in the Blank

29. Elevated, aldosterone
30. Pituitary gland
31. Short-loop negative feedback
32. Exopthalmos, thyroid hormone
33. Diaphysis, epiphysis
34. Dwarfism
35. Hormones
36. Melanocyte-stimulating hormone
37. Endocrine system
38. Paracrines
39. Steroids
40. 4
41. Permissiveness
42. Antagonism

43. Hypothalamus, anterior pituitary
44. Type II Diabetes Mellitus
45. Anabolism
46. Islets of Langerhans
47. Glycogenesis
48. Lamellae
49. Calcitonin
50. Stored
51. Thyroglobulin
52. Grave's disease

Matching

53. s
54. b
55. k
56. r, j
57. d
58. c
59. r
60. r
61. l
62. n
63. e
64. n
65. q
66. j
67. o
68. i
69. n
70. e
71. f
72. s
73. j, r
74. p
75. c
76. m
77. n
78. e
79. a
80. n
81. g
82. c
83. s
84. n
85. b
86. e
87. a
88. h

89. g
90. e
91. e
92. h

Multiple Choice

93. c
94. b
95. e
96. d
97. e
98. d
99. a
100. e
101. c
102. c

Modified Multiple Choice

103. a
104. a
105. b
106. b
107. c
108. a
109. a
110. a
111. a
112. a
113. a
114. b
115. b
116. a
117. a
118. c
119. b
120. b
121. b
122. a
123. a
124. a
125. c
126. c
127. a
128. a
129. c
130. c
131. a
132. a
133. a

134. b
135. a
136. a
137. b
138. a

Chapter 18: The Reproductive System

True/False

1. False—Gametogenesis is the formation of sperm and eggs.
2. True
3. True
4. False—Somatic cells are diploid.
5. True
6. True
7. True
8. False—in the tubule fluid.
9. True
10. False—Leydig cells secrete testosterone.
11. False—46 doubled chromosomes.
12. True
13. False—Sertoli cells, not Leydig cells.
14. True
15. True
16. False—LH, not FSH.
17. True
18. False—The prostate, not the bulbourethral glands.
19. True
20. True
21. False—Estrogen, not progesterone.
22. False—similar to ECF
23. True
24. False—Follicle, not corpus luteum.
25. False—FSH stimulates
26. True
27. False—FSH, not LH.
28. True
29. True
30. True

Fill in the Blank

31. Menopause
32. Granulosa cells, primary follicle
33. Ovulation
34. Follicular cells
35. Estradiol

36. Myometrium, endometrium
37. Aromatase
38. ZP3
39. Ampulla
40. Seminal vesicles
41. Resolution
42. Erectile tissue
43. Polyspermy
44. Orgasmic platform
45. Gametes
46. Somatic cells
47. Autosomal chromosomes
48. DHEA
49. Cryptorchidism
50. Leydig cells, interstitial cells
51. Spermeogenesis
52. A head, an acrosome, a midpiece, a tail
53. Blood-testes barrier
54. Luteinizing hormone (LH), follicle-stimulating hormone (FSH)
55. Blastocyst
56. Decidua
57. Chorion
58. HCG (human chorionic gonadotropin)
59. 38
60. Oxytocin
61. Involution
62. Colostrum

Matching

63. a, b
64. d
65. b
66. c
67. d
68. b
69. c
70. e

Multiple Choice

71.   b
72.   c
73.   d
74.   e
75.   b
76.   b
77.   e
78.   e
79.   b
80.   a
81.   c
82.   c
83.   e

Modified Multiple Choice

84.   c
85.   c
86.   b
87.   a
88.   b
89.   b
90.   n
91.   h
92.   o
93.   a
94.   e
95.   g
96.   t
97.   c
98.   b
99.   u
100.   d
101.   r
102.   d
103.   m
104.   q
105.   j
106.   s
107.   u
108.   l
109.   m
110.   p
111.   o
112.   i
113.   p
114.   n
115.   u